THERE'S NO SUCH THING AS FREE SPEECH

There's No Such Thing as Free Speech
and it's a Good Thing, Too

STANLEY FISH

New York   Oxford
OXFORD UNIVERSITY PRESS
1994

323, 443

B. & T.

25.00

8/94

Oxford University Press

Oxford   New York   Toronto
Delhi   Bombay   Calcutta   Madras   Karachi
Kuala Lumpur   Singapore   Hong Kong   Tokyo
Nairobi   Dar es Salaam   Cape Town
Melbourne   Auckland   Madrid

and associated companies in
Berlin   Ibadan

Published by Oxford University Press, Inc.,
200 Madison Avenue, New York, New York 10016

Oxford is a registered trademark of Oxford University Press

Library of Congress Cataloging-in-Publication Data
Fish, Stanley Eugene.
There's no such thing as free speech /
Stanley Fish.
p.   cm.   Includes index.
ISBN 0-19-508018-1
1. Freedom of speech.   2. Freedom of speech—United States.
3. Academic freedom.   4. Academic freedom—United States.
I. Title.   Z657.F5   1994   323.44'3'0973—dc20   93-15347

2 4 6 8 9 7 5 3 1

Printed in the United States of America
on acid-free paper

For Jane
and to the memory of
Max Fish and Henry Parry,
two honest men

Preface

When she was the chair of the National Endowment for the Humanities, Lynne Cheney used to delight in pointing out that her views on the curriculum, literary value, and the canon were shared not only by fellow conservatives, but by liberals like Arthur Schlesinger and the occasional Marxist like Eugene Genovese. In Cheney's eyes that broad-based consensus was a clear indication that she was simply voicing common sense in the face of a radical, nihilistic assault on the entire fabric of Western Civilization. Although she never quite put it this way, the message was unmistakable: if persons so far apart on the political spectrum share a distaste for an agenda—call it multiculturalism, postmodernism, deconstruction, or political correctness—that agenda is beyond the pale and should be rejected by all reasonable people.

The matter, however, is a bit more complicated than that, for conservative ideologues and liberal stalwarts recoil from the same thing not because they are similarly clear-sighted but because they are committed to a similar structure of thought, although they would fill it out in different ways. It is a structure that opposes the essential to the accidental and enjoins us to be vigilant lest the latter overwhelm the former. A Lynne Cheney and a Eugene Genovese might disagree as to what the essential is and where the danger to its flourishing lies, but they will agree on the obligation to protect it and they will join forces against anyone who argues, as I do, that the essential is a rhetorical category whose shape varies with the contingencies of history and circumstance. In a debate about the curriculum, for example, there might be arguments about which texts will produce the desired result—the formation of a virtuous citizenry or a commonwealth free of oppression—but all parties will believe in the project and will reject any suggestion that there is nothing *general* to say about the relationship between pedagogical choices and preferred outcomes and that finally one can only respond to the question provisionally, by determining what might work at this time, in this place, and for this momentary purpose. This refusal to come down on one side or the other of an issue posed in absolute terms will always be heard as some kind of relativism or solipsism or nihilism and as the subordination of value to political expediency, but in the chapters that follow I assert that it is nothing of the kind and that values, rather than being opposed to political commitment, grow only in its soil and wither in the arid atmosphere of bodiless abstraction, whether that abstraction is named reason, merit, fairness, or procedural neutrality. The upshot of this is not, as some would have it, that anything goes or that words have no meaning, but that the line between what is permitted and what is to be spurned is always being drawn and redrawn and that structures of constraint are simultaneously always in place and always subject to revision if the times call for it and resources are up to it. Neither the defender of the status quo nor the proponent of radical change will find much comfort in these pages, which issue no clarion calls and recommend (if they recommend anything) only that we resist overheated and overdramatic characterizations of our situation, whether they come from the left or the right.

I would expect that for many readers the most distressing thing about these essays will be the skepticism with which they view the invocation of high-sounding words and phrases like "reason," "merit," "fairness," "neutrality," "free speech," "color blind," "level playing field," and "tolerance." My argument is that when such words and phrases are invoked, it is almost always as part of an effort to deprive moral and legal problems of their histories so that merely formal calculations can then be performed on phenomena that have been flattened out and no longer have their real-world shape. An exemplary (it is a bad example) instance of this practice has just been provided for us by the Supreme Court in its recent (June 28, 1993) decision that the creation by the North Carolina legislature of two black majority districts may be unconstitutional because it smacks too much of "race consciousness." "It is unsettling," observes Justice O'Connor in her majority opinion, "how closely the North Carolina plan resembles the most egregious racial gerrymanders of the past." But the resemblance that strikes Justice O'Connor so forcibly only emerges if one has forgotten or bracketed out everything about the past that makes the present an intelligible (and moral) response to it. Specifically, one must forget that past redistricting practices were devised with the intention of disenfranchising an already disadvantaged minority; discrimination was not a byproduct of the policy, but its goal, and a goal whose achievement can be measured by the fact (noted by Justice White in his eloquent dissent) that it is only now, as a result of the plan under attack, that "the State has sent its *first* black representatives since Reconstruction to the United States Congress."

This is to say, the goal of the North Carolina policy is to bring African Americans into the political process, and the idea that by doing so a white majority that remains a majority "in a disproportionate number of congressional districts" will have been unduly deprived of its influence on that process is, as White forthrightly says, "a fiction." It is a fiction that can only be maintained if one removes the present fact situation from its context and inserts it into the game I call "moral algebra," a game that is played by fixing on an abstract quality and declaring all practices that display or fail to display that quality equivalent. In this case Justice O'Connor plays the game by substituting for the question "What is the purpose of the Voting Rights Act and how does the North Carolina Plan square with that purpose?" the question "Does the North Carolina Plan exhibit race consciousness?" When this second question has yielded an affirmative answer (and given the fact of the Voting Rights Act, *any* plan developed under its aegis would be race conscious, for absent race consciousness there would be no Voting Rights Act) one can then declare, with an apparently straight face, that the North Carolina legislature's effort to undo the effects of past discrimination "closely . . . resembles the most egregious racial gerrymanders of the past."

My point has been made with an admirable conciseness by a lead editorial in *The New York Times:*

[Justice O'Connor's] opinion wraps the legal assault on the Voting Rights Act in a noble language, proclaiming "the goal of a political system in which race no longer matters." Thus she steps over the fact that we have a political history—and a body of franchise law—in which race has always mattered. Indeed, since the Civil War, the struggle to achieve a healthy race consciousness in our politics has been an ennobling part of our system.

Only a willful disregard of that history would allow a majority opinion that says the equal protection clause of the 14th amendment—passed to extend political rights to blacks—must be read to protect a white majority with no history of discrimination. [June 30, 1993; p. A14]

It is the willful disregard of history that is the object of critique in the pages that follow.

Along the way more people helped me than I could possibly name but I would be remiss if I did not publicly thank Miriam Angress, Stanley Blair, Lisa Haarlander, Howard Horwitz, Elizabeth Maguire, Amy Roberts, Jessica Ryan, Susan Ryman, Ruth Sandweiss, Barbara Herrnstein Smith, and William Van Alstyne.

Durham, North Carolina   S. F.
August, 1993

CONTENTS

THERE'S NO SUCH THING AS FREE SPEECH

1

INTRODUCTION:
"THAT'S NOT FAIR"

The Case for the Boss's Nephew

In the movie version of *How to Succeed in Business without Really Trying,*
Robert Morse, playing a Protestant version of Budd Schulberg's Sammy
Glick, insinuates himself into the firm of his heart's desire and is assigned
to the mailroom, where he finds one other co-worker who just happens to
be the boss's nephew. It turns out that one of the two will be elevated to
the position of mailroom head, and the junior executive who is to make the
decision announces solemnly, "I've been told to choose the new head of
the mailroom on merit alone." To this the boss's nephew immediately responds, "That's not fair."

The audience, of course, is supposed to laugh at the inappropriateness of
invoking fairness as a protest against the introduction of a regime of merit,
but the boss's nephew is perfectly serious, and if we step back from the
scene (as we are not encouraged to do) and look at it from *his* perspective,
the point he wants to make comes into focus. He has been proceeding under
the expectation that *as* the boss's nephew he would have a leg up on any
rewards that are to be distributed, and he is complaining that the rules of
the game are suddenly being changed in the seventh inning. It is not so
much that he is disputing the introduction of "merit" as a standard; rather,
he is saying that according to the understanding that prevailed before this
unhappy moment, it was a *part* of his merit that he was related to the head
of the company. In effect he is reacting just as the elder son in a hereditary
monarchy might react were he to be told just before the king died that from
now on we're going to do it differently and hold elections. In both cases
the disappointment is more than personal; it extends to the overturning of a
whole way of life, complete with a tradition, a set of expectations, obliga-

3

tory routines, normative procedures, in-place hierarchies, and so on. And when a new way of life is forcefully introduced, the sense of betrayal is cosmic, and the terms in which it is expressed are abstract and universal: "That's not fair," to which the boss's nephew might have added, had he thought of it, "That's not merit."

The moral is one that is drawn in many of the essays in this book: while notions like "merit" and "fairness" are always presented as if their meanings were perspicuous to anyone no matter what his or her political affiliation, educational experience, ethnic tradition, gender, class, institutional history, etc., in fact "merit" and "fairness" (and other related terms) will have different meanings in relation to different assumptions and background conditions. That something is fair or meritorious is not determined above the fray but within the fray; the words do not mark out an area quarantined from the pull of contending partisan agendas; they are among the prizes that are claimed when one political agenda is so firmly established that its vision of the way things should be is normative and can go without saying. At such moments of political victory (which can last a moment or a millennium), the agent who lives entirely within the reigning paradigm will look at a practice and say, as the boss's nephew says, "That's not fair," in perfect confidence that his judgment will be as obvious to his hearers as it is to him.

Of course in the scene from *How to Succeed in Business without Really Trying,* the confidence of the speaker, intact only a few seconds ago, is already shaken as he speaks. What he has experienced (and I am aware of how much weight I am placing on so slender a reed) is nothing less than the intrusion into his world of another world whose presuppositions have been no part of his prior experience and seem to him, as they push themselves into his life, to be bizarre. On a much larger and socially significant scale, this is exactly what has happened in recent years to many in the academy who have been discombobulated by the double whammy of hearing strange doctrines promulgated in bewildering proliferation and seeing them win favor with the very constituencies (students, administrators, journal editors, funding agencies, professional societies) that had so long smiled in their direction. "Do you hear what they're saying? Words have no intrinsic meaning, values are relative, rationality is a social construct, everything is political, every reading is a misreading." One must imagine these and similar pronouncements as they are processed by two different audiences: the one audience, of which I myself would be a member, hears them as commonplaces, old hat, long ago digested, and understood in ways that lead to none of the dire consequences usually predicted; the other audience, made up largely, though not exclusively, of older academics with track records to savor and defend, hears only those consequences and declares that if such things are allowed to be said or even thought, civilization will collapse and chaos will come again. The fear is absurd in the overgeneral terms of its usual expression—ideas, especially ideas from so esoteric a field as literary theory, do not generate such large effects so easily—but the fear is a real one for those who recognize in the rising popularity of new slogans and mantras ("the personal is the political," "interpretation cannot be avoided or constrained," "there is nothing outside the text") a challenge to the very foundation of their professional and personal lives. A few years ago I saw Tom Wolfe asked by a TV interviewer what he thought about deconstruction. Wolfe answered by recalling a conversation with Margaret Atwood, to whom he had put the same question. "All you have to know Tom," she replied, "is that it's bad for *us.*" That's what the boss's nephew knows the minute he hears the word "merit." He is being told that the game is up and he is now playing a new game for which he has not been trained.

By using the game analogy I do not mean to trivialize his discomfort or poke fun at his cry of unfairness. I merely want to stress the extent of the change that is now overtaking him. A game is not simply a collection of discrete actions casually brought together in a demarcated field. A game— be it baseball or literary criticism or philosophy—is a little universe in which each action acquires its significance and value only in relationship to an underlying sense of purpose that at once defines the enterprise and gives shape to the events that occur within it. It only matters that you should be able to hit a horsehide-covered spheroid with a stick if there is understood to be an activity called "scoring runs" that is crucial to the even more basic activity of "winning." It only matters that you should be able to give evidence for your reading of a poem if there is understood to be an activity called "literary interpretation" whose point is to get the poem's meaning right. If that underlying sense of purpose or point is taken away or replaced by another, the result will not simply be locally disconcerting but globally disorienting. All of a sudden you literally do not know what to do because everything you have been taught to do—often by a mode of instruction that has been indirect, a matter of cultural conditioning—has been declared obsolete, or without value, or even evil.

When this happens, when a life project such as being the boss's nephew or arriving at the right reading of a poem is rudely and powerfully challenged by an alternative project, all parties immediately claim to occupy the moral high ground and declare that "it" isn't fair. The boss's nephew says "It isn't fair" because in the world he has always inhabited a family affiliation is a mark of virtue and therefore a true indicator of how he should be treated; when he is treated that way, people are being fair, and when he is treated in another way (as he now will be), he is the victim of unfairness.

To this the Robert Morse character could reply that "true fairness" requires the setting aside of family affiliation as an "extrinsic" factor that obscures the genuine standard of "merit" now being redefined as the basis of the new definition of fairness. (The process is wholly and inevitably circular.) One would not be able to adjudicate the dispute because the opposing arguments rest upon radically opposed presuppositions about what is desirable, possible, and even real. These same presuppositions will generate the polemical stances of the parties. The boss's nephew (I am imagining an expansion of the scene more philosophical than the tone of the piece allows) would accuse the newcomer of undermining traditional family values (I write this in the months before the 1992 presidential election); the newcomer would announce that he came to expose the hollowness of the old order, of a dessicated ancien régime undergirded by little more than moldy custom and unearned privilege.

The nice thing about *How to Succeed in Business without Really Trying* is that neither ethic is given the moral advantage over the other. Although the introduction of "merit" as a standard would seem to award the palm to the Horatio Alger figure, especially in the context of the American valorization of entrepreneurial self-improvement, the brazen opportunism he displays makes him an uneasy object of approval. The broad comedy of his rapid rise in the corporate structure renders all the talk about "merit" somewhat suspect; and indeed, it soon becomes apparent that what "merit" amounts to in this brave new world is a better (i.e., more effective) kind of talk. Within the changed rules, producing "merit talk" is going to get you further than you would get by producing family affiliation talk. Let me hasten to add, lest the point be misunderstood, that the inadvisability in the new regime of engaging in family affiliation talk does not mean that such talk is devoid of or unrelated to merit, only that in a bold political stroke one party has managed to *define* merit so that the concept is congruent with the assumptions underlying its own practices; and insofar as that definition comes to be regarded as normative, practices informed by other assumptions—for example, the assumptions underlying primogeniture and hereditary entitlements—will be seen as without merit or subversive of merit. (This understanding of "merit" corresponds to its use in the law, in which the statement "the argument is without merit" means that the argument does not fit within the currently acceptable ways of posing questions and giving answers; that is, the law recognizes that merit is not an abstract, independent standard but one that follows from the traditions and practices of a community whose presuppositions are not at the moment the object of scrutiny or skepticism.)

Getting hold of the concept of merit and stamping it with your own brand is a good strategy, and it is a strategy that has been used to great advantage in recent years by those who range themselves against multiculturalism, feminism, ethnic studies, gender studies, campus speech codes, deconstruction, literary theory, and popular culture. The list may seem puzzlingly heterogeneous until one realizes that each of its items can be characterized as a *special interest* and therefore as insufficiently general to serve as the basis of a program, or a course, or a standard of judgment. You know the arguments: feminists have a particular axe to grind, and therefore their accounts and evaluations are tainted by bias and partisanship; works of popular culture appeal to the untutored tastes of regional or lower-class audiences and are therefore not the proper objects of serious study; multiculturalism indiscriminately honors the norms of every culture and is therefore without a higher standard that would allow one to distinguish between good and bad or between good and better. In each case the claims of the "ism" are declared to be narrow, political rather than genuinely aesthetic or rational, and as "merely" political claims they weigh little against the large claims made in the name of "truths that pass beyond time and circumstance; truths that, transcending accidents of class, race and gender, speak to us all."

I quote from Lynne Cheney's pamphlet *Humanities in America: A Report to the President, the Congress, and the American People* (1988). Cheney's ringing declaration is altogether typical of the statements that issue from her side of the street (of course she wouldn't see it as a "side" at all but as just the broad and common way), and it conveniently highlights both the appeals of the position (again, she wouldn't see it as a "position") she espouses and the problems that lurk not below its surface but on its surface. First one must ask a double question: What are these truths and how and *by whom* are they to be identified? The question takes this form because no one in the field is aligning himself or herself with falsities; everyone insists on the truth of his or her assertions; and since the assertions of different would-be truth tellers conflict, there must be a mechanism for determining which truths are transcendent and which are tied to (and tainted by) time, circumstance, and the accidents of class, race, and gender; otherwise Cheney's call to transcendence is without operative (as opposed to hortatory) force. Classically, that mechanism is found in something called Reason, but notoriously, what is a reason for you may not be a reason for me and may even seem irrational, that is, incompatible with the principles that ground my perception and judgment. Reasons do not confirm or shore up your faith; they are extensions of your faith and are reasons *for you* because of what you already believe at a level so fundamental that it is not (at least while you are in the grip of belief) available for self-conscious scrutiny. It would seem, then, that no one is or could be capable of making the necessary determination (the determination of which proffered truths are the genuinely transcendent ones) because everyone is so enmeshed in time and circumstance that only circumstantial and timely (i.e., historically bounded) truths will be experienced as perspicuous.

Moreover, not only is there no one who could spot a transcendent truth if it happened to pass through the neighborhood, but it is difficult even to say what one would be like. Of course we would know what it would *not* be like; it would not speak to any particular condition, or be identified with any historical production, or be formulated in the terms of any national, ethnic, racial, economic, or class traditions. In short, it would not be clothed in any of the guises that would render it available to the darkened glasses of mortal—that is, temporally limited—man. It is difficult not to conclude either (a) that there are no such truths, or (and this is my preferred alternative) (b) that while there are such truths, they could only be known from a god's-eye view. Since none of us occupies that view (because none of us is a god), the truths any of us find compelling will all be partial, which is to say they will all be political. It follows that when Cheney invokes transcendent truths and urges that they be the basis of our decisions, the invocation has no determinate content (for neither she nor anyone else could say with certainty what the transcendent truth *is*) and could not be the basis of a plan of action.

Nevertheless, even though the gesture of invoking transcendence is an empty one, it can be quite full politically. The unavailability of any transcendent truths to invoke can be an advantage to someone who has the skill and resources to advance the truth he or she prefers to a place of honor that has no "natural" inhabitants. One can reduce the strategy to a formula: first detach your agenda from its partisan origins, from its history, and then present it as a universal imperative, as a call to moral arms so perspicuous that only the irrational or the godless (two categories often conflated) could refuse it. You can do this in many ways, but one way, tried and true, is to appropriate a vocabulary that is already an honored one and then "spin" it so that it will generate the conclusions—the marching orders—that are the content of your politics. This is precisely what has been done, and done brilliantly by the neoconservative participants in the recent culture wars. Perhaps their most stunning success has been the production (in fact a reproduction), packaging, and distribution of the term "political correctness." The phrase is a wonderfully concise indictment that says that a group of unscrupulous persons is trying to impose its views on our campus populations rather than upholding views that reflect the biases of no group because they are common to everyone. It is these commonly shared views, we are told, that are really correct, while the views of feminists, multiculturalists, Afrocentrists, and the like are merely politically correct, correct only from the perspective of those who espouse them.

One sees why the strategy has been so successful: it puts the other side on the defensive by assigning it a description it will feel obliged to refuse. That, however, is the trap; if you respond (as the president of Stanford did in a debate with William Bennett) by declaring that you too are committed to the disinterested search for truth and eschew politics, you will be playing the game on the other fellow's field, and you will be giving yourself the herculean task (doable but a public relations disaster) of explaining why feminist, African American and gay concerns are universal. The better response (now beginning to be made) would be to acknowledge the political implications of a revised curriculum but point out that any alternative curriculum—say, a diet of exclusively Western or European texts—would be no less politically invested, only differently so. In this way you not only refuse the label of "politically correct," you refuse the game of which it is a part by denying the game's central premise, which is that *any* party to the dispute could occupy a position above or beyond politics. Political correctness, the practice of making judgments from the vantage point of challengeable convictions, is not the name of a deviant behavior but of the behavior that everyone necessarily practices. Debates between opposing parties can never be characterized as debates between political correctness and something else, but between competing versions of political correctness.

At least one member of the anti–political correctness crowd knows this and has been willing to say so in public. In the *Wall Street Journal* of December 26, 1991, Robert Kelner chastised his fellow conservatives for what he called "our phony war on political correctness." The war is phony, he says, because those who wage it bill themselves as nonpolitical or apolitical whereas in fact "many conservatives *want* a politically correct campus"; it's just that right now on many campuses the politics and its correctness go in directions they dislike. Kelner is distressed not only because he thinks the strategy dishonest but because he sees that by pursuing it his colleagues on the right forfeit the opportunity to foreground their agenda, since by the logic of their accusations having an agenda is something they cannot acknowledge. Were they more honest, they would admit, like Kelner, that what distresses them about the educational scene today is not that it is informed by politics but that it is informed by politics of the wrong kind. ("We sought—and still seek—ascendancy.") Any such admission, however, would require giving up the stronger claims encoded in words and phrases like "merit based," "neutral," "objective," and "fair," for they would now be seen to be parasitic on the very interests—political interests—to which they are polemically opposed. The battle lines would have to be redrawn so that the opposition was no longer between fairness and bias but between the sense of fairness that follows from one form of bias— one understanding of what is appropriate, desirable, and real—and the sense of fairness that follows from another.

Isn't this relativism? The question always raises itself whenever someone argues that judgments as to what is fair or meritorious flow from local contexts rather than from the identification of transcendent or general standards. The question seems urgent because of the fear that if judgments are relative to particular contexts and there is no context of contexts—no source of authority so compelling that everyone, no matter what his or her history, education, political situation, etc., defers to it—then there is no way to tell the difference between right and wrong, to prefer one argument to another, to distinguish Shakespeare from graffiti. This fear imagines human agents as standing apart from all contexts of judgment and faced with the task of identifying the right and true ones. But as I argue in "The Common Touch," none of us is ever in that "originary" position, unattached to any normative assumptions and waiting for external guidance; rather, we are always and already embedded in one or more practices whose norms, rules, and aspirations we have internalized, and therefore we are not only capable of making distinctions and passing judgments but cannot refrain from doing so. When I read a passage from *Paradise Lost* or read someone else's analysis of that same passage, I do not proceed in two stages, first processing the text at a basic level and then sorting it out into patterns and meanings in accordance with some rule or calculus. Rather, the act of sorting is coextensive with the act of reading; the very processing of the text is driven by my preknowledge of what kind of text it is as derived from my experience as a professional interpreter, and when the processing is done, I am not faced with the task of making decisions about how the text is organized or what it means because just those decisions have been the content of my performance. The fact that one is not in possession of (or possessed by) standards that transcend the site of one's practice is in no way disabling; one can still pose questions, give answers, meet objections, and discern error, all without moving outside the circle of one's professional or disciplinary competences.

This, however, does not end the worry about relativism, which is merely transferred to another level; for there will be other readers of *Paradise Lost* as well educated as I who will by virtue of alternative forms of preknowledge, alternative expectations and assumptions, perform quite differently and come up with different answers to questions like "What does it mean?" Isn't it here that the absence of independent standards becomes disabling, since in the event of competing and contradictory accounts of the matter one has nowhere to go? Well, yes and no. It is true that there is nowhere to go, no locus of judgment to which disputants can appeal for an authoritative announcement. But this doesn't mean that they must throw up their hands or toss the dice; it means that they must argue, thrash it out, present bodies of evidence to one another and to relevant audiences, try to change one another's mind. To be sure, the process is not guided by any unchallengeable authority, but authority, not unchallengeable but temporarily regnant, is what is fashioned in the course of it. That is to say, authority does not *preside* over the debate from a position outside it but is the prize for which the debaters vie.

Of course what I am describing is an idealized form of debate in which everyone's cards are on the table and the stakes are frankly acknowledged. In many debates, however, one side has recourse to the tactics decried by Kelner; it seizes the high ground by laying claim to a certain charged vocabulary and using it not to further argument but to shut it down in a fit of moral posturing. That is what the junior executive does when he grandly announces to the boss's nephew that from now on he will be evaluated on the basis of merit alone. The possibility that merit is itself a contested concept, or that in some regimes it might include the very qualities (of consanguinity) that are now being ruled out of court, is not considered. The nephew is never given a chance to reply because the case is declared open and shut, and, not surprisingly, his attempted protest ("that's not fair") is never given a serious hearing.

Meritscam

It is the thesis of the essays in part 1 of this book that the antimulticulturalist, anti–affirmative action, antifeminist, antigay, antiethnic studies backlash has proceeded in just the manner I have described, and it is the effort of these chapters to deprive the backlashers of a vocabulary to which no one has an exclusive claim. In "The Common Touch," "common" is the target word, and the argument is that its invocation always masks a *special* agenda that would pass itself off as the agenda of everyone (that is, of no one). In the five essays that follow, all written for a series of debates between Dinesh D'Souza and me, the same argument is pursued in a number of directions. In "The Empire Strikes Back," I analyze the very political attack on "political correctness," which, I contend, is a specter largely fabricated by those who urge us to take up arms against it. In "Reverse Racism," I examine the logic by means of which the neoconservative polemic acquires its force, and find that the basic move is to abstract persons and issues from the flow of history (this is not only a tactic; it is an article of neoconservative faith) so that real-world issues can be reduced to problems in a moral algebra: any policy that takes race into consideration is equivalent to any other policy that takes race into consideration, Nazis equivalent to Israeli hard-liners, Ku Klux Klanners equivalent to those who favor minority set-asides. In the making of such equivalencies, differences in power, motive, and morality just don't count; they are declared "accidental'' or ''extrinsic,'' and any attention paid to them is quickly labeled special pleading or flagrantly political thinking.

In ''You Can Only Fight Discrimination with Discrimination,'' I continue the analysis of the affirmative action backlash and point out that it assumes not only the desirability but also the possibility of a form of government or institutional arrangement that is fair to everyone because, for example, it is ''color-blind'' or presided over by a purely formal mechanism (e.g., SAT scores). The assumption, in other words, is that discrimination—the favoring of some groups over others—is a deviant practice and that the appropriate response to it is simply to eliminate it. I challenge that assumption by observing that whenever a policy of ''fairness'' or ''merit'' is put into place, those values will have been defined in ways that could be challenged by parties whose concerns were not uppermost in the minds of the policy's drafters. A policy that was fair to everyone could be devised only if everyone's interests and perspectives were the same; but if everyone were the same—believed the same things, envisioned the same future as a realization of the same hopes brought to fruition by the same agreed-upon means—there would be no problem because there would be no politics (politics after all is unthinkable apart from difference and faction), and the question of fairness would never arise. It arises because of the irreducible differences that make it impossible for anyone ever to *think up,* never mind implement, a policy that is universally equitable. Discrimination is not a deviant practice; it is the practice everyone is always and already engaged in. And when its particular effects are overturned by a plan devised specifically to remove them, that same plan will inevitably produce *new* discriminatory effects felt by persons whose interests are, for the moment, being slighted. You can only fight discrimination—practices that disadvantage some groups—with discrimination—practices that disadvantage some other groups.

This does not mean that all discriminatory practices are equal; all it means is that one cannot condemn a practice just for being discriminatory (since there are none that are not). Rather, one must consider the effects of a practice and attempt to calculate as best as one can the costs of either allowing it to flourish or moving to curtail it. Any such calculation will involve taking into account the history into which a new practice (always advertising itself as a reform) would insert itself; and in ''Bad Company'' I make a start at uncovering the historical antecedents and affiliations of the antimulticulturalists, who usually present themselves as the children of Reason rather than as the inheritors of a tradition that might be critically examined. What I find is a straight line between contemporary hostility to black studies, ethnic studies, gay and lesbian studies, and so on and the anti-immigration, anti-Catholic, frankly racist writings of the late nineteenth and early twentieth century—by and large the same fears, the same scapegoats, the same rhetoric. The only difference is that in 1992, "national unity" or "cultural coherence" or "harmonious citizenry" takes the place of outright appeals to racial prejudice; but although the language is softer and spoken more often than not by persons (like Arthur Schlesinger) innocent of consciously malign motives, it still serves racist ends in relation to which it stands as a code. "Speaking in Code" is the title and subject of the final essay written for the debates with D'Souza. The tone of this piece is more severe than that of its predecessors, for its target is not only the practice of sending coded messages but also the complicity, often witting, of those who receive them. The essay is a direct challenge to those of its readers (and they will be many) who will resonate positively to invocations of "fairness," "merit," "common values," "race neutrality," and the sayings of Martin Luther King, Jr. The charge is that by genuflecting in the presence of these culturally sacrosanct formulations, credulous audiences fail to inquire into the agendas of those who traffic in them; and the further charge is that for some the failure is deliberate, for it enables them to hold up clean hands (see, I'm just for equality of opportunity, what could be wrong with that?) while participating, at a safe remove, in dirty work.

I assume that many will find this argument offensive, and those who do will probably be even more offended by the two essays on free speech, which in my account joins "fairness," "merit," "color-blind," etc., as a concept that today bears more ideological baggage than its champions ever acknowledge. Recently Frederick Schauer, a noted First Amendment scholar, said the unsayable when he suggested that standard free-speech arguments have all the earmarks of an ideology because it is assumed in the society that counterarguments are dangerous and must be rejected by all right-thinking persons. There is, observes Schauer, "little free inquiry about free inquiry and little free speech about free speech" in the current climate ("The First Amendment as Ideology," *William and Mary Law Review* 33, no. 3 [1992]: -856). It is a paradox that an orthodoxy of tolerance is *in*tolerant of those "who have less protective rather than more protective views about freedom of speech" (863).

In fact it is not a paradox at all if your view of free speech is consequential or instrumental, that is, if you favor the largest measure of toleration because you believe that the proliferation of speech "will do us more good than harm over the long run" (Ronald Dworkin, "The Coming Battles over Free Speech," *New York Review of Books,* June 11, 1992, 56). Any such view will require that you specify the "good" whose protection or emergence will be promoted by a regime of free speech; but once such a good has been specified—be it the discovery of truth, or the realization of individual cognitive potential, or the facilitation of democratic process (the three most popular candidates put forward in the literature)—it becomes possible to argue that a particular form of speech, rather than contributing to its realization, will undermine and subvert it. This is so because in a consequentialist argument freedom of speech is not identical with the good but is in the service of the good; it is not a prime but a subordinate value, and when its claims conflict with those of its superior, it must give way. What this means is that insofar as you hold to a consequentialist view of free speech—insofar as you have an answer to the question "What is free speech *for?*"—you are already committed to finding in a particular situation that speech with certain undesirable effects should not be tolerated; and what that means in turn is that *there is no such thing as free speech,* because from the very start your sense of just how free speech should be is shadowed by your identification of, and obligation to, the good in whose name acts of speech are to be justified. "Free speech" always means for consequentialists "free speech so long as it furthers rather than subverts our core values"; and when an exception to a free-speech policy is made, it is not an anomaly or an afterthought but a continuation of the logic that has ruled the policy from the beginning.

It would seem, then, that strong free-speech proponents must take a nonconsequentialist, noninstrumentalist position, one in which freedom of speech is not subordinate to some other value or tied to the calculation of empirical effects but is asserted and honored simply for itself. "A non-consequentialist justification, says Kent Greenawalt, "is one which claims that something about a particular practice is right or wrong independent of the consequences" ("Free Speech Justifications, *Columbia Law Review* 89, no. 1 [January 1989]: 128). Ronald Dworkin puts some flesh on these bones when he declares that "freedom of speech is valuable, not just in virtue of its consequences but because it is an essential and 'constitutive' feature of a just political society that government treat all its adult members, except those who are incompetent, as responsible moral agents" (56).

The trouble, however, with a nonconsequentialist position is that no one can maintain it because it is always sliding over into consequentialism. Dworkin is a case in point. The doctrine of moral responsibility, he says, prevents government from withholding speech "from us on the ground that we are not fit to hear and consider it," on the ground that it will do some of us harm. Yet his own formulation contains two qualifications on the right of everyone to hear everything, the young ("'all its adult members'") and the incompetent, and these qualifications only make sense if he believes that the young and the incompetent are not capable of making good use of the speech that freely comes their way; but that means that free speech is envisioned as having a point external to itself—something like the furthering of rational deliberation—and that, it hardly seems necessary to say, is a consequentialist position. Dworkin gives the game away again when he acknowledges that his nonconsequentialist justification rests on values that "may be overridden in special cases; in deciding, for example, how far military information may be censored" (57). By failing to formulate a non-consequentialist justification that does not traffic in the calculation of likely effects (such as the effect of harming the young and the incompetent or the effect of weakening the nation's security), Dworkin establishes, against his own intention, that *any* coherent understanding of so-called free speech will be consequentialist.

This is a conclusion that Dworkin and other free-speech liberals are loath to reach because it amounts to saying that any understanding of free speech will be political; for in order to form a consequentialist position, you must choose some consequences—some vision of the way you want the world to be in the future—above others, and that choice, which will also be a choice of the "special cases" you are willing to recognize, will inevitably be opposed by those who would prefer other consequences and recognize other exceptions. Short of an absolutely absolutist position (which no one holds or defends), a line must be drawn between protected speech and speech that might in some circumstances be regulated, and that line will always reflect a *political* decision to indemnify some kinds of verbal behavior and devalue others. To be sure, the line will always be presented as if the political considerations were all on one side and the considerations of principle on the other, but the mark of politics will be on both sides.

The point is obvious if the line is drawn between so-called high-value and low-value speech, for someone must decide what is high and what is low, and someone else, were he or she in power, would decide otherwise; but the point holds even when the line is specifically drawn to rule out political considerations, as in the distinction between content regulation (regulation based on preferred and dispreferred ideas) and time-manner-place regulation. Supposedly the latter is indifferent to ideas and is merely a matter of maintaining order (e.g., by preventing sound trucks from going through neighborhoods in a manner that would disturb residents), but as the many challenges to such regulations show, their effect is almost always either to maximize the speech opportunities of certain favored groups (those who can afford to take off work or purchase amplifying equipment) or to minimize the speech opportunities of less favored groups (picketers, the homeless, the poor, neo-Nazis). One can of course contend that such patterns are merely the unintended consequences of a purely formal rule, but the rule—regulate time, manner, and place, not substance—cannot give content to its own terms, and the act of giving them content, of deciding what is and is not a matter of substance, will itself be substantive and will vary with agents differently positioned in the economic and social structures of society.

The moral is the one I draw in "There's No Such Thing as Free Speech, and It's a Good Thing, Too" and "Jerry Falwell's Mother": The First Amendment is not a self-declaring statement and will assume the form given to it by powerful and authoritative interpreters. And the moral that follows from that one is that the First Amendment does not in and of itself (finally a meaningless phrase) direct a politics but will display the political "spin" of whatever group has its hand on the interpretative machinery. "Free speech" is thus just like "fairness" and "merit"—rather than a concept that sits above the fray, monitoring its progress and keeping the combatants honest, it is right there in the middle of the fray, an object of contest that will enable those who capture it to parade their virtue at the easy expense of their opponents: we're for fairness and you are for biased judgment; we're for merit and you are for special interests; we're for objectivity and you are playing politics; we're for free speech and you are for censorship and ideological tyranny. It is a wonderful (not here a word of approbation) strategy, and if it is pursued as successfully as it has been in recent years by the neoconservatives, the result is to place the opposition in the difficult position of having not only to respond to arguments but to dispute the very vocabulary in which the issues have come to be framed, a vocabulary which, because it occupies the rhetorical high ground, stigmatizes counterarguments ("you mean you're against fairness?") even before they are heard.

That vocabulary is by and large the vocabulary of liberalism, and it is the structure of liberal thought that is my target in every one of these essays. Liberal thought begins in the acknowledgment that faction, difference, and point of view are irreducible; but the liberal strategy is to devise (or attempt to devise) procedural mechanisms that are neutral with respect to point of view and therefore can serve to frame partisan debates in a nonpartisan manner. I put the matter in this way so as to point up what seems to me an obvious contradiction: on the one hand, a strong acknowledgment of the unavailability of a transcendent perspective of the kind provided by traditional Christianity (against whose dogmas liberalism defines itself), and on the other, a faith (curious word to associate with liberalism) in the capacity of partial (in two senses) human intelligences to put aside their partialities and hew to a standard that transcends them.

The names by which that standard goes are the familiar ones we have already met: "fairness," "merit," "neutrality," to which we can add "the marketplace of ideas" as it operates in free-speech polemics. The marketplace of ideas is supposed to regulate in a purely formal way the contest between conflicting agendas; "purely formal" means "without regard to content," because the marketplace (sometimes called the forum of public discourse) leans in no particular ideological direction. It works, we are told, only to assure that each party will get its turn at bat. In fact, however, the marketplace has to be *set up*—its form does not exist in nature—and since the way in which it is to be set up will often be a matter of dispute, decisions about the very shape of the marketplace will involve just the ideological considerations it is meant to hold at bay. As Robert Post has put it in a scrupulous analysis, the fact that the designing of the marketplace of ideas will itself involve choices "between competing value judgments" leads to "the startling conclusion" that "the boundaries of a discourse defined by its liberation from ideological conformity will themselves be defined by reference to ideological presuppositions" ("The Constitutional Concept of Public Discourse," *Harvard Law Review* 103, no. 3 [January 1990]: 683). Post's analysis of the marketplace of ideas exactly parallels my analysis of "fairness," "merit," etc.; in each instance a term or a concept or a mechanism that is supposed to protect us from the infection of politics is itself infected in its very constitution (here both a noun and a verb).

This applies also to "Reason," the most popular and prestigious of the names given to the nonexistent center of liberal thought. Historically Reason emerges as the preferred alternative to the enthusiasms that fueled theological disputes, and, ultimately, religious wars, in the centuries before the Enlightenment. Religious stances are by definition dogmatic, a word that is usually heard as an accusation (one more indication of how thoroughly the ideology of Reason has made its way), but a word that, properly understood, identifies the unyielding commitment at the heart of any theology. But that is just the trouble according to the prophets of Reason: it is because they are unyielding that religious arguments are productive only of heated conflict; rather than submitting themselves to the cool light of rational inquiry, they maintain themselves obdurately (i.e., dogmatically) and leave those whose stand on them in a state of ideological intransigence. Reason, in contrast, privileges no ideology, sanctifies no point of view, but invites all positions to the seminar table, where each will be given the chance to be heard and assessed.

The appeal of this picture of Reason's operations is undeniable, and it is the same appeal that renders the marketplace of ideas and the notion of "fairness" so attractive: it promises release from ideological gridlock by providing a means of adjudication—the law of Reason—which, because it is hostage to no ideology, enables us to test the coherence of any ideology. It all sounds fine as long as you remain on the level of abstraction and do not descend to particular cases; for in a particular case, a real-life situation that requires you both to act and to justify your course of action, the imperative to Reason takes the form of giving *reasons,* and there is a huge and fatal disparity between the claim of Reason to be independent of ideology and the pedigree of any of the reasons you might think to give. This is so because the reasons you think to give will always be a function of the personal and institutional history that has brought you to a moment of dispute. If, for example, you ask me to give reasons for my reading of a poem, I would surely be able to do so; but the reasons I then proceeded to give would rest on assumptions—about the nature of poetry, the practice of literary interpretation, the difference between poetry and other forms of discourse or the absence of any such difference, and a thousand other things— that gave to those reasons the intelligibility and force they would have for *me* and for those whose intellectual and institutional history was commensurate with mine. Others would not hear them as reasons at all, at least not as *good* ones, but would hear them as mistakes and perhaps, were the dispute heated enough, as instances of *ir*rationality. At that moment the appeal to Reason will have run its course and produced the kind of partisan impasse from which Reason supposedly offers us an escape.

Why not, then, one might ask, go deeper (in the wonderful world of rationality accidents are always being peeled away in order that essences might be revealed) and put those underlying assumptions on the table so that *they* could be scrutinized and assessed? To this suggestion I would pose a simple question: if you propose to examine and assess assumptions, what will you examine and assess them *with*? And the answer is that you will examine and assess them with forms of thought that themselves rest on underlying assumptions. At any level, the tools of rational analysis will be vulnerable to the very deconstruction they claim to perform. You can never go deep enough, for no matter how deep you go, you will find reasons whose perspicuity is a function of just those factors—institutional history, personal education, political and religious affiliations—from which Reason supposedly stands apart. (There is no slide to nihilism or relativism implied here; "going deep" is an analytical action performed by philosophers and metacritics in the privacy of their seminars; outside those seminars they, like the rest of us, move quite nicely on the very ground they have deconstructed.)

What this means is that whenever Reason is successfully invoked, whenever its invocation stops the argument and wins the day, the result will be a victory not for Reason but for the party that has managed (either by persuasion or intimidation or legerdemain) to get the reasons that flow from its agenda identified with Reason as a general category, and thereby to identify the reasons of its opponents as obviously *un*reasonable. Like "fairness," "merit," and "free speech," Reason is a political entity, and never more so than when its claim is to have transcended politics. The liberal dream of a purely formal mechanism of adjudication, one whose authority rests on a pedigree of procedure rather than content, will always have more content than it can acknowledge, even to itself, *especially* to itself. And if this is the case, it is too little to say that liberalism doesn't work; one should say rather what I say in the essay of the same name: liberalism, bereft of its formal center, doesn't exist.

What You Say Is What You Get

"Liberalism Doesn't Exist" is the last of the essays in part 1 of this book, and it sums up the argument of the rest: the key terms invoked by neoconservative polemicists in the recent "culture wars," terms that come to us wearing the label "apolitical"—"common values," "fairness," "merit," "color-blind," "free speech," "Reason"—are in fact the ideologically charged constructions of a decidedly political agenda. I make the point not in order to level an accusation but to remove the sting of accusation from the word *politics* and redefine it as a synonym for what everyone inevitably does. The essays in the second half of the book pretty much take the unavailability of nonpolitical modes of being as a given and go on to the next question: What follows? What does this mean for the way we live? My answer is "not very much," and it is an answer that will be distressing both to the forces on the intellectual left and to their opponents. The right will find it distressing because they won't believe it and will refuse to take any comfort from it. Surely, they will insist, a life everywhere impinged on by politics will be a life devoid of morality, decency, and good order. Those on the left will find it distressing because in their view the insight that words like *merit, fairness,* and *Reason* always clothe a political agenda should generate a wariness and a skepticism about truth claims that will lead to a kinder, gentler mode of interacting with our fellows; for now that we have unmasked the pretensions of an absolutist vocabulary, should we not abandon that vocabulary and substitute for judgments of right and wrong, true and false, softer formulas ("I see your point," "I respect what you have to say") that will at once reflect and contribute to a growing tolerance of diverse points of view? The result, we are promised, will be a brave new world in which no one enslaves and tyrannizes anyone else and everyone's opinion is accorded a respectful hearing.

Although this vision of an ever more pluralistic future seems diametrically opposed to the neoconservative insistence on identifying and hewing to the one true way, the two are in fact mirror images of each other because they both require the same (impossible) act of will. That is, they require us to separate ourselves from our currently held convictions so that we might either ally ourselves with an ahistorical truth (the imperative of the right) or float free of all truths and remain open to innumerable voices (the dream of at least one part of the left). To be sure, there is a difference between resolving to transcend our beliefs in order to embrace a truth beyond belief and resolving to loosen the grip of our beliefs so that we will not be led by any of them to perform invidious acts of judgment and exclusion; but in both cases it is the resolve that is crucial and raises the crucial question: With what does one either transcend belief or loosen its hold? The answer can only be with a part of the mind that is itself not already occupied by belief, some aspect of the self that stands to the side of commitments and affiliations. It is an answer that makes sense if you hold a strong view of Reason or fairness or merit such that these or similar entities provide a space free of partisan entanglements; but if you hold the reverse view and insist, as I have been doing in these pages, that Reason, fairness, and merit only come in partisan shapes, it is simply a contradiction to promote as an agenda an escape from entanglements, since by your own argument there is nowhere to escape *to* and no means by which this escape-without-destination might be effected. The knowledge that what moves you is a historically limited picture of the way things are or should be does not render that picture less compelling, for it could only do so if you were holding out for a picture that was *not* historically limited, and if you were to be doing that, you would have become indistinguishable from your essentialist opponent. Paradoxical as it may seem, those who are persuaded by the arguments of this chapter can do nothing with them—can derive from them no program or set of marching orders—without being unfaithful to them. The most you can do is employ them as I have here, as a means of denying to those on the other side the high ground of purity so essential to their polemic; but once that has been done (and I urge that it be done unceasingly), you cannot claim a new purity that has somehow arisen, like a phoenix, from the ashes of absolutist discourse.

This is the hardest of lessons for the cultural and intellectual left, whose members want very much to think that what they take to be their epistemological sophistication gives them an advantage over their adversaries and makes their reasons different *in kind* from the reasons of those who retain a faith in objectivity. But this is a flat-out misreading of the lesson antifoundationalism preaches, for if all arguments are inevitably intermixed with policy and therefore challengeable, no argument, even the argument that all arguments are inevitably intermixed with policy, can claim an epistemological superiority that would give its proponents an advantage independent of the hard work of presenting evidence, elaborating analogies, marshaling authorities, and so on. It is *because* all arguments owe their force to contingent historical factors that no meta-argument can make contingency a matter either of suspicion or of celebration; contingency is a given and can count neither for nor against an argument; any argument must still make its way by the same routes that were available before contingency was recognized as a general condition.

This conclusion will seem counterintuitive both to those who fear that a strong historicism will deprive our practices of their normative justifications and to those who hope that a strong historicism will deprive our practices of their normative justifications. The essays in part 2 undermine both the hope and the fear by arguing that a strong historicism leaves our practices precisely where it found them, resting on the bottom of their own histories. I begin with the law, which has always foregrounded norms and mechanisms of justification but has lately been the object of a critique that declares it to be without either. I refer to the Critical Legal Studies movement, a left-leaning segment of the legal academy whose members argue that legal reasoning is not a formal mechanism for determining outcomes in a neutral fashion but is rather a ramshackle ad hoc affair whose ill-fitting joints are soldered together by suspect rhetorical gestures, leaps of illogic, and special pleadings tricked up as general rules, all in the service of a decidedly partisan agenda that wants to wrap itself in the mantle and majesty of THE LAW.

This critique has an obvious relationship to the position I have been elaborating in these pages, and in many ways the first two parts of "The Law Wishes to Have a Formal Existence" unfold in the manner of a standard CLS analysis. In the first part of the essay, I examine the parol evidence rule, a device designed to constrain the scope of interpretation in contract cases, and conclude that the rule necessarily fails of its purpose because it demands a restraint (do not look beyond the four corners of the written contract) no one could possibly exercise. In the second part, I turn to "consideration," a formal requirement of contract formulation intended to prevent judges from substituting their own terms for the terms agreed on by the contracting parties. As the case history shows, however, judges have no difficulty recharacterizing the shape of consideration so that it supports the conclusion they wish to reach. Indeed, directions for performing this interpretive feat are built into contract doctrine itself, which turns out in my analysis to be as porous and plastic as poetry or abstract art. Nevertheless, in the third part, I refuse the conclusion to which my examples seem to be pointing, the CLS-style conclusion that the law is a sham or an elitist conspiracy, and assert instead that these very features of the law, even though they are in tension with the law's "official story," are what enables the law to perform its task, the task of advertising its actions as following faithfully from general principles of justice, due process, impartiality, and so on while at the same time tailoring and remaking those principles in accordance with the pressures exerted by present-day exigencies. The law, in short, is always in the business of *constructing* the foundations on which it claims to rest and in the business too of effacing all signs of that construction so that its outcomes can be described as the end products of an inexorable and rule-based necessity.

By regarding the rhetorical and legerdemain of the law as an achievement (as an "amazing trick") rather than as the matter of scandal, I forestall the call to reform that often accompanies CLS analyses. In these analyses, the direction of reform is said to follow from the debunking of objectivist claims: since we now see that the vocabulary of legal reasoning is circular, question begging, and endlessly manipulable, we should either abandon it for something more answerable to the demands of justice or employ it "under erasure," that is, with a tentativeness and reserve in keeping with our knowledge of its lack of an independent justification. But if we were to abandon the vocabulary of law for something else, then we would have that something else and we would no longer have law; and as for retaining the vocabulary while at every moment questioning its legitimacy, the result would not be an improved legal practice but the obstruction of legal practice, since in the process the goals of legal practice—the rendering of decisions and the giving of remedies—would have been forever deferred.

In making this argument I want to drive a wedge between the project of subjecting the law to a philosophical or metacritical analysis (what Peter Goodrich calls "reading" the law) and the activity of simply (or not so simply) doing law. I want to say that it is not the case, as many assume, that the first underlies or is a prerequisite of the second. Rather, the two are different practices—the practice of lawyering or judging and the practice of reflecting on lawyering or judging—and the price of confusing them is the loss of one's grasp of what makes the law distinctive, what makes it law and not something else. Here I touch upon the issue that dominates the next four chapters, the issue of disciplinary autonomy. In recent years it has become fashionable to question the autonomy of disciplines on the grounds that they always borrow terms, concepts, data, methods, and a great deal else from one another and therefore cannot be said to be identical with themselves in any strong sense. (The point is a Derridean one.) This, however, is to mistake the nature of autonomy, which is not a matter of refraining from commerce but of stamping whatever is imported or appropriated with a proprietary imprint. While it is true that disciplines do not originate much of what appears in their operations, it is not the materials they traffic in that makes for their distinctiveness, but the underlying purpose or point in the context of which those materials acquire a disciplinary intelligibility. Autonomy, I explain in "Play of Surfaces," *requires* the incorporation of foreign elements, which once incorporated—seen in the light of the discipline's underlying point or purpose—are no longer foreign. Autonomy is a social and political achievement (rather than something initially given), and it can only maintain itself by reconfiguring itself in the face of the challenges history puts in its way.

This means that disciplinary autonomy is not stable, and it will seem paradoxical to some that instability could be the source of identity, that the law, for example, could be continually in flux and still be said to be the law. The paradox vanishes, however, when one realizes that the law (or literary criticism) does not remain what it is because its every detail survives the passing of time, but because in the wake of change society still looks to it for the performance of a particular task. That is what survives, a distinctiveness that rests not on an essential difference but on a difference in the sense of purpose (to secure justice or to interpret poems) informing those who think of themselves as lawyers or literary critics. Even when a practice is racked by debate, it is nevertheless itself so long as what is being debated is the best way to do the job the culture continues to assign to it.

The recharacterization of disciplinary autonomy as a matter of self-modifying practice rather than of unchanging essences is in the tradition of pragmatism as it is described in "Almost Pragmatism." In the pragmatist vision cores and (temporary) essences are made, not found, and what makes them are the verbal and rhetorical resources that a realist epistemology devalues. If, as Richard Rorty puts it, "there is no way to think about . . . the world . . . except by using language" (*Consequences of Pragmatism*, [Minneapolis, 1982], xix), the world that appears to us will be a function of that language, and we cannot alter the one without altering the other. Since the features of the legal (or any other) landscape come into view by virtue of the descriptive terms certified practitioners unreflectively employ, the removal of those terms will not clean up that landscape but depopulate it, or repopulate it with the terms, and therefore with the entities, of some other enterprise. Insofar as you value the job being done by a particular enterprise—be it law, literary criticism, or anything else—it behooves you to retain and strengthen the vocabulary that marks its distinctiveness; for you can not get rid of the vocabulary without depriving yourself of the resources (including resources of *action*) it makes available. In a pragmatist world what you say is what you get.

It follows from this analysis that the narrowness of disciplinary vocabularies is in fact their strength, and it is a strength that is dissipated to the point of enfeeblement in the project I call "interdisciplinarity." In "Being Interdisciplinary Is So Very Hard to Do," I distinguish between interdisciplinary work—the practice of borrowing materials from fields other than your own—and interdisciplinarity—the hope that by moving from field to field and back again you can enlarge the boundaries of your consciousness and become a more clear-sighted human being. The first, I argue, is just business as usual and involves no metaphysical claim; the second depends on such a claim, the claim that one can by an act of the will see through the limitations of one's professional practice and come to engage in it without being confined within its imperatives. What is promised by the gospel of interdisciplinarity is a kind of double vision: with one eye you perform as a literary critic or a judge or a physicist; with the other you focus on the concerns and values that are shut out by the totalizing impulse of the discursive forms you are employing. By inhabiting a mental space more capacious than that allowed by any particular discipline, you defeat the tendency of disciplinary vocabularies to suppress and remove from memory everything that is inconvenient to their narrow assumptions and purposes.

My response to this is simple. First of all, there is no such mental space: thinking and the actions that follow from thinking are only possible and conceivable within some demarcated field of reference. The blurring of boundaries does not improve thought but incapacitates it, except when the blurring is a sign that *new* boundaries have been established within which thought and action can go on in a different but still perspective-dependent way. Second (and this is the same point from a slightly different angle), it is only by *not* keeping everything in mind, by *not* remembering all that a particular vocabulary (the only kind of vocabulary there is) would exclude, that thoughts can form and generate performance. This is what H. M. Collins means when he declares that "the objects of science are made by hiding their social origins" (*Changing Order: Replication and Induction in Scientific Practice* [Chicago, 1992], 188). Collins is discussing the relationship between scientific practice proper—what scientists do in their labs— and the sociology of science—the excavation of the underlying social conditions that make scientific practice possible. His point is that in order to engage in either of these projects one must rigorously exclude—that is, forget—the imperatives of the other: "Science—the study of an apparently external world—is constructed by not doing the sort of thing that sociology of scientific knowledge does to science." Critics of Thomas Kuhn think it a mark against him that his work is not taken seriously by scientists, but as Collins makes clear, one cannot *be* a scientist and take Kuhn seriously, "for science would not make sense as an institution unless it were normally the case that acting scientifically meant acting as though the sociology of science were not true" (190). Acting scientifically means acting on the assumption of a determinate nature waiting to be described by a neutral observation language; acting sociologically means acting on the assumption that nature is socially constructed by the very speech acts of which it is supposedly the cause. Everything about the two practices—their respective facts, discovery procedures, mechanisms of justification, and so on—is different, and the attempt to unite them will result only in confusion and a loss of focus.

Focus cannot be expanded—made to take in more things; it can only be adjusted with the resulting (and inevitable) gains and losses: some things seen more clearly, other blurred or obliterated. The hope that focus can be expanded and that as a consequence we might become less bound to the perspectives provided by our personal and institutional histories is the left's version of the right's desire to (re)institute the authority of "truths that pass beyond time and circumstance." One party (the Lynne Cheneys of this world) seeks its escape in a heaven of rational forms as (supposedly) embodied in a list of great books. The other party (the various apostles of interdisciplinarity) seeks its escape in a liberal utopia of enlarged sympathies and nonjudgmental (i.e., ever more tolerant) mental processes. Each for different reasons (finally not all that different) rejects the narrowness of "merely" disciplinary work and tries to move away from it, either by simply (ha!) rising above it or by endlessly complicating it. In either case, however, the escape is illusory. The one party "escapes" only to a previous state of the temporality it scorns when it identifies the period of its own youth as the Golden Age. The other "escapes" to ambitious projects of reconciliation and cooperation, projects that in fact reconcile nothing and sacrifice the real advantages of local intelligibility to the empty dream of nonexclusionary ways of knowing. (The programs of rationality and of toleration, while often opposed, are manifestations of the same wish to detach the mind from its historical choices.)

In some of its versions, the New Historicism is such a project. New Historicists assert, in Louis Montrose's words, "the textuality of history, the historicity of texts" and thereby refuse the claims of either historians or literary critics to be engaged in a discrete activity with its own laws and sphere of operation. On the one hand, the argument goes, any access to the past is mediated by the contestable vocabulary of description a historian chooses to employ; therefore the "facts" of the resulting narrative are facts only in relation to that vocabulary, and the historian stands on no more firm a ground than does the author (or interpreter) of a lyric poem. But on the other hand, the poet does not create his artifact out of whole cloth while ranging freely in the zodiac of his own wit; rather, he is situated amidst a network of material practices—of government, commerce, education, manners—that constrain his efforts and (to a large extent) determine their significance. A lyric poem is as much a social and historical fact as it is a fact of the "imagination" (a concept that tends to drop out in New Historicist formulations).

I find this analysis persuasive, but I do not find persuasive the conclusion often drawn from it, that works of literature "so called" should be understood in terms of the larger network of social forces rather than in the terms internal to a Platonizing aesthetics; for if one focuses not on the literary text as it presents itself in the field and history of aesthetic productions but on the cultural text within which it emerges, the literary object rather than being made more perspicuous will have been made to disappear. Just as the sociologist of science asks questions the scientist can not afford to ask (because they would deny him the object central to his discipline), so does the sociologist of literature ask questions—In whose interest is it that this poem should appear in 1595? In what agenda of economic exploitation does the country-house poem participate?—whose answers turn him away from those very features that mark an object as poetic. A New Historicist might reply by declaring that there is no distinct "poetic" realm and that what is and is not poetic is a function of historical forces; and I would counterreply that while it is true that distinctiveness of the "poetic" is a historical rather than an essential fact, it is nevertheless a fact and one whose implications for analysis and description cannot be ignored, especially by persons who proclaim the bottom line status of historical formations. If poets who take up their pen in 1595 do so with their heads full of Virgil, Theocritus, Aristotle, Cicero, Petrarch, Ariosto, Dante, Chaucer, and Lydgate, any assessment of their achievements must begin with and stay close to those materials and to the tradition (as historical as any) that has produced them. To be sure, one can step back at any time from that tradition and replace its questions with the questions of some other project, political, economic, theological, whatever; but the exercise will not demonstrate that poetry is really "about" any of those things, only that it is always possible to interrogate an artifact in ways that remove it from the context of its primary intelligibility.

But why would anyone want to do that? The answer is to be found in the ambitions of the New Historicism as I have described them in "The Young and the Restless." The chief ambition is to insert literary productions and commentaries on them into the mainstream of political life, where they can help alter the basic structures of society. The reasoning is that you can accomplish this goal by demonstrating that literary productions have always and already been implicated in political agendas, that, for example, the country-house poem, far from being the pleasantly idyllic representation of rural life, was an important component of a program of agrarian exploitation by the ruling class. The reasoning fails, however, on precisely political grounds; for it is a fact of political life that recharacterizations of literary works, including new accounts of their genesis, have no impact beyond the circles in which such recharacterizations are fungible currency. A member of an English department may well have his or her mind changed by a revisionary analysis of the pastoral, but a member of the general public will not even understand what is at stake, will not know what the pastoral *is* (or used to be), and will certainly not be moved to alter his or her views on the economy, or abortion, or affirmative action, by what has recently been argued in the pages of *Representations* or *Signs*. The truth, sad or happy as you may happen to find it, has been declared with flinty brevity by Evan Watkins. Literary study, he says, "occupies a marginal position in the larger, organizing apparatuses of cultural production and circulation within the dominant formation" (*Work Time: English Departments and the Circulation of Cultural Value* [Stanford, 1989], 271). No revision of the internal structure of literary study will change that position or bring literary critics into the "dominant formation."

Indeed, it is worse than that. Those who conflate and confuse literary with political work end up doing neither well. The literary critic who has his or her eye on the possible connections between textual features and issues in the body politic will lose the focus provided by the routine practices of his or her discipline and will gain only the license to be irresponsible. This, I argue, in "Milton's Career and the Career of Theory," is exactly what has happened in that corner of Milton studies that has responded to the New Historicist imperatives. In the essays I examine, the rush to find large significances in small poetical moments leaves those moments underscribed and overtheorized. Although the New Historicist polemic emphasizes the virtue of hard archival work, New Historicist assumptions permit interpreters to get away with doing almost no work at all of the kind that would result in persuasive arguments as opposed to discrete, ad hoc speculations. Persuasive arguments are arguments that can be seen as advancing a project whose goals are clearly articulated. Persuasive arguments are disciplinary arguments.

I have been making disciplinary arguments for thirty years, and I pay tribute to them and (I must confess) to myself in "Milton, Thou Shouldst Be Living at This Hour." To some extent I was given permission by the occasion, an evening on which I was the recipient of the highest honor conferred by the Milton Society of America. The moment was very personal and very professional, and it was obvious that I had long since become incapable of distinguishing one from the other. That indeed was the point of the address, and it is therefore perfectly continuous with my defense and praise of labor in the disciplinary vineyards.

"The Unbearable Ugliness of Volvos" offers a darker view of those same groves, one that foregrounds the perverse psychology by which academics transmute self-inflicted pain into treasure. I begin by asking why it is that so many academics seem devoted to an automobile that is at once ungainly and expensive, and I develop the answer into a dyspeptic analysis of academic practices. For a time it seems that the ruling spirit of the performance is cynicism, but in the end cynicism is put in the service of a frankly political appeal of just the kind I have so often belittled. I tell the members of my audience (gathered for the annual meeting of the English Institute) what they already know, that they are the targets of a concerted attack against the humanities. I urge them (albeit indirectly as befits a literary person) to forgo the pleasures of humiliation and fight back; and I predict that in the election year to come the battle will be intensified (as indeed it proved to be under the banner of family values and the assault on the "cultural elite"). The posturing of the piece is shameless, and the shamelessness is doubled (here tripled) when I flaunt it instead of apologizing for it. I haven't the slightest idea if what I did that day had any effect, but as I say in the closing sentence, it is something I have always wanted to do.

I cannot end this introduction without making one final observation that returns us to the perspective of the essays in part 1. Late in "Milton's Career and the Career of Theory," I fault the New Historicism for its inattention to the demands of evidence and argument and accuse it of being overly "enamored of its political correctness." When I made that judgment in 1989, "political correctness" had not yet become a code phrase for the supposed sins of the academic world. My use of it was simply descriptive of what I took to be the misplaced emphasis of a criticism that had strayed from its proper (professional) course. That is to say, my use of the phrase was innocent. Those were the days.

PART I

2

THE COMMON TOUCH, OR, ONE SIZE FITS ALL

When Robert Penn Warren died in the fall of 1989 the *New York Times* printed a lengthy and admiring obituary in which it was noted, among other things, that Warren was greatly influenced by John Crowe Ransom. No doubt the Ransom influence extended to many matters, but the *Times* chose to highlight only one. Ransom, it informed us with authoritative solemnity, "once pointed out the impoverishment of modern life and the handicap to a writer in the destruction of commonly held myths that had been the heritage of the Western world" (16 September 1989, p. 11). What is curious is not that Ransom's observation (if it is his; there is no reference) is remarked, but that it has almost no relationship to what the rest of the article goes on to say about Warren, and indeed it is only by way of an obviously strained transition that the journalist is able to return to his or her putative subject.

If this is a puzzle, it is a small one that registers only on someone like me who is watching for signs. The puzzle becomes less puzzling and the signs more foregrounded when one turns to another vehicle of high-middlebrow journalism, *The New Yorker* magazine, and finds in the 21 August 1989 issue two reviews that appear on consecutive pages. The first, by Brad Leithauser, is a review of the fiftieth anniversary reissue of John Steinbeck's *The Grapes of Wrath*. The tone is set from the very first sentence when Leithauser reminds us that Steinbeck quarried his titles and epigraphs from the Bible, hymnals, Shakespeare, Milton, Burns, and Blake. This practice is immediately read as a "signal of the grandeur [Steinbeck] self-consciously aspired to," and we are not surprised by the subsequent

Previously printed in Darryl J. Glass and Barbara Herrnstein Smith, eds., *The Politics of Liberal Education* (Durham, N.C.: Duke University Press, 1992). Reprinted with permission of the publisher.

judgment that this grandeur was seldom if ever achieved. More than a hint of what is to come is provided when the list of Steinbeck's strengths is barely distinguishable from the list of his weaknesses: on the one hand, "sympathy for the disenfranchised, moral urgency, narrative propulsion," on the other, "repetitiveness, simplistic politics, sentimentality."[1] Only the asymmetry of the order in the two lists prevents us from seeing that sympathy for the disenfranchised (strength) is the same as sentimentality (weakness), narrative propulsion the same as repetitiveness, moral urgency the same as simplistic politics.

As it turns out, *any* politics is regarded as simplistic by this reviewer, who tells us, first, that one could aptly describe *The Grapes of Wrath* as a novel about the homeless and, second, that nevertheless it is "old fashioned, especially in its very willingness to tender solutions for the social problems it documents." The apparent paradox of downgrading the book because it is both up-to-date and out-of-date is resolved when these two categories are subsumed under the larger category of "popular fiction," ever up-to-date and therefore soon dated: ". . . how at home the novel would seem on a current *Times* best-seller list, with a blurb reading something like 'Three generations of a dispossessed Oklahoma family head west toward hope.' " In the context of this judgment the fact that "the book remains one of the best selling American novels of all time" reads as an indictment. Mixing faint praise with genteel condescension, Leithauser moves in the following pages to a final assessment: Steinbeck leaves us with "a regretful sense . . . of how much better a writer he might have been"; as it is, his book only "occasionally offers one of the rarest and most gratifying pleasures that literature opens up to us, . . . that little miracle of transformation by which . . . a stick figure becomes an Everyman."[2] This last word says it all: a work rises to the stature of serious literature only if it transcends the local concerns that inspire its author.

If Leithauser's assumptions remain implicit, they are explicitly proclaimed in the book reviewed on the next page, Robert Alter's *Pleasures of Reading in an Ideological Age*. Alter's thesis is in fact adumbrated in his title: the true pleasures of reading—the pleasures of contact with a "transhistorical human community" through the medium of "the common stuff of our human existence"—are opposed to "the explicit ideological commitments" of "political systems."[3] The thesis is warmly approved by the anonymous reviewer, who, with an irony that escapes him or her, calls it "timely": "This timely study takes issue with contemporary schools of literary criticism that maintain that the literary canon is an instrument of domination, or that literary works are no different from other verbal communications, or that a literary text has no meaning."[4]

It is tempting to linger on the inaccuracy of this characterization of the contemporary critical scene, to point out, for example, that a relationship to particular political agendas is only one of the things predicated of the canon by advanced critics, or that typically theorists do not deny the difference of literary works but inquire into the production and revision of that difference by social and political forces, or that it is Alter who deprives literary texts of their meaning by regarding their local contexts of reference as discardable and beside the literary point. But I will resist the temptation and focus instead on the relationship between the two apparently dissimilar reviews, one dismissive of popular or middlebrow culture, the other taking potshots at the alleged absurdities of the high academy. What links them is their joint affirmation of a supposed *common* ground in relation to which popular culture and contemporary literary theory are alike passing fads, deviations from the main path. Both political causes and scholarly agendas constitute mere fashion, and as George Steiner (the author of another book I shall be examining) puts it (he borrows from Leopardi), "Fashion is the mother of Death,"[5] by which he means that what is fashionable—of local and temporary urgency—takes our attention away from what abides, what is central, what is common. So far is Alter from being fashionable that, as the reviewer notes, he declines even to *debate* "Marxism, feminism, structuralism, or deconstruction." As we shall see, this strategy (replicated by the reviewer) is typical of those who oppose the common to the merely ideological, and one can understand why: if what you are affirming is basic to human experience, argument is unnecessary; for the deficiencies of these exotic programs will be obvious to anyone with, as we say, a bit of "common sense." The review of Alter's book (shorter than my analysis of it) ends with the double praise of the author's "lucid and moderate way" and of literature for preserving "from generation to generation" the "diversity, passions and playfulness" of "human beings."[6] The curious—even bizarre—word in this encomium is "diversity." What does it mean? What could it mean? How do you celebrate diversity in a context that affirms the common so strongly?

These and other related questions are raised often in the twelve brief essays that appear in the summer 1989 issue of the *National Forum*. The agenda is set in the introduction by editor Stephen White when he affirms "the notion that we need consensus on some of the things we need to know." He is followed by William Bennett, former Secretary of Education, who laments "the disappearance of a common curriculum in many of the nation's colleges and universities, and the resulting failure of many students to acquire . . . even a rudimentary knowledge of the civilization of which they are both products and heirs." A few pages later Lynne Cheney, head of the National Endowment for the Humanities, offers praise of Columbia University's core curriculum with its "remarkable stability"; Chester Finn,

former Assistant Secretary in the Department of Education, concisely sum-
marizes the group concern when he declares that "the foremost job of formal education is to teach our children—all of them—about those things we
have in common." From another perspective Elizabeth Fox-Genovese wor-
ries that the "extreme claims of feminism . . . risk undermining any aspiration to common standards and a common culture, including a common
ideal of justice," and she concludes by warning that "without some sem-
blance of a collective culture and of common ideals, we are left without a
common basis from which to defend the claims of the individual against
oppression."[7]

No doubt this celebration of the common is intended to be reassuring and
even benign, but to these ears at least, phrases like "collective culture"
and "common curriculum" have a disconcerting sound, for they suggest
the imposing of special and indeed *un*common standards on the very persons whose commonness is supposedly being affirmed. Here we meet in
the context of educational policy the familiar problem at the heart of liberal
politics: how to reconcile the exercise of authority with the very values—
freedom, tolerance, diversity—supposedly protected by that authority. By
and large, the contributors to this issue of the *National Forum* evade the
problem by assuming its solution in an unexamined invocation of the
"common"; that is, in order to get their agenda off the ground they assume
that the *content* of the category "common" is uncontroversial; and once
this assumption is in place, the claim to be taking account of, and even
honoring, diversity can be asserted with apparent coherence. The deep *in*-
coherence of the claim is embedded in a sentence I have already quoted:
"The foremost job of formal education is to teach our children—all of
them—about those things we have in common." The question, suppressed
by this formulation, is "when?" That is, when do "we" have these things
in common? The present tense "have" suggests, indeed insists, that the
common is *already* ours, but if that were so there would be no "job" for
formal education to do. The sentence fudges the real relationship between
its components: education comes first; the common, or rather *some* common, is its product. Our children—"all of them"—do not begin with shared
perspectives; they are to be brought to the perspectives common to some of
us by a process in which the perspectives they may have shared, had in
common, with others of us are either expunged or marginalized.

There is nothing necessarily scandalous about this; the normative and
therefore coercive force of education is, or should be, a given. The point
is worth making only because Finn and others assume that the category of
the common is uncontroversial, a matter of what everyone easily sees, of
what is so obvious and perspicuous that only the most skewed perceptions
could miss it. This assumption is not argued for but merely asserted in

words like "have" in "have in common," for a reason I have already noted: to argue for the common would be to acknowledge that it is *arguable*, a matter of dispute, and as such incapable of serving as the self-evident baseline in relation to which supposedly uncommon views are identified and stigmatized. In place of argument polemicists like Finn render judgments that (in their view) only madmen would gainsay: "Democracy is not just different from totalitarianism; it is better. Freedom of expression beats censorship all hollow, and freedom of worship is preferable both to Inquisitions and to state-enforced atheism."[8] Many of us would agree with these judgments; nevertheless they are not so *indisputably* true as Finn implies. First, the terms are loaded—labels like "totalitarianism" and "censorship" are not the names of practices one is submitting for judgment but of practices that have already been judged; second, even if the terms are not challenged, and censors or inquisitors are willing to identify themselves as such, it would still be possible to debate the value of the actions they perform. Military establishments (including the establishment of the United States) will often insist, and with reason, that censorship is necessary if the nation's defenses are to be maintained; and in times of war inquisitional techniques are engaged in even by democracies. On the other side, democracy has often been derided (in our era, in our country) as mob rule, and in the minds of some feminists, freedom of expression licenses violence against women in the form of pornography.

I am not endorsing any of these positions, but merely pointing out that they can and have been put forward, and put forward by people who are not obviously eccentric or insane. Nor am I denying the possibility of judgment, the possibility of deciding, with William Bennett, that "some forms of government are better than others,"[9] but merely observing that any judgment one might make in that direction is disputable, and disputable by persons no less well *educated* than you or I. The moral is concisely stated by R. T. Smith, another contributor to the *National Forum,* who began, he reports, as a believer in "the test of time," but now realizes that "he had been camping on embattled ground all along."[10] What I want to say is that it is *all* embattled ground and no less so when it is labeled as "common" ground, as something firmly under everyone's feet. Someone who says to you, "This is *our* common ground," is really saying, "This is *my* common ground, the substratum of assumptions and values that produces *my* judgments, and it should be yours, too."

Now this is not an impossible proposition; it could easily happen that you were persuaded to exchange your assumptions and values for someone else's; but in that event, the ground you would then stand on would be no more common, in the sense of being shared by everyone, than the ground you had left behind. The common, in short, is a contested category; its content will vary with the varying perspectives of those who assert it. When Finn declares that the "primary subject matter of public schooling" is "the commonalities, not the differences," he thinks that the difference between the two is easy to tell; and indeed it is, from the particular point of view that Finn (or any particular person) now occupies; but from another point of view the difference will be different; that is, what is the same or common will itself be different and what is different from the differing commonalities will be different. I multiply the differences in such a dizzying way in order once again to make my only point: *it is difference all the way down;* difference cannot be managed by measuring it against the common because the shape of the common is itself differential.

Indeed, the common is differential—relative to the context or interpretive community within which its shape is specified—even when its supposed basis is difference itself. The identification of difference as the common ground on which we all (should) stand might seem to be a powerful and coherent response to the arguments of Finn, Bennett, and Cheney, but it is a response that falls into the very error it would correct. Consider, for example, the essay authored by Gerald Graff and William Cain, "Peace Plan for the Canon Wars," originally published in *The Nation* and reprinted in the *National Forum*. Graff and Cain are expanding on Graff's now familiar directive, "teach the conflicts," by which he means "structuring the curriculum around . . . conflicts." Under current pedagogical conditions, Graff and Cain complain, students see only the preferred unifying project of particular instructors; they are thus "spared the unseemly sight of their teachers washing their dirty linen in public," but the price they pay for this "peace and quiet" is intellectual sterility.[11]

Cain and Graff want us to abandon this form of academic gentility and put conflict at the center of the curriculum. But if conflict is made into a structural principle, its very nature is domesticated; rather than being the manifestation of difference, conflict becomes the theater in which difference is displayed and stage-managed. Once a line has been drawn around difference, it ceases to be what it is—the remainder that escapes the drawing of any line, no matter how generous—and becomes just another topic in the syllabus. By making difference into a new "common" ground Cain and Graff succeed only in evading the lesson of its irreducibility. Strange as it may seem, the effect of bringing difference into the spotlight front and center is to obscure its operation, to hide the fact that the perspective from which one thinks to spy difference is itself challengeable, partisan, conflictual, differential.

Difference is evaded in a different but related way by Betty Jean Craige when, in apparent contrast to Graff and Cain, she counsels "holistic" rather than "oppositional" thinking. "Holohumanists," she tells us, will abandon the dualism of objective versus political points of view, and in the context of a full acknowledgment of the relativity of value, will "teach students to recognize and welcome cultural difference." To that end she urges a curriculum that "would foster a tolerance—and, ideally, an appreciation—of cultural beliefs and behaviors different from our own." [12] This seems admirable, but one must ask from what perspective this recognition of cultural difference will occur, or (it is the same question) from what position the tolerances one is supposed to foster will be identified? Only two answers to these questions are possible and they both subvert Craige's project. If, on the one hand, the perspective from which cultural differences are to be recognized is a perspective shared by everyone—is no *perspective* at all— one would then be claiming for it precisely the objectivity and universality Craige declares "not possible"; and if, on the other hand, the perspective is, in fact, a perspective—shared by some but not by all—the tolerances it encourages one to foster would not be recognized *as* tolerances by the inhabitants of other perspectives. Tolerance, in short, is not an independent value but a context-specific one, and therefore to exercise it is not to avoid oppositional activity but to engage in it. Graff and Cain want to privilege conflict; Craige wants to privilege tolerance. Each fails to see that conflict and tolerance cannot be privileged—made into platforms from which one can confidently and unproblematically speak—without turning them into the kind of normative and transcendental standards to which they are putatively opposed.

In the end, the difference between those who, like Bennett, Cheney, and Finn, would subordinate difference in the name of the common and those who, like Cain, Graff, and Craige, would acknowledge difference by making it into a central value is only apparent, a difference, finally, between different ways of managing (or attempting to manage) difference. (This is not to say that distinctions between these positions cannot be made on other grounds—pragmatic, institutional, humane—and on those grounds the Cain-Graff-Craige position is decidedly more attractive to me.)

The point may become clearer if we substitute for difference—a word that by now must be inducing something akin to seasickness—the notion of the political. It is politics or, as it is sometimes put, ideology that the right-establishment wishes to control, even eliminate; and it is politics that the left-challengers urge us to acknowledge and affirm. But both agendas fail in the same way: on the one side, the control or elimination of politics requires a vantage point which is not itself political, and the impossibility of specifying what is common without provoking a dispute demonstrates that no such vantage point is available; on the other side, the affirmation of politics implies that politics is something one can either refuse or embrace, and that in turn implies a moment when politics is an *option* rather than a name for the condition—the condition of difference—that one can never escape. Those who think they can *choose* politics are no less evading the fact of the political—the fact that point of view and perspectivity are irreducible features of consciousness and action—than those who think they can bracket politics. Politics can neither be avoided nor positively embraced; these impossible alternatives are superficially different ways of *grasping* the political, of holding it in one's hand, whereas properly understood, the political—the inescapability of partisan, angled seeing—is what always and already grasps us.

If we are always in the grasp of the political and can never move either to the high ground of moral certainty or to the even higher ground of a universal tolerance (universals in any direction are what the operation of difference renders unavailable), then the only question is, by what form of the political shall we be grasped? An answer to this question will not be found in the arguments of the opposing sides of the educational debate because those arguments, as we have seen, rest on claims (the claim either to have identified the exclusively normative set of values or to have moved beyond exclusion altogether) that cannot be sustained. It is time therefore to seek an answer elsewhere, not in the nuances of philosophy but in the more material consequences that will follow from the triumph of either agenda. The parties to the debate are telling us, in effect, what will be good for us, and it seems reasonable to ask if the conception of the good they offer is one we would like to embrace.

We will be helped in this inquiry by the growing number of what I call "ethicist" books. The list is an ever-lengthening one, and I can only pause here to rehearse some of the authors (notably male) and their titles: Allan Bloom, *The Closing of the American Mind,* Robert Alter, *The Pleasures of Reading in an Ideological Age,* Wayne Booth, *The Company We Keep,* George Steiner, *Real Presences,* Peter Shaw, *The War against the Intellect,* John Silber, *Straight Shooting,* Page Smith, *Killing the Spirit,* Roger Kimball, *Tenured Radicals,* Charles Sykes, *ProfScam,* James Atlas, *The Book Wars,* Frank Kermode, *An Appetite for Poetry,* Dinesh D'Souza, *Illiberal Education,* Alvin Kernan, *The Death of Literature,* Robert Bork, *The Tempting of America,* Bernard Bergonzi, *Exploding English,* Peter Washington, *Fraud: Literary Theory and the End of English.* The genre is not a new one (it begins as early as the Book of Jeremiah, chapter 2, verse 7: "ye . . . made mine heritage an abomination"), and the story it tells is always a story of loss, the loss of a time when common values were acknowledged and affirmed by everyone. The villains vary (the devil, the

Antichrist, foreigners, immigrants, Jews) but in this latest version of the genre they are moral relativists in general and literary theorists in particular.

The reasoning is simple: by teaching that all norms and standards are specific, contingent, historically produced, and potentially revisable, literary theorists and poststructuralist thinkers, it is said, undo the foundations on which any truly ethical action might be based. After all, the argument continues, if ethical judgments are disputable, and none can ever be grounded in anything firmer than the local conditions of practice, the act of judgment is rendered meaningless and trivial. The response, given many times by many persons but apparently not readily taken in, is that since those who are embedded in local practices—of literary criticism, law, education, or anything else—are "naturally" heirs of the norms and standards built into those practices, they can never be without (in two senses) norms and standards and are thus always acting in value-laden and judgmental ways simply by being competent actors in their workplaces. The poststructuralist characterization of the normative as a local rather than a transcendental realm, far from rendering ethical judgment impossible, renders it inevitable and inescapable. Antifoundationalist thought, properly understood, is not an assault on ethics but an account of the conditions—textual and revisable, to be sure—within which moments of ethical choice are always and *genuinely* emerging; it is only if ethical norms existed *elsewhere* that there would be a chance of missing them, but if they are always and already where you are they cannot be avoided. The counterresponse—that such moments are *not* genuine because they are not rooted in standards and norms that are independently and objectively established—fails when one realizes that were such standards to exist somewhere—in the mind of God or in the totality of the universe—there would be no one capable of recognizing or responding to them. After all, none of us lives in the mind of God or in the totality of the universe; rather, we all live in specific places demarcated in their configurations and in their possibilities for action (including ethical action) by transient, partial, shifting, and contingent understandings of what is and what should be.

But I fear I have slid back into philosophy again so soon after announcing my intention to remove the discussion from its precincts. Let me return the focus to where I promised it would be, on the relationship between the ethicist books and the specification by the educational conservatives of a common good as embodied in a common culture. The relationship is pretty much a straightforward one that proceeds via the story of loss so characteristic of this tradition: if a common cultural heritage has been lost, it must be the case that we have also lost or lost sight of the perspective from which that heritage could be identified and validated; we have lost the perspective

of normative ethics, the set of principles that at once underwrites and requires the common. The imperative, then, is to recover and preserve those
principles and that is why, in this argument, the maintenance of the traditional literary canon is such a priority; for the canon (again in this argument) is the repository, the ark, of those principles, not only containing
them but extending them in the effects it has on its readers.

The process by which this extension occurs is sketched out for us in
Robert Alter's book. The first stage rests on a familiar formalist thesis:
literary language, unlike ordinary language, escapes the confines of particular, local, historical referents; "densely layered and multi-directional," it
provokes "multifarious connections and . . . interpretations," which work
against the appeal of "explicit ideological commitments." [13] What literary
language works *for* is evidently more difficult to say; Alter attributes to it
the very general capacity of "address[ing] reality," and, after acknowledging that " 'reality' is a notoriously slippery term," defines it none too firmly
"as an umbrella for the underlying aspects of our being in the world,"
which in turn is glossed as "the common stuff of our human existence." [14]
At this point the line of argument becomes clear: the reader who engages
with the language of literature (as opposed to the language of advertising
or popular culture) will be immersed in "the common," and by virtue of
that immersion will become more and more like it. "Literary texts," Alter
tells us, "invite a special mode of attention" [15]—the mode of attending to
transhistorical aspects of being—and if the invitation is taken up, that mode
will be the mode of *our* existence; we will ourselves be "special"—that is,
not special, but universal—in the very same way literature is.

All of this assumes that "literature" is not a descriptive but a normative
category, a label applied in approbation and honor and a label withheld
from inferior productions like Steinbeck's *The Grapes of Wrath* that do not
display the proper transcendental qualities. Much of Alter's book is devoted
to picking out the works that in his view deserve the label and then presenting them as "the Canon." Once this is done, the paradigm is complete:
books made out of transcontextual language point to transcontextual values
and thus make their serious readers into transcontextual beings informed by
those same (common) values. The entire process is nicely encapsulated in
a single sentence from Wayne Booth's *The Company We Keep,* another of
the ethicist tracts and in many ways the best of them. Speaking of genuine
works of literature like *King Lear, Don Quixote, Bleak House, War and
Peace,* etc., Booth says, "When I read [them] . . . I meet in their authors
friends who demonstrate their friendship not only in the range and depth
and intensity of pleasure they offer, . . . but finally in the irresistible invitation they extend to live during these moments a richer and fuller life
than I could manage on my own." [16]

We have all had the experience Booth describes (although perhaps with books different from those he would invoke), but the relation of such "moments" to the strong ethicist argument is problematical. If the truly great works are those that irresistibly invite us to live a richer and fuller life, why have so many readers of the books Alter and Booth list managed to resist the invitation and gone on from reading Shakespeare and Goethe to acts of incredible cruelty? And why, conversely, have readers of works made of supposedly inferior stuff—works of popular and even "low" culture—been moved by their reading to acts of great altruism and service? And why, if the works of the Western canon are the repository of common (in the sense of universal) ethical values, has there been so much argument about the values to be found in them, including the argument that they have nothing especially ethical to teach at all? If what these works "irresistibly" convey to us and convey us to is so common, why is it so much in dispute? And given the dispute, how do we know when the ethical direction we spy in a book is the right one? It would seem that in order to answer these questions one must *already* be in the state of ethical perfection to which the canon is supposed to bring one, which suggests the superfluousness or at least causal irrelevance of the canon to the very values it is said to produce.

At this point an ethicist might respond that the experience of literature does not point us to any particular ethics but to the realm of the ethical in general, and I would respond in turn that the realm of the ethical in general is either empty or full of some contestable set of values that someone or some group wishes to pass off as the general. The point is the same one I have made before: like the "common," the category of the "ethical"—which is, after all, a particular instance of the common—is continually in dispute; it cannot serve as an antidote to politics and ideology, as Alter and others want it to, because it will be, in any form that makes human sense, a political and ideological construction. Another way to put this is to say that the trouble with ethicist arguments is that they don't have an opponent; there are no *non*ethicists against whom the ethical critic can position himself, for in order to *be* a nonethicist one would have to stand free of any local network of beliefs, assumptions, purposes, obligations, etc., and such a standing free—necessary both to the moment of pure ethical choice and to the possibility of choosing *against* ethics—is not an option for human beings. Everyone is always and already an ethicist, enacting value in every activity, including the activity of reading.[17] The only question is, which of the many possible ethicisms—ethical stances—should one affirm?

In short, the ethicists are not *the* ethicists, in the sense of being the sole proprietors of a moral vision in a world of shameless relativists; rather, they are the purveyors of a *particular* moral vision that must make its way in the face of competition from other moral visions that come attached to texts no less inherently worthy than the texts recommended by Booth and Alter. In the absence of an exclusive claim to the moral high ground and without the supporting prop of a set of obviously sacred books, the ethicists are left only with the force of their example as displayed in their writings. Is it an example we wish to follow? Is their conception of the common good so compelling that we would wish to embrace it?

On the evidence of their own moral performances, the answer, I think, must be "no." A favorite ploy of these authors is to quote out of context and often without attribution passages or sentences that are presented as self-evident demonstrations of the foolishness, turpitude, stupidity, and absurdity of their opponents. This practice is then justified *ethically,* as when Alter says of examples of what seem to him to be obvious atrocities, "for reasons of simple decency I will not cite the sources or the authors' names."[18] Here is a moment in which the claim of an ethicist to be presenting us with a common moral vision (the word "common" makes many appearances in Alter's book) intersects with the singularity and dubiety of the moral vision informing his practice. In what sense is it "decent" to declare the sentence of another writer "repellent" without in any way acknowledging the large project (whose configurations would include the predecessors to whom he or she was responding) of which the sentence is a part? In what sense is it "decent" to call someone names at the very moment one is depriving the victim of the name he or she has been forthright enough to sign? No doubt Alter would reply that he merely wishes to save his nameless authors from the embarrassment they would suffer were they fully identified; one would think, however, that they might prefer to represent themselves rather than be *mis*represented by a critic who uses the notion of "decency" in order to mask and excuse a form of behavior that many will find *morally* objectionable. (I will not even speak here to the obvious point that Alter's preferred authors—Milton, Henry James, Fielding—would fare no better than his chosen targets were they to be subjected to his procedures.)

The fact that Alter's scholarly practices are ethically questionable does not distinguish him from any number of writers on either side of the debate; but since it is precisely his claim to *be* distinguishable from his less responsible colleagues (who remain nameless), the point is worth making, for it raises in a local, particular form the question I have been asking: do we want such persons to be our *moral* guides? The same question can be put to another self-proclaimed ethicist, Professor Anne Barbeau Gardiner of the John Jay College of Criminal Justice. Writing in the fall 1989 issue of the *ADE Bulletin,* Professor Gardiner is concerned with celebrating the "power of a literary text to convey . . . an ethical point"; "great literary texts of the past are . . . civilization building," she says, and offers as her example a class's experience with Dante's *Inferno,* a must on any ethicist's list.

What Professor Gardiner did was describe a number of twentieth-century moral actors—such as the person "who never leaves home without consulting a horoscope"—and then ask her students to assign them to the appropriate circle of hell. She calls the exercise "The 1980s through Dante's Eyes: Name the Circle, Ditch, and Punishment Reserved for These Wrongdoers," and reports happily that "students had no trouble exercising their moral imagination and, after putting themselves in Dante's place, assigning an appropriate level of punishment to each offender." [19] Putting aside for the moment the injustice here done to Dante, whose "place" is surely more capacious than Gardiner imagines, one is at least uneasy at this display of moral insensitivity in the name of the moral qualities supposedly inculcated by great texts. Are we ready to follow the ethical example of a professor who declares herself pleased to "see that the students enjoyed passing moral judgment"? Is this the training we wish our students to receive? Are these the ethics that underlie our common cultural tradition?

Let me be clear. I would not deny Professor Gardiner her perspective, either on Dante or on the goals and purposes of education. My quarrel is with the assumption, in her writing and in the writings of others, that a few people (self-selected) have a privileged access to common or universal ethical values and that they have this access in part because, unlike their less clear-sighted colleagues, they have identified and bonded with the small set of texts that embodies those values. What the ethicists are saying to us is, "We (not you) are educated, we (not you) are sensitive; therefore you should listen to us." The question again is, should we? and again I look for an answer in the example offered us in the performance of a high ethicist, Martha C. Nussbaum, David Benedict Professor of Classics, Professor of Philosophy, and Professor of Comparative Literature at Brown University. It would seem that if anyone has the proper ethicist credentials for serving as our moral exemplar, it is Professor Nussbaum, who has for some time been conversing with Aeschylus and Virgil (presumably in the original Greek and Latin), engaging strenuously with Plato, Aristotle, Hume, Kant, and others, with ample time, apparently, for serious study of Stendhal, Henry James, and Proust.

The piece of Professor Nussbaum's I have in mind is an admiring review of Wayne Booth's *The Company We Keep*. Nussbaum approvingly rehearses the book's arguments and quotes, among other excerpts, the sentence I have already cited in which Booth pays tribute to literary works that "enable" readers "to live . . . a richer and fuller life than they could manage on their own." [20] Later she offers a discrimination that presumably exemplifies the fuller and richer life *she* is now able to live. Her concern is the difference (slighted, she says, by Booth) between the way one treats a book and the way one treats "a real live person." "Sometimes," she says, "people feel the need for complete numbing distraction, distraction so complete that it blots out all stress and worry." "Consider," she continues,

> two people in search of such undemanding release. The one hires a prostitute and indulges in an evening of casual sex. The other buys a Dick Francis novel and lies on the couch all evening reading. There must be, I think, a huge moral difference between these people . . . (I say this as someone who reads in just this way whenever I finish writing a paper . . .). The person who hires a prostitute is seeking relief by using another human being; he or she engages in a transaction that debases both a person and an intimate activity. The person who reads Dick Francis is not, I believe, doing any harm to anyone. Surely she is not *exploiting* the writer; indeed, she is treating Francis exactly as he would wish, in a not undignified business transaction.[21]

There is so much wrong with this that it is difficult to know where to begin. Even the most attractive aspect of the passage—the acknowledgment (in the parenthesis) that Nussbaum is talking about herself and is her own chief example—breathes self-promotion and self-dramatization: "Here I am, at this moment, writing a paper that engages the full, rich, and moral me, the very paper you are no doubt admiring me for, at this moment." As for the *argument* of the passage, problems arise at every juncture. Is the moral difference between the two persons Nussbaum imagines as clear as she thinks and in the direction she assumes? Could she be unaware that there is at least an argument for prostitution as a feminist praxis as well as a historical analysis in which the prostitute is the scapegoat category for everything a society cannot bear to confront? And, in any case, will the prostitute be pleased to know that Nussbaum has chosen against him or her and for an evening with Dick Francis? And what of Mr. Francis? Will he be pleased to be the winner of a contest framed in these terms? Does he write, do you think, in order to provide thoughtless distraction for wearily heroic academics? Would he, if given the choice, want Nussbaum for a reader? Would he want to be consumed by a moral being so rarified that she simply discounts whatever serious intentions animate his rather subtle explorations of one aspect of the British character? And what of those of Mr. Francis's readers who are so foolish and morally deficient as to derive from his books something more positive than "numbing distraction," who are not teased out of, but into, thought? And what, too, of all those reviewers who were apparently moved by reading Mr. Francis (perhaps in bed) to speculations about the venerable (even canonical) tradition in which his work appears? And what of the entire world of popular culture, both inside and outside the academy, a world here dismissed without so much as a backward glance? This list of questions could be continued indefinitely, but the answers are less important than the fact that they can be posed with some force, for they speak not merely to Nussbaum's judgment on particular matters of literary culture but to the issue of judgment in general and therefore to Nussbaum's fitness for the role she and her fellow ethicists would play in our society.

One could reply that, after all, there is little danger that the reins of government or even of educational policy will in fact be handed over to Alter, Gardiner, and Nussbaum, and that therefore the thinness of their moral vision is not a matter of great concern. Unfortunately, however, that same vision, similarly thin but politically robust, is now being enacted in the edicts and rulings of highly placed government officials; I am thinking of documents like Lynne Cheney's *Humanities in America: A Report to the President, the Congress, and the American People.* Here, for example, is a key statement from that publication: "The humanities are about more than politics, about more than social power. What gives them their abiding worth are truths that pass beyond time and circumstance; truths that, transcending accidents of class, race and gender, speak to us all." [22] Note that this is a politics that dares not speak its name, but speaks (typically and disingenuously) in the name of the common sense "we" all share. Note also the dismissal of "social power" at the very moment it is being exercised by the head of a socially powerful agency in the form of a publication funded, printed, and mailed by the federal government. This political sleight of hand does considerable work, including the work of eliding questions one might otherwise be moved to ask, questions like "who gets to *say* which are the works that embody 'abiding worth'?" and "who are 'us all'?" The pamphlet itself is an answer: Cheney and her associates will get to say, and what they say will have all the impact and force provided by their relationship to that same federal government. It is also obvious who will *not* get to say, those for whom matters of class, race, ethnicity, and gender are of paramount importance and abiding concern, that is, those who are poor, black, Hispanic, Asian, female, gay, etc.

This is not a characterization that Cheney and company would accept. No polemic issuing from this establishment would be complete without paying lip service to diversity; but more often than not, the recognition is, in itself, an act of marginalization, as in this recommendation near the end of Cheney's pamphlet: "Undergraduates should study texts of Western civilization and should learn how the ideals and practices of our society have evolved. Students should also be encouraged to learn about other cultures." [23] The number of questions begged here is large and familiar: what about the students for whom the word "our" in "our society" is problematical, students for whom the culture made up of texts from fifth-century

Athens, first-century Rome, and sixteenth-century England is decidedly "other"? What happens to *their* concerns, perspectives, values, heritages? The answer is that they must learn the "ideals and practices" identified by Cheney. Of course, neither they nor their classmates are *barred* from learning other things; indeed they are "encouraged" to learn them, presumably in the odd or spare moment stipulated in the fifty-hour proposal. After all, secondary and inessential matters must wait their time, even if it never comes.

The work done by Cheney's "also"—"should also be encouraged"—is done elsewhere by what I call the "and Alice Walker" move, performed when you make up a long list of great texts from Plato to T. S. Eliot and then say, "and Alice Walker," thus testifying to your commitment to diversity. Bennett prefers Martin Luther King, Jr., as his token[24] and now has taken to adding Coretta Scott King and Corazon Aquino, a new odd couple pressed into the service of displaying the capaciousness of the former secretary's mind. And Cheney now has kind words (at the end of a paragraph) for Susan B. Anthony and Frederick Douglass. Can Jackie Robinson be far behind?

When pressed on the point that such addenda are obvious afterthoughts, sops thrown to a notion of diversity that is being dishonored in the breach, conservative educators will typically reply, with Chester Finn, that we can't help it if "for a long time [white males] were the principal folks on the planet with the learning, the leisure, and the resources to create literature."[25] This argument, made at length in what the *National Review* (13 October 1989) advertised with melancholy glee as Sidney Hook's "last article" (presumably we are to regard this as deathbed testimony), rests on an assumption that will not survive scrutiny: the assumption that in a world of diverse cultures and subcultures, only those powerful enough to produce their own propaganda in the form of histories, curricula, reading lists, and poet laureateships were creating literature, singing songs, fashioning traditions, having thoughts. The fact that we now know little of the cultures that have been swept into one of history's dustbins hardly seems reason to dismiss them as obviously inferior, unless the reason is the one often given by victors: we won, you lost, now keep quiet! When Hook exclaims, "Of course the culture of the past was created by the elites of the past!" and asks, "Who else could have created it in a time when literacy . . . was the monopoly of the elite?"[26] he doesn't really want to hear an answer; that is, he doesn't want to take seriously his own word "created," for to do so would be to consider the possibility that the creation of a cultural history—of a great tradition—depends on the marginalization and suppression of other traditions and indeed of other elitisms which are now referred to as minor, or ethnic, or regional, or popular, or vulgar, or primitive, when they are referred to at all. Hook, like Finn, wants to take these dismissive labels as natural, even as he acknowledges that he and others of his party have applied them in the act of establishing their own taste as the taste "common to us all." In the end the argument amounts to no more than saying, "We know what's good; what's good is what people like us produce and appreciate, and we are going to see to it that everything else is excluded."

There is a word for this politics and it is "undemocratic"; not because it resists the demand for proportional representation—egalitarianism and democracy are not the same thing—but because it would arrest the play of democratic forces in order to reify as transcendent a particular and *uncom-*mon stage in cultural history. This politics would stop time and elevate one set of tastes to a position of privilege from which it could then label all other tastes as vulgar, inessential, and corrupting. Nowhere is the rationale for this political program (which always portrays itself as apolitical and even antipolitical) more clearly in evidence than in George Steiner's *Real Presences.* The first third of Steiner's book is an extended critique of the secondary or parasitic—that which is either a deviation from or unthinking imitation of the genuine, the immediate, the primary, the authentic. Against the endlessly resourceful discourse of the parasitic Steiner opposes "that which the heart knows," for example, that Tolstoy was offering an aberrant judgment when he proclaimed *"King Lear* to be 'beneath serious criticism.' "[27] Of course, the problem (illustrated by Tolstoy) is that not all hearts know the same thing; the problem, once again, is difference. But Steiner knows how to deal with it: the heart that doesn't know or that knows the wrong thing must be disregarded and disenfranchised. "Given a free vote," Steiner laments, "the bulk of humankind will choose football, the soap opera or bingo over Aeschylus." Steiner's strategy is simply *not* to give the bulk of humankind a vote and to refuse it on the basis of a principle that is enunciated as clearly as one might wish: "Democracy is fundamentally at odds with the canonical."[28] In fact, democracy is at odds with the canonic only if the canonic is viewed as a category established for all time; but if canons, in the company of the standards and norms that underwrite them, are always emerging and reemerging in response to historical needs and contingencies, then democracy is simply a name for the canon-making process. It is a process that Steiner (along with Cheney, Bennett, and Finn) would stop in its tracks by declaring, on no obvious authority whatsoever, that after a certain date (it varies from 1490 to 1940) nothing is to be admitted into the curriculum. This is clearly the implication of Steiner's scornful account of recent developments at an unnamed university (we can assume that it is Harvard):

The bellwether of American universities assigns to its "core curriculum," this is to say, to its minimal requirements for literacy, a course on black women novelists of the early 1980s. [No "and Alice Walker" for him!] Poets, novelists, choreographers, painters of the most passing or derivative interests, are made the object of seminars and dissertations, of undergraduate lectures and post-doctoral research. The axioms of the transcendent . . . axioms which this essay seeks to clarify—are invested in the overnight.[29]

The moral is obvious and it is drawn: these inferior authors and their texts must be "ruled out of court." Here, in contrast to the cautious timidity of Cheney's core curriculum with its cosmetic concessions to other cultures, races, genders, is the real thing, the authoritative imposition of one group's very particular tastes in the name of the common and the transcendental.

The question is, do we really want it? and if we are to trust the media (a dubious proposition), the answer would seem to be a ringing "yes"! Since 1989, when the *National Forum* published its "canon issue," the questions it raised have been debated with increasing urgency, but the shape of the debate has changed markedly because as things are now only one position is fully represented. In the *National Forum,* pieces by defenders of the traditional curriculum alternate with pieces by advocates of women's studies, minority studies, literary theory, etc.; but in the innumerable essays that have appeared in our major magazines and newspapers in the past two years, these latter groups are given no substantial voice and are introduced only as the excoriated and ridiculed other. Whatever the truth about the relative strength of the warring forces on college campuses, the conservative backlash has certainly won the media battle, so much so that a reader of the 18 February 1991 issue of the *New Republic* (once a respected liberal journal) would have to believe that there was no one on the "multiculturalist" side except a small band of radical-left crazies who had either never encountered or failed to absorb the rigors of the Western tradition. Along with this story, the *New Republic* tells another, contradictory story of a massive subversion which captured the villages and citadels as we slept, leaving only a few brave, beleaguered, powerless souls (like William Bennett, Lynne Cheney, Hilton Kramer, and William Simon) to mount a counterinsurgency. Here is the classically fissured shape of paranoid thought, in which absolute power and absolute vulnerability are simultaneously declared.

But tempting though the pleasures of diagnosis may be, I will forgo them and concentrate on what is for me the significant feature of the *New Republic*'s call to arms: it continues to present the issues in philosophical terms, as a battle between rationality and some darkly terrible alternative. The editor is correct when he observes that "no generation goes by without a 'crisis' in the humanities."[30] In his view, however, these recurring crises trace out a pattern in which the forces of evil reappear in every age to challenge the forces of good: "The most common cause of these recurrent crises has been the demand that the university conform to one orthodoxy or another." My analysis would be less apocalyptic, and I would rewrite his sentence to read: "The most common cause of these recurrent crises has been the resistance of an orthodoxy long in power to orthodoxies just now emerging and experiencing a new strength." Although the attack on multi-culturalism and diversity is almost always made in terms of principle, behind it lies something much less grand and more human, a feeling by many older academics that the world they entered so many years ago has changed in ways they find threatening. One gets a glimpse of this feeling in a poignant sentence written by a university teacher in *Academic Questions,* the journal of the National Association of Scholars: "In the more than thirty years that I have been teaching . . . I have observed an increasing drift away from the kind of intellectual-cum-moral consensus I found to obtain . . . when I began my teaching career in 1954."[31] One does not know whether to lament or rejoice for this writer. On the one hand, it is more than thirty years and he is still in the game; on the other, it is not the game he was trained to play, and we can suppose that the very students who should now be presenting him with a festschrift are instead calling into question the assumptions that have guided his work all these years.

The personal distress registered here is real and should not be slighted or diminished (a version of it awaits us all, unless we die young), but neither should personal distress be allowed to read itself as evidence of the triumph of politics over reason. Although it has become fashionable to characterize the debate in these terms, a more accurate characterization would see it as a contest between one kind of politics—derisively called "political correctness"—and another politics—which I would call "political disappointment" or even "political envy." In short, there is politics (and within the differing political frameworks, reason) on both sides, and, as even a cursory reading of Gerald Graff's *Professing Literature* will show, it was ever thus. It has always been the case that the orthodoxy productive of one generation's sense of "the common" (that is, "common sense") is challenged by the concerns and emphases of the next generation; and it has also always been the case that the older generation can only hear the challenge as the emergence of irrationality, the abandoning of standards, the beginning of the end. To the person experiencing it as a diminution of his or her professional authority and influence, change can only be seen as the work of the devil.

I would suggest that we see it simply as change or, to use an even more basic term, as history; for then the arguments against it might begin to seem less heroic than quixotic, deployed not in the service of reason and the American way but as part of a (doomed) effort to stay the replacement of one historically produced "common ground" by its inevitable successor. The question then becomes not, "Will standards and values be subverted by the introduction of new materials and methodologies?" but, "What will be the likely effects of trying to say 'no' to history?" The first question makes sense only if standards and values exist in a realm apart from history where they define the obligations and responsibilities of historically situated actors; but if values and standards are themselves historical products, fashioned and refashioned in the crucible of discussion and debate, there is no danger of their being subverted because they are always and already being transformed.

Transformation, however, is always a painful process, at least for those persons (and I am often one of them) who want things to stay the same, and one can understand the Juvenalian laments ("we are going to hell in a hand-basket") even if one is not moved to echo them. In the end, however, I prefer the quieter tones of pragmatic inquiry: what is to be gained or lost in our everyday lives as students and teachers by either welcoming or rejecting various new emphases and methodologies urged on us by various constituencies? Unlike questions posed in the timeless language of philosophical abstractions, this is a question one can answer. If we harken to those who speak in the name of diversity (and I say again that I myself resist the invocation of diversity as a principle, as a new theology), the result will be more subject matter, more avenues of research, more attention to neglected and marginalized areas of our society, more opportunities to cross cultural, ethnic, and gender lines, more work, in short, for academics. If, on the other hand, we harken to those who would hold back the tide and defend the beachhead won thirty-five or fifty years ago, the result will be more rules, more exclusionary mechanisms, more hoops to jump through, more invidious distinctions, more opportunities to be demeaning and be demeaned, more bureaucracy, more control. I know what I like. In the words of the old song, how about you?

Preface To Chapters 3 Through 7

Chapters 3 through 7 were written for a series of debates between Dinesh D'Souza and me that began in September 1991 and ended in March 1992. The debates were sponsored by the student organizations of five campuses, the University of South Florida, Northern Illinois University, Pennsylvania State University, Erie-Behrend, the University of Alabama, and Center College, Kentucky. The format, as devised by the two of us, was simple: initial presentations of fifteen to eighteen minutes, followed by five minutes of rebuttal, followed by up to an hour and one-half of audience questions, and concluded by a three-minute closing statement. The audiences were uniformly large and animated, lining up patiently at two microphones placed on either side of the hall. Some carried copies of D'Souza's Illiberal Education, the book that more than any other fueled the controversies that made us into an attraction. For the most part D'Souza quarried his presentation from chapters of his book, skillfully weaving together some of the anecdotes that give his polemic its undoubted power. I had no book (this is it), and so I wrote an essay for each occasion. The first essay, "The Empire Strikes Back," attempted a general overview of the "culture wars" with a particular emphasis on the situation (barely a situation at all) at Duke University. Subsequent essays were prompted by the question-and-answer period that took up the bulk of the evenings. It became apparent at South Florida that members of the audience were agitated more by affirmative action and the issue of "fairness" than by any other topic, and accordingly I decided to meet the "reverse racism" objection head-on. At Penn State, Mr. D'Souza drew an approving response when he declared, "You can't fight discrimination with discrimination," and on the spot I produced the beginning of a counterargument that became the full essay I read at Alabama. The last piece, "Speaking in Code," was written in the knowledge that it would indeed be the last (there was no dearth of invitations; it just seemed to me that the series had run its natural course), and I let out all the stops and allowed myself a harsher tone than I would have otherwise employed.

However harsh the accents either of us fell into on stage, our personal interactions were unfailingly cordial. We dined together, traveled together, and played tennis whenever we could. (When his serve was on—and it was on more often than I liked—Mr. D'Souza would always win.) After the formal sessions we would continue the conversation in a bar or restaurant, and on one occasion, when we ended up in a Hardee's (the only place open in town), a few students wandered in, and for twenty minutes or so we did the whole thing over again. In May I danced happily at his wedding, and we have since appeared together on a panel discussing First Amendment questions about which we are pretty much in agreement. (In fact, the areas of agreement between us are wider than one might have expected.) Neither of us, I think, changed the other's mind on the issues we debated, but it is fair to say that both of us sharpened our arguments in the course of the rapid-fire thrust and parry that characterized our exchanges. I am always asked, "Who won?" By our reckoning, which could surely be disputed, the debates ended in a draw: two each and a tie. What cannot be disputed, because it was reported to us at every turn, is that the campus communities won as they always will when important questions are taken up by serious and informed opponents. It was short-lived, but it was a great show.

3

The Empire Strikes Back

I appear before you today by virtue of a mistake made by central casting that has tapped me for the role of ardent academic leftist, proponent of multiculturalism, and standard-bearer of the politically correct. Unfortunately, my qualifications for this assignment are so slight as to be nonexistent. First of all I am, as you can see, a 53-year-old white male. More important, I have for the past thirty years taught only traditional texts written by canonical male authors of the ultracanonical English Renaissance— John Milton, John Donne, Edmund Spenser, George Herbert, Francis Bacon, Ben Jonson, Andrew Marvell. When not writing on these classical authors, I have in recent years addressed a number of issues in literary and legal theory, and I think it fair to say that I have come out on the "right" end of the spectrum every time, arguing against the liberationist claims often associated with deconstruction and some versions of feminism, against the political pretensions of the New Historicism, against the utopian vision of interdisciplinarity, against the revisionary program of the Critical Legal Studies movement, the left wing of the legal academy. Why then have the media so consistently mischaracterized my position and misreported my views? The question is a real one, but it is of interest only to a few, perhaps only to my mother. The larger question is "What is going on? What developments in the classroom and in the pages of various academic journals have stirred up such a commotion?"

It is easier to say what is *not* going on. Notwithstanding reports to the contrary, there is no evidence that either Shakespeare or those who teach him have been run out of the academy by an intolerant coalition of Marxists, rabid feminists, godless deconstructionists, and diseased gays. Noting media reporting of "a fight over the political or cultural content of classroom learning and of speaker presentations," the 1991 survey of *Campus*

Trends (compiled by the American Council on Education) concludes that despite "anecdotal accounts" "problems are not widespread." Ninety percent of all institutions report no "controversies over the political or cultural content of remarks made by invited speakers." Ninety-seven percent of the same institutions report no "controversies over course texts or over information presented in the classroom." At Duke, often cited as a hotbed of political correctness, recent months have seen appearances by George Will (as commencement speaker), Lynne Cheney, Charles Sykes, and Dinesh D'Souza, all without incident; Duke's "radical" English department runs an average of six Shakespeare sections a semester, and two springs ago ninety-five undergraduates filled two Milton courses to overflowing. The English major at this same "decadent" institution requires a major author course in Chaucer, Shakespeare, Spenser, Milton, and Pope, a specified number of courses in literature before 1800, and an introductory course in the techniques of literary analysis, New Critical style. There are no multiculturalist requirements (perhaps there should be), no seminars in sensitivity training, no harassment of instructors presenting traditional courses in traditional ways (if there were, I would be one of those harassed). A reporter from a nearby city made an independent count and found that of 135 courses given last year by the department, 123 would have been familiar to anyone attending college in the past thirty-five years, while only 13 dealt with such outré subjects as film studies, black studies, gender studies, gay studies, and literary theory. In this respect Duke is representative of its sister campuses. Dinesh D'Souza reports a professor at Penn State as saying, "I would bet that . . . *The Color Purple* is taught in more English departments today than all of Shakespeare's plays combined." He would lose that bet, and he would lose it if the field of reference were the nation's high schools, where, as another survey has recently indicated, Shakespeare is alive and well and massively studied.

Is there, then, nothing going on that deserves the label "political correctness"? Before I answer the question, it might be good to examine the label and inquire into what it implies. First of all, it implies the introduction of politics into an area (often called the life of the mind) where politics doesn't belong; and second, it implies that this intrusion of politics is itself politically organized, the result of design and coordinated activity. In *that* sense of politics, however, all of the action is to be found on the side of those who are yelling "political correctness" (as they once yelled "Communist sympathizers") rather than on the side of those at whom the epithet is hurled. It is the neoconservative forces on and off campus (and, as we shall see, more off than on) that operate an efficient network of semistudent organizations, nonofficial semistudent newspapers, and nonlocal faculty action groups. It is the neoconservatives who intrude themselves into other people's classes and demand the removal of courses and programs put in place by regular university procedures; it is the neoconservatives who generalize a few tired incidents into an assertion of wholesale crisis and then feed the public's appetite for crisis with the help of a cadre of well-placed and largely ignorant journalists. And, above all, it is the neoconservatives who are enabled in these activities by *massive* infusions of outside funding from a familiar list of far-right foundations, think tanks, and individuals. In the past two years the National Association of Scholars (a successor to the infamous Accuracy in Academia) has received $425,000 from two of those foundations alone; and the *Dartmouth Review*—the flagship of yellow journalism, academic style—has received $300,000 from the Olin Foundation in the past decade.

Some apologists have attempted to portray the *Review* and similar publications as no different from other student organizations, but the chairman of Dartmouth's Board of Trustees, George Munroe, gives the lie to this description when he asks in a *Wall Street Journal* piece, "How many typical student newspapers exist virtually without advertisers or paid subscribers[?] . . . How many student journalists are rewarded by their *national* benefactors with prestigious jobs in government, public policy institutes and national media? How many typical student papers have members of their boards [none of them alumni of the college] who, when a crisis arises, fly in from Washington to hold a press conference?" Munroe's point is mine; operations like the *Dartmouth Review* and the National Association of Scholars are the local habitations of a national partisan agenda, one that has stipulated as its goal the control and alteration of the American campus, down to its smallest detail. This agenda does not originate on the campus; rather, it uses the campus as a giant backdrop, a nation-size outdoor movie screen, on which is projected an allegory of moral conflict, good against evil, all the roles assigned by off-campus agents who interpret their own screenplay for a credulous audience. That audience is not primarily an academic one— I would guess that less than 10 percent of faculty, students, and administrators in any college or university are in sympathy with the chicken-little message—but is primarily made up of something called the "concerned public," which, in combination with alumni ready to believe that things have gone downhill since they sat in the ivied halls thirty years ago, acts as a booming chorus for a song that has been taught at long distance. (Of the thousands who complain of "political correctness," very few—and here I include journalists—have taken the trouble to verify their complaints by actually visiting a campus or sitting in on a class or reading a book.)

Now don't get me wrong. There is nothing illegal or even immoral about this; concerted organization for partisan ends is the American way. My only point is that it is the way of politics—itself a perfectly honorable word despite the fact that it has been turned into an *accusation* by the very people who are so assiduously practicing it. What we have here, then, is not, as has been advertised, a brave resistance to politics by the representatives of an apolitical rationality but rather an argument between two forms of politics, or if you prefer, two forms of political correctness. Once this point is clear and the polemical use of the term has been blunted, we can rephrase the question I raised earlier: Is there nothing going on in the academy today that might raise the ire of the politically correct who live on the right? The answer is "yes"; there have indeed been developments in the past twenty years that would be distressing to persons invested in traditional modes of intellectual inquiry. Some of these developments are demographic and therefore not a matter of anyone's design. Before World War II both the college population and the ranks of the professoriate were drawn from a pool made up largely of middle-class and upper-middle-class males, many of whom would have been destined in an earlier age for a life in the ministry. In the years following the war the G.I. Bill of Rights brought many to the university who would not have thought of going before, and some of those made their way to graduate school and then into the classroom as instructors. A short time later the percentage of women in higher education began to rise at virtually the same moment that a strong feminist movement was making its way into every corner of American life. Add to these developments the impetus of the civil rights movement, new patterns of immigration following the Immigration Act of 1965, and the reawakening of interest in ethnic origins and traditions stimulated in part by the television program "Roots," and it would be surprising had there *not* been significant changes in the materials making their way into the curriculum and onto the reading lists of our colleges and universities.

This, however, is only half the story, for these demographic and social changes coincided with a change in the *intellectual* configuration of scholarly inquiry. I refer to what has been called the interpretive turn, or the turn to language. It involves a reversal of the relationship that was traditionally held to obtain between descriptive vocabularies and their objects. The usual and common sense assumption is that objects are prior and therefore at once constrain and judge the descriptions made of them. Language is said to be subordinate to and in the service of the world of fact. But in recent years language has been promoted to a constitutive role and declared by theorists of various stripes (poststructuralists, postmodernists, feminists, Bakhtinians, New Historicists, Lacanians, among others) to bring facts into being rather than simply report on them. No longer is it taken for granted that poems come first and interpretations of them second, or that historical events come first and historical accounts of those events come second, or that molecules and quarks come first and scientists' models of molecules and quarks come second; in discipline after discipline the reverse argument has been powerfully made, the argument that the vocabulary a practitioner finds ready to hand—the vocabulary that precedes his or her entrance into the practice and constitutes its prism—limits, and by limiting shapes what can be seen. If the categories available to you as a literary critic are lyric, drama, epic, and novel, you will see something identified as a literary work as one of these and the details of the work—from large structural patterns to the smallest stylistic feature—will be *produced* by your sense of what is appropriate to that category. And if the language of historical description is informed by concepts of progress, decline, consolidation, and dispersion, historical inquiry will *produce* events that display those characteristics rather than the very different characteristics that might emerge in the wake of an alternative descriptive vocabulary (e.g., the decentered and decentering vocabulary of postmodernism).

Merely to state this view is to see the problems it presents to "traditional" thinking: notions of objectivity, accuracy, verisimilitude no longer provide the comfort and guidance they once did, for they are now not absolute judgments, but judgments relative to differing and competing vocabularies or paradigms; and a whole host of distinctions—between fact and value, norm and deviation, reason and rhetoric, center and periphery, truth and politics—become, if not untenable, at least *disputable* in any of their proffered forms. Not surprisingly, the elaboration of these notions by deconstructionists, feminists, sociologists, cultural anthropologists, and others met with resistance, and some resisted not by engaging with the new arguments point by point but by dismissing them as obviously nonsensical and betting that in a short time they would fade away like a bad dream. It was a losing bet; if there is now no vigorous discussion of deconstruction in the academy, it is because its lessons have been absorbed and its formulations—the irreducibility of difference, the priority of the signifier over the signified, the social construction of the self—have been canonized; and if poststructuralism has given way to postmodernism as the new all-purpose term, it is because the implications of the first term are now being extended far beyond the realm of aesthetics and philosophy to the very texture of everyday life.

In short, the revolution, if that is the word, has succeeded and passed through several stages of revision; but meanwhile those who drew back from it in horror and closed their eyes in the hope that when they looked again everything would be as it was remained in the academy, where they became the equivalent of White Russians, convinced that history had taken a perversely bad turn and that weak-minded students had been misled by false prophets and corrupt teachers. And like White Russians, they waited, waited for the moment when the power and prestige that had been taken from them would be returned, waited for the moment when the honors, chairs, and salaries misbestowed on less deserving others would be restored and everything would be as it had been promised in 1959 when they received their degrees. It was in this posture—made up of equal parts of disappointment and millennial expectation—that they were found by William Simon, William Bennett, Lynne Cheney, Charles Sykes, Roger Kimball, Hilton Kramer, Nat Hentoff, Jonathan Yardley, Dorothy Rabinowitz, Dinesh D'Souza, not themselves full-time academics, but bringing a message a bypassed generation of academics wanted to hear: we will help you reclaim your legacy; we will expose the profscam by which your birthright was denied you; we will dislodge the radicals whom you foolishly tenured; we will put you in charge of reopening the American mind; we will put an end to the politics of race and sex on campus; we will put those women and blacks and gays in their proper places, at your feet.

Let me not be misunderstood. There are no simple villains here. One cannot fault the new generation of scholars and students for having responded with enthusiasm to the emergence of methodologies and materials that were unknown (or at least unappreciated) thirty years ago; and one cannot blame those who entered the academy thirty years ago for feeling discombobulated and dispossessed by developments they could not have possibly predicted when they signed on for the magical mystery tour; and one cannot even criticize the soldiers of the new and old right for seeing an opportunity and seizing it, for having the skill (which I can only envy) to mobilize a powerful coalition of disgruntled professors, nostalgic alumni, antiacademic journalists, concerned parents, and suspicious citizens in the face of a threat they have brilliantly fabricated. What has been lost in the presentation of the situation—what is always lost—is its complexity; and it is not hard to see why. Complexity does not play well in Peoria or anywhere else. People don't want to listen to detailed historical accounts of academic controversies; don't want to read thick sociological analyses of the various groups that populate the campus landscape; don't want to examine the thousands of printed college catalogs and course descriptions in order to determine what is really being taught; don't want to consider the relationship of changes in university life to changes in the life of the society as a whole; don't want to think about Columbus, sexual harassment in the workplace, date rape, the war against smoking, the dilemmas of the AIDS crisis, the spiritual malaise of modern life, the budget deficit, the trade deficit, the plight of the homeless, the media explosion, the information revolution, the interplay between educational and political policy, the impossibility any longer of separating the life of the mind from the life of the legislature and the marketplace.

In the face of this complexity the press, understandably, has told a single story; it has in fact told the story of there being but a single story, the story of subversive youths and ethnics and nihilists who are at once a lunatic fringe and a threat to a strangely endangered center, a story of epistemological evil emerging from some unfathomable impulse to destroy and lay waste, a story of reason under siege, of the decline of culture, of the abandonment of standards, of the triumph of barbarism. It's great theater, and one can only anticipate the movie that should have been made by Frank Capra starring Jimmy Stewart and ending with everyone singing "Auld Lang Syne." The alternative is not a particularly appealing one, since it involves the telling of the story of multiple stories, and that in turn involves painstaking analyses of phenomena that refuse to assume a simple shape. Nevertheless it is the alternative we need, and my hope for today is that I can persuade you first to consider it and then, perhaps, to embrace it.

4

REVERSE RACISM, OR, HOW THE POT GOT TO CALL THE KETTLE BLACK

I take my text from George Bush, who, in an address to the United Nations on September 23, 1991, said this of the U.N. resolution equating Zionism with racism: "Zionism . . . is the idea that led to the creation of a home for the Jewish people. . . . And to equate Zionism with the intolerable sin of racism is to twist history and forget the terrible plight of Jews in World War II and indeed throughout history" (*New York Times,* September 24, 1991, A6). What happened in World War II was that 6 million Jews were exterminated by persons who regarded them as racially inferior and a danger to Aryan purity. What happened after World War II was that the survivors of that Holocaust established a Jewish state, that is, a state centered on Jewish history, Jewish values, and Jewish traditions, in short, a Jewocentric state. What President Bush objects to is the logical sleight of hand by which these two actions are declared equivalent because they are both expressions of racial exclusiveness. Ignored, as Bush says, is the *historical* difference between them, the difference between a program of genocide and the determination of those who escaped it to establish a community in which they would be the makers, not the victims, of the laws.

It is only by thinking of racism as something that occurs principally in the mind, a falling away from proper notions of universal equality, that the desire of a victimized and terrorized people to band together can be declared to be morally the same as the actions of their would-be executioners. It is only when the actions of the two groups are detached from the historical conditions of their emergence and given a purely abstract description that they can be made interchangeable. What President Bush is saying to the United Nations is "Look, the Nazis' conviction of racial superiority generated a policy of systematic genocide; the Jewish experience of centuries of persecution in almost every country on earth generated a desire for

60

a homeland of their own; if you manage somehow to convince yourself that these are the same, it is you, not the Zionists, who are morally confused, and the reason you are morally confused is that you have forgotten history.''

What I want to say, following Bush's reasoning, is that a similar forgetting of history has in recent years allowed some persons to argue, and argue persuasively, that affirmative action is reverse racism. The very phrase ''reverse racism'' contains the argument in exactly the form the president objects to: it was once the case in this country that whites set themselves apart from blacks and claimed privileges for themselves while denying them to others; now, on the basis of race, blacks are claiming special status and reserving for themselves privileges they deny to others; isn't one as bad as the other? The answer is ''no,'' and one can see why by imagining that it is not 1991 but 1955 and that we are in a town in the South. No doubt that town would contain two more or less distinct communities, one white and one black, and no doubt in each community there would be a ready store of dismissive epithets, ridiculing stories, self-serving folk myths, and expressions of plain hatred, all directed at the other community, and all based in racial hostility. Yet it would be bizarre to regard their respective racisms—if that is the word—as equivalent, for the hostility of one group stems not from any wrong done to it but from the wrongs it is able to *inflict* by virtue of its power to deprive citizens of their voting rights, to limit access to an educational institution, to prevent entry into the economy except at the lowest and most menial levels, and to force members of the stigmatized group to ride in the back of the bus; the hostility of the other group is the result of these actions, and while hostility and racial anger are unhappy facts wherever they are found, there is certainly a distinction to be made between the ideological hostility of the oppressor and the experience-based hostility of those who have been oppressed.

Not to make that distinction is, in George Bush's words, to twist history and forget the terrible plight of Afro-Americans, not simply in World War II but in the more than two hundred years of this country's existence. Moreover, it is further to twist history to equate the efforts to remedy that plight with the actions that produced it. Those efforts, designed to redress the imbalances caused by long-standing discrimination, are called affirmative action, and it is a travesty of reasoning to argue that affirmative action, which gives preferential treatment to disadvantaged minorities as part of a plan to achieve social equality, is no different from the policies that created the disadvantages in the first place. Reverse racism is a cogent description of affirmative action only if one considers the virus of racism to be morally and medically indistinguishable from the therapy we apply to it. A virus is an invasion of the body's equilibrium, and so is an antibiotic; but we do not equate the two and decline to fight the disease because the medicine we employ is disruptive of normal functionings. Strong illness, strong remedy—the formula is as appropriate to the health of the body politic as it is to the body proper.

At this point someone will always say, "But two wrongs don't make a right; if it was wrong to treat blacks unfairly, it is wrong to give blacks preference and thereby treat whites unfairly." But this objection is just another version of the forgetting and rewriting of history. The work is done by the adverb "unfairly," which suggests two more or less equal parties, one of whom has been unjustly penalized by an incompetent umpire or official scorer. But the initial condition of equality in relation to which the prep-school virtue of fairness might be an appropriate yardstick has never existed. Blacks have not simply been treated unfairly; they have been subjected first to decades of slavery, then to decades of second-class citizenship, massive legalized discrimination, economic persecution, educational deprivation, and cultural stigmatization; they have been killed, beaten, raped, bought, sold, excluded, exploited, shamed, and scorned for a very long time. The word *unfair* is hardly an adequate description of their experience, and the belated gift of "fairness" in the form of a resolution no longer to discriminate against them legally is hardly an adequate remedy for the deep disadvantages that a prior and *massive* discrimination has produced. When the deck is stacked against you in more ways than you can even count, it is small consolation to hear that you are now free to enter the game and take your chances.

The same insincerity and hollowness of promise infect another formula that is popular with the anti–affirmative action crowd, the formula of the level playing field. Here the argument usually takes the form of saying, "It is undemocratic to give one class of citizens advantages at the expense of other citizens; the truly democratic way is to have a level playing field to which everyone has access and where everyone has a fair and equal chance to succeed on the basis of his or her merit." Fine words, but they conceal the true facts of the situation as it has been given to us by history: the playing field is already tilted in favor of those by whom and for whom it was constructed in the first place; if the requirements for entry are tailored to the cultural experiences of the mainstream majority, if the skills that make for success are nurtured by institutions and cultural practices from which the disadvantaged minority has been systematically excluded, if the language and ways of comporting oneself that identify a player as "one of us" are alien to the lives minorities are forced to live, then words like "fair" and "equal" are cruel jokes, for what they promote and celebrate is an institutionalized unfairness and a perpetuated inequality. The playing field is already rigged, and the resistance to altering it by the mechanisms of affirmative action is in fact a determination to make sure that the present imbalances are continued as long as possible.

One way of rigging the field is the SAT, or Scholastic Aptitude Test, administered by a *private* agency, the College Board. This test figures prominently in Dinesh D'Souza's *Illiberal Education,* in which one finds many examples of white or Asian students denied admission to colleges and universities even though their SAT scores were higher than some of those— often African Americans—who were admitted to the same institution. This, says D'Souza, is evidence that as a result of affirmative action policies colleges and universities tend "to depreciate the importance of merit criteria in admissions." D'Souza's assumption—and it is one that many would share—is that what the test measures is *native intelligence,* the raw brain power persons have independently of their social, racial, ethnic, or economic status; it is, we are told, a test of *merit,* with merit understood as a quality objectively determined in the same way body temperature can be objectively determined by a thermometer.

In fact, however, the test is nothing of the kind. First of all, statistical studies have established that test scores are calibrated to income and zip codes; students coming from families with incomes of over $50,000 will perform on the average of 25 percent better than students coming from families with incomes of $12,000, a differential of about a hundred points in both the math and verbal sections of the test. Second, it has been demonstrated again and again that scores vary in relation to cultural background; the test's questions assume (without stating) a certain uniformity in educational experience and life-style and penalize those who, for any number of reasons, have had a different experience and lived different kinds of lives. As one commentator puts it, "What exactly can ETS mean by 'ability' if one of the differences between having it and not having it is the difference between knowing and not knowing about oarsmen and regattas or about polo and mallets," all of which have made their appearance on the tests? In short, what is being measured by the SAT is not absolutes like native ability or merit but accidents like birth, social position, access to libraries, the opportunity to take vacations or to take tennis lessons.

This conclusion is supported by a third point, the presence of the not-so-small industry of SAT coaching. The most successful of these, called the Princeton Review Course, reports that after a brief two- or three-week period of intense tutoring, students are able to improve their score by an average of 185 points, and 30 percent of the students improved by a measure of 250 points or better. Moreover, what is taught in this and similar courses is not content or subject matter but test-taking skills, skills that are themselves not innate or universal but specific to a particular cultural environment, say the environment of those whose families spend Sunday morning doing the *New York Times* crossword puzzles. Of course it is these same families that can afford to send younger members to the Princeton Review Course at a fee of from $500 to $1,000 as reported in 1985, and no doubt much higher now. "Most of our kids are wealthy," says the director of the course. "Those are the kids who have an advantage to begin with and we're moving them up another level." So much for "native intelligence" independent of cultural or socioeconomic variables.

The fact that test scores can be pumped up artificially by a cram course that teaches more tricks than content may in part explain my fourth point, which is that there is not much of a relationship between test scores and academic performance. It has been shown, for example, that the "correlation between SAT scores and college grades . . . is lower than the correlation between height and weight; in other words you would have a better chance of predicting a person's height by looking at his weight than you would of predicting his freshman grades by looking only at his SAT scores" (David Owen, *None of the Above: Behind the Myth of Scholastic Aptitude* [Boston, 1985], p. 207). Even the SAT literature itself acknowledges that "your high school record is probably the best evidence of your preparation for college," and one must add that there is more than a little evidence that your grade point average is no better a predictor than your score. Everywhere you look in the SAT story, the claims of fairness, objectivity, and neutrality fall away, to be replaced by suspicions of specialized measures, unfair advantages, and unimpressive statistics.

It is against this background that my fifth point, which in isolation might have a questionable force, takes on a special and even explanatory resonance: the principal devisor of the test was an out and out racist. In 1923, Carl Campbell Brigham published a book called *A Study of American Intelligence,* in which he declared among other things that we face in America "a possibility of racial admixture . . . infinitely worse than that faced by any European country today, for we are incorporating the negro into our racial stock, while all of Europe is comparatively free of this taint." "The really important steps," he continued, "are those looking toward the prevention of the continued propagation of defective strains in the present population." Two years later Brigham became the College Board's director of testing and instituted a test based on another famous racist text, Madison Grant's *The Passing of the Great Race,* which divided American society into four distinct racial strains, with Nordic, blue-eyed blond people at the pinnacle and the American Negro, whose arrival in America Brigham described as "the most sinister development in the history of this continent," at the bottom. So here is the great SAT test, devised by a racist in order to confirm racist assumptions, measuring not native ability but cultural advantage, an uncertain indicator of performance, an indicator of very little except of what money and social privilege can buy. And it is in the name of this faultless mechanism that we are asked to reject affirmative action and reaffirm "the importance of merit criteria in admissions."

Nevertheless, the case against affirmative action is not yet done; there is at least one more card to play, and it is a strong one. Granted that the playing field is not level and that access to it is reserved for an already advantaged elite, it still remains true that the disadvantages suffered by others are not racial—at least not in 1991—but socioeconomic; therefore, shouldn't it be the case, as D'Souza urges, that "universities should retain their policies of preferential treatment, but alter their criteria of application from race to socioeconomic disadvantage" (251) and thus avoid the unfairness of current policies that reward middle-class or affluent blacks at the expense of poor whites? One answer to this question is given by D'Souza himself when he acknowledges that the overlap between minority groups and the poor is very large, a point underscored recently by Secretary of Education Lamar Alexander when he said, in response to a question about funds targeted for black colleges, "98% of race specific scholarships do not involve constitutional problems"—by which he meant, I take it, that in 98 percent of the cases race-specific scholarships were also scholarships to the economically disadvantaged, to the poor.

Still, there is that other 2 percent, those nonpoor, middle-class, economically favored blacks who are receiving special attention on the basis of disadvantages they do not experience. What about them? The force of the question depends on the assumption that in this day and age race could not possibly be a seriously disadvantaging fact for those who are otherwise well positioned in the society. But the lie to this assumption was given dramatically in a recent broadcast of the ABC program "Prime Time Live." In a stunning twenty-five-minute segment, the reporters and a camera crew followed two young men of equal education, cultural sophistication, level of apparent affluence, etc., around St. Louis, a city where neither was known. The two differed only in a single respect: one was white, the other black; but that small difference turned out to mean everything. In a series of encounters with shoe salesmen, record store and bank employees, rental agents, landlords, employment agencies, taxicab drivers, and ordinary fellow citizens, the black member of the pair was either ignored or given a special and suspicious attention; was asked to pay more for the same goods or come up with a larger down payment for the same car; was turned away as a prospective tenant; rejected as a prospective taxicab fare; treated with contempt and irritation by clerks, bureaucrats, and city officials; and in every way possible made to feel second-class, unwanted, and inferior.

The inescapable conclusion was that alike though they may be in every other way, the blackness of one of these young men meant that he would

lead a significantly different and lesser life than that of his white counterpart; he would be less well housed and at greater expense; he would pay
more for services and products when and if he were given the opportunity
to purchase them; he would have difficulty establishing credit; the first emo-
tions he would inspire on the part of those he met would be distrust and
fear; his abilities would be discounted even before he had a chance to display them; and, above all, the treatment he received from minute to minute
would chip away at his self-esteem and self-confidence with consequences
that most of us could not even imagine. As the young man in question said
at the conclusion of the broadcast, "You walk down the street with a suit
and tie and it doesn't matter. Someone will make determinations about you
that affect the quality of your life."

Of course, those same determinations are being made quite early on by
kindergarten teachers, grade school principals, high school guidance coun-
selors, and the like with results that cut across socioeconomic lines and
place young black men and women in the ranks of the disadvantaged no
matter what the bank accounts of their parents happen to show. Racism is
a cultural fact, and while its effects may to some extent be diminished by
socioeconomic variables, those effects will still be sufficiently great to warrant the nation's attention and thereby to warrant the continuation of affirmative action policies. This is true even of the field thought to be dominated by blacks and often cited as evidence of the equal opportunities society
now affords them. I refer, of course, to professional athletics; but national
self-congratulation on this score might pause in the face of a few facts,
such as the fact that a minuscule number of Afro-Americans ever receive a
paycheck from a professional team; such as the fact that even though there
are 1,611 daily newspapers reporting on the exploits of black athletes, there
are only five full-time black sports columnists; such as the fact that, despite
repeated pledges and resolutions, major league teams have not managed to
put blacks or Hispanics in executive positions. The commissioner of major
league baseball has declared himself frustrated, disappointed, and at a loss.
Maybe he should consider affirmative action.

When all is said and done, however, there is one objection to affirmative
action that is unanswerable in its own terms, and that is the objection of
the individual who says, "Why me?" Sure, there is a history of massive
discrimination, and I acknowledge that the damage done has not been removed by changes in the law; I understand that the tests by which qualifications are determined are slanted in the direction of certain cultural and
educational experiences, and I agree that it would be good for the country
if more of its citizenry were given the opportunities now afforded a relative
few. But why me? I didn't own slaves; I didn't vote to keep people on the
back of the bus; I didn't turn water hoses on civil rights marchers; why,

then, should I be the one who doesn't get the job or who doesn't get the scholarship or who gets bumped back to the waiting list?

I sympathize with this feeling, if only because in a small way I have had the experience that produces it. Last year I was nominated for an administrative post at a large university. Early signs were encouraging, but after an interval I received official notice that I would not be included in the next level of consideration, and subsequently I was told, unofficially, that at some point there had been a decision to look only in the direction of women and minorities. Although I was disappointed, I did not conclude that the situation was "unfair," because it was obvious that the policy was not directed at me—at no point in the proceedings did someone say, "Let's find a way to rule out Stanley Fish"; nor was it directed even at persons of my race and gender—it was no part of the policy to disenfranchise white males. Rather, the policy was driven by other considerations, and it was only as a by-product of those considerations—not as the main goal—that white males like me were dispreferred. What were those considerations? I was not on the inside of the process, but it isn't difficult to guess: the institution in question is an urban university with a high percentage of minority students and a very low percentage of minority faculty and an even lower percentage of minority administrators; given those circumstances, it made perfect sense to focus on women and minority candidates, and it was within that sense, and not as the result of prejudice, that my whiteness and maleness became disqualifications.

I can hear the objection in advance: "What's the difference? Unfair is unfair; you didn't get the job; you didn't even get on the short list." The difference is not in the outcome but in the ways of thinking that led up to the outcome; it is the difference between an "unfairness" that befalls one as the unintended effect of a policy rationally conceived and an unfairness that is pursued as an end in itself. It is the difference between the awful unfairness of Nazi extermination camps and the unfairness to Palestinian Arabs that arises from, but is not the chief purpose of, the founding of a Jewish state.

The point is not a difficult one, but it is difficult to *see* when the "unfairness scenarios" are presented, as Mr. D'Souza and others always tend to, as simple contrasts between two decontextualized persons who emerge from nowhere to contend for a job or a place in a freshman class. Here, Mr. D'Souza typically says, is student A; he has a board score of 1300; and here is student B; her board score is only 1200, yet she is admitted and A is rejected. Is that fair? Given the minimal information provided, the answer is of course "No"; but if you expand your horizons and consider fairness in relation to the cultural and institutional histories that have brought the two students to this point, histories that weigh on them even if they are not their authors, histories that include the facts about SAT scores I have already rehearsed, then both the question and the answer suddenly grow more complicated.

The sleight-of-hand logic that first abstracts events from history and then assesses them from behind a veil of willed ignorance gains some of its plausibility from another key word in the anti–affirmative action lexicon. That word is "individual," as in "the American way is to focus on the rights of individuals rather than groups." Now individual and individualism have been honorable words in the American political vocabulary, and they have often been well employed in the fight against various tyrannies. But like any other word or concept, individualism can be perverted and twisted to serve ends the opposite of those it originally served, and this is what has happened when, in the name of individualism, indeed in the name of individual rights, millions of individuals are enjoined from redressing historically documented wrongs. How is this managed? Largely in the same way that the invocation of fairness is used to legitimize an institutionalized inequality. First you say, in the most solemn of tones, that the protection of individual rights is the chief obligation of society and its institutions; and then you define individuals as souls sent into the world with equal entitlements as guaranteed either by their Creator or by the Constitution; and then you pretend that nothing has happened to them since they stepped onto the world's stage; and then you say of these carefully denatured souls that they will all be treated in the same way, irrespective of any of the differences that history has produced. Bizarre as it may seem, individualism in this argument turns out to mean that everyone is or should be the *same,* and a recent letter to the *Atlantic* magazine makes the point with a brilliant (if inadvertent) clarity. "I believe," the correspondent says, "it is time to stop insisting to black students that they are different. It is time to let them get on with their studies and with living as individuals in society, *like everyone else*" (emphasis mine) (July 1992, 8). It is fitting (if merely serendipitous) that the man who thinks that being an individual requires you to be indistinguishable from your fellows is named Smith. (How about a new movie, *Mr. Smith Says No to Multiculturalism?*) This dismissal of individual difference in the name of the individual would be funny were its consequences not so serious; for it is the mechanism by which imbalances and inequities suffered by millions of people through no fault of their own can be sanitized and even celebrated as the natural workings of unfettered democracy.

Individualism, fairness, merit—these three words are continually in the mouths of our up-to-date, newly respectable bigots who have learned that they need not put on a white hood or bar access to the ballot box in order to secure their ends; rather, they need only clothe themselves in a vocabulary emptied of its historical content and made into the justification for attitudes and policies they would not acknowledge were they frankly named. So skillful have these new bigots become in appropriating vocabularies and symbols thought to be the property of their natural opponents that they can often represent themselves as the preservers of the values they are subverting. Recently a poster has appeared on a college campus that says in large letters STOP APARTHEID; it is only after you have absorbed the message that you find out that the apartheid in question involves affirmative action, minority scholarships, black dormitories and fraternities, etc. The equation of these mild attempts to afford a disadvantaged minority educational opportunities, along with such rights as voluntary association, with the totally repressive mechanisms of the South African state is a particularly egregious instance of the funny-money logic decried by George Bush, the logic by which the victims of racism become accused of racism the moment the tide turns ever so slightly in their favor. I don't know about you, but I prefer my bigotry straight. I would rather hear someone say, "I really don't believe that blacks or women or Arabs or gays should be enfranchised in every corner of our society. I really am opposed to the Equal Rights Amendment and the Civil Rights Bill of 1964 and the Immigration Act of the following year, and the various laws against discrimination that are multiplying at this very moment. I believe in the right of Anglo-Saxon American males, or in the right of those who are willing to comport themselves as Anglo-Saxon American males, to rule and to guide; and I want to institute policies that ensure the continuation of this natural prerogative." Of course, very few people are saying such things these days; instead they are prating on about fairness, merit, and individuality, and under cover of those once honorable words, they excite the fears of mainstream Americans and assure them that if they act on those fears, it will only be out of the very highest motives. You are some of those Americans, and it is your fears that are being appealed to. I hope you resist.

5

You Can Only Fight
Discrimination with Discrimination

On previous occasions, Mr. D'Souza has spoken first and I have followed. Today we reverse that order, but I find myself still thinking in terms of a response, since it is Mr. D'Souza's accusations that provide the backdrop for evenings such as this. Accordingly, I have decided to use some of my time to rehearse *his* points, if only so that I can feel comfortable in what has become my natural position. In the next hour or two you will hear Mr. D'Souza make most, if not all, of the following arguments:

—While the world slept, left-leaning activists have taken over the curriculum of our colleges, discarding time-honored texts in favor of works whose only recommendation is that they were written by blacks, women, Tahitians, or homosexuals. This program of subversion has been carried out in the name of something called "multiculturalism," which might better be called "oppression studies" or "victim studies."

—This decline in academic standards has gone hand in hand with a decline in admission standards, as students are no longer admitted on the basis of objective criteria but on the basis of race, gender, Third World status, and sexual orientation.

—Those who approve of this policy call it "affirmative action," but what is really being affirmed is the replacement of reason as a guideline by the skewed guideline of preferential treatment, and the result is unfairness for everyone: unfairness for the white male students whose earned credentials are set aside for specious reasons; unfairness for the minorities whose accomplishments will forever be tainted by the suspicion that they were the beneficiaries of unwarranted preference; and unfairness for a society left wondering if its doctors and lawyers and teachers are truly qualified or simply the inferior products of a bad program of social engineering.

—And, worst of all, the entire educational experience has been *politicized* by an intolerant orthodoxy that seeks to impose its views by pushing through so-called curricular reform, by putting in place campus speech codes that stifle discussion and chill dissent, and by intimidating both students and professors who are harassed for speaking their minds and for not toeing the party line as laid down by multiculturalists and gay activists.

—And the remedy for all of this? We must return to standards that transcend the limiting categories of race, gender, and sexual orientation; universities must stop rigging the rules and repledge themselves to the neutral, nonpartisan principles that will assure fairness for everyone, principles based on merit rather than on the skewed arguments of special interests. Only in this way will we be able to restore to our educational system the ethic of hard work and honest competition that is so basic to the democratic way of life.

Each of these points deserves a detailed response, but in the time allotted to me, I can only make a beginning.

First, as a matter of *fact,* the picture of a campus world seized by a radically politicized left professoriate that has trashed the traditional curriculum and terrorized its ideological opponents in ways reminiscent of Senator McCarthy is simply unsupported by the evidence. In a recent cover story, *Time* magazine, no friend to progressive trends in the academy, declared this highly polemical picture of academic life "flatly absurd." It asserted,

> the comparison to McCarthyism could only be made by people who either don't know or don't wish to remember what the senator from Wisconsin and his pals actually did . . . the firing of campus professors in mid-career, the inquisitions by the House Un-American Activities Committee on the content of libraries and courses, the campus loyalty oaths, the whole sordid atmosphere of persecution, betrayal and paranoia. The number of conservative academics fired by the left thought police is, by contrast, zero. (Robert Hughes, Feb. 3, 1992, p. 46)

Time's judgment is backed up by every survey that has been conducted, by surveys that reveal a remarkable stability in the curriculum, by surveys that reveal a professoriate still from 80 to 90 percent male and white, by surveys that show 97 percent of colleges reporting *no* undue pressure on conservative scholars and teachers. This general finding can be further substantiated by the facts about the English department at Duke, which has been offered by Mr. D'Souza and others as the very symbol of what has gone wrong; I say flatly that there is no relationship whatsoever between the media characterization of that department and the reality of its day-to-day life, and I am prepared to back up that statement with massive documentation.

Nor is it the case that affirmative action policies have eroded standards and drastically diluted the quality of the student body on our campuses. The truth is that the vast majority of affirmative action programs operate from a pool made up of conventionally qualified students and that the adjustments for minority or disadvantaged applicants are then made *within* that pool and not, as is sometimes implied, by walking outside and taking the first ten unqualified youths you happen to meet. The University of California at Berkeley, one of Mr. D'Souza's prime exhibits, is a good example. By law, only the top 12.5 percent of graduating high school students are U.C. eligible; the pool from which the university selects its students is thus made up of uniformly qualified applicants; 60 percent of those admitted from this U.C.-eligible pool are admitted on the basis of numerical test scores and grades; it is only after this "mechanical" measure has done its work that the admissions officers begin to take into account other factors such as special talents, athletic ability, disabled status, underrepresented minority status, socioeconomic disadvantage, rural background, and students seeking reentry. A significant number of those admitted under these categories are white and Asian; more important, all are qualified by virtue of being in the top 12.5 percent. To be sure, at this level admission may be denied to someone whose raw test scores are higher than those of someone else who has been admitted, but this only reflects the fact that test scores, while an important measure of what is desirable in a campus community, are not the only measure, and that Berkeley, like every other university in the country, uses the admissions process as a way of shaping a student body whose diversity will itself be an important factor in the educational experience. Not only is this strategy a generally accepted one and one given the blessing of the Supreme Court in the famous *Bakke* case (which acknowledges diversity as a legitimate goal), it is a necessity, since of the more than 21,300 who applied for entry to the class of 1990, 5,800 had straight-A averages and were competing for only 3,500 available spots.

When those 3,500 spots are filled, it will always be possible to point to unsuccessful applicants who, by at least one measure, rank higher than some of those who have been more fortunate. But while this result is properly the cause of personal disappointment, it cannot, I think, be characterized as an injustice, as it could be if the rejected applicants had been *singled out* for unequal treatment. What justice requires is not that each case be decided on its own merits (whatever that would mean) but that any decision reached in a particular case follow from some rationally defensible set of considerations. It would be reprehensible if the decision were flatly arbitrary, traceable to no logic of selection whatsoever, but so long as there *is* a logic of selection, and one that could be justified with reference to a moral and educational vision, the individual inequities it yields will in an important sense be principled. This does not mean, of course, that the principles are ones everybody will acknowledge, for there will always be rival visions that bring with them alternative logics of selection. But whatever vision informs the process, the logic of selection it entails will produce inequities that will be the cause of personal disappointments in different populations. Those who suffer such disappointments are less victims, as they would be were they robbed or mugged, than they are casualties, as they would be if they were let go by a company that was restructuring its work force according to some master plan.

But why not devise plans and policies that have no casualties, policies that are fair to everyone? The answer is that there is not and could not be such a policy because fairness is itself a contestable concept and will be differently defined depending on what assumptions inform those who brandish it as a measure. Fairness for everyone would be possible only if everyone's interests were the same, if everyone were in agreement as to what baseline considerations must be in place for a procedure to be labeled "fair." But if that were the case, the question of fairness would never be raised. It is raised precisely because everyone's interests are *not* the same, and since different interests will generate different notions of fairness (the debate between those who call for equality of access and those who call for equality of opportunity is an example), any regime of fairness will always be *un*fair in the eyes of those for whom it was not designed. A change in design will not produce less unfairness but unfairness differently directed. The amount of unfairness in the world can never be eliminated or even diminished; it can only be redistributed as in the course of political struggle one angled formulation of what it means to be equitable gives way to another.

That is why, as the title of this chapter asserts, you can only fight discrimination with discrimination. The usual wisdom is to proclaim exactly the reverse and require remedies for discrimination to be rigorously nondiscriminatory in their turn. If an admissions policy is found to be discriminatory, one corrects it according to this argument, not by tilting the balance in the other direction, but by purging it of any tendency to take considerations of race, gender, ethnic origin, economic class, and so on into account. What remains will be a policy responsive only to considerations of merit. As compelling as this reasoning may seem, however, it founders on the fact that before a policy of such purity can even be conceived (never mind implemented), one would have to specify exactly what a consideration of merit *is,* and any specification one comes up with will be challengeable and (potentially) controversial. Merit, like fairness, is a contested concept; as a measure it is itself derived from a particular view of what is worthy and indispensable, and because that view is partial—that is what it means to be a view—so is the measure of merit it generates.

Suppose, for example, a decision between two job candidates were to turn in part on the fact that one is a woman. Wouldn't that be a case in which merit will have been compromised in favor of extraneous factors? My answer is "no." Merit will not have been compromised but reconceived in relation to an alternative measure; for if it is your judgment (as it is mine) that the presence of women in a department makes a difference— in the number and kinds of voices heard, in the perspectives that enrich the conversation, in the general feel of the workplace—and if you value that difference and put it into play by factoring gender into your decision, you will be acting in accordance with your notion of merit, a notion that will seem distressing and even bizarre to those for whom the measure of merit is a quantity of publications, or familiarity with literary theory, or, in some places, ignorance of literary theory. Once the measure of merit relative to a particular practice has been established (if only temporarily), those who would have been judged meritorious under another measure will believe, and believe correctly, that they have been the victims of discrimination. Any regime of merit, like any regime of fairness, will be a form of discrimination; moreover, any effort to redress the effects of discrimination will only reinstitute discrimination as its unsought-for but inevitable by-product.

The demand that discrimination be eliminated entirely is finally the demand that we live outside (or above or to the side of) the varied and conflicting perspectives that give to each of us a world saturated with goods, goals, aspirations, and obligations. It is the demand that we no longer be human beings—beings defined by partiality—but become as gods, beings who know no particular time or place. This is the dream not only of philosophy but of theology (in relation to whose assumptions it at least makes sense), but until we are the beneficiaries of a revelation or of a god who descends to begin his reign on earth, it must remain just that, a dream, and we will continue to be confined within the traditions and histories that generate our differing senses of what is true and good and worth dying for. To put it another way, each of us lives in a narrative, a story in which we are at once characters and the tellers. No one's story is the *whole* story, and in the various lights shed by our various stories, different truths will seem self-evident and different courses of action will seem obviously called for. Those we now criticize as racists, those who in the nineteenth century and for the first sixty years of the twentieth argued for second-class citizenship and segregated facilities and limited access to the ballot box, did not think of themselves as evil persons pursuing evil policies; they thought of themselves as *right,* and from the vantage point of the story they were living and telling—a story I find unpersuasive and repellent—they were. In the years since 1960, that story has become less and less compelling to more and more people, which means not that its limitations have been transcended but that another story, with its own limitations, has become more compelling. The effect of telling that newer story has not been to eliminate partiality but to alter its shape, so that while the old story strongly recognized and validated some facts and de-emphasized some others, the new story recognizes and validates a different set of facts and in so doing necessarily slights facts to which the inhabitants of alternative stories cling for dear life.

The conclusion is perhaps distressing—especially if you are holding out for a vision rooted in no story but in the Whole Truth as seen by the eyes of God—but it is inevitable: alternative stories are alternative vehicles of discrimination, alternative narratives in which some interests are slighted at the expense of others. No agenda operates (or can even be conceived) that does not privilege some concerns and turn a hostile or blind eye to some others, and what follows from that conclusion is the even more distressing conclusion that you can only fight discrimination—dislodge one story whose telling has consequences you don't like—by discriminating, that is, by putting another story whose consequences someone else won't like in its place.

And where does that leave us? Just where we have always been, debating various agendas, each of which pursues goals that exclude or de-emphasize the goals of its rivals and none of which can legitimately claim to be more fair or more objective or more neutral than any other. Now, realizing that no agenda can make good on that claim (if only because those disadvantaged by it, those left out by its vision, will always cry "foul") is not in and of itself a helpful insight. It doesn't lead to anything positive. But it can have a salutary negative effect; for by identifying as nonfruitful the path of inquiry that seeks to determine which of two or more schemes of organization is the more fair or the more objective, it shifts our attention away from the realm of abstract moral calculation and into the realm of particularized history, where questions are asked in a context and not in a vacuum.

Consider, for example, the vexed issue of hate speech on campus. It is often said that the logic of speech codes would require the disciplining of those who spoke scornfully of whites and males and even Nazis as well as the disciplining of those who spoke scornfully of blacks, women, and gays; for after all, if the rule is that one should not discriminate, are not all acts of discrimination equal? The answer, I think, is *no,* because discrimination is not a problem in logic but a problem in historical fact, and it is a fact about discrimination that it is usually practiced by the powerful at the expense of the relatively powerless. The point has been made by Thomas Grey, professor of law at Stanford and a principal author of that university's code. Grey acknowledges that under the provisions of the Stanford code, calling a black student "nigger" would constitute harassment, but calling a white student "white trash" would not. The reasoning is that since in our society whiteness is the norm, not only statistically (and that, of course, may change) but more importantly in the sense that normative values are understood to be derived from a white Anglo-Saxon history, "there are *no* epithets in this society at this time that are 'commonly understood' to convey hatred and contempt for whites *as such*." This is so because the common understanding has been fashioned by and for whites, and therefore any epithet denigrating them would be "commonly" regarded as a mistake, something not to be taken seriously. In contrast, insults directed at traditionally persecuted or disadvantaged groups

> draw their capacity to impose the characteristic civil-rights injury to "hearts and minds" from the fact that they turn the whole socially and historically inculcated weight of . . . prejudices upon their victim. Each hatemonger who invokes each of these terms summons a vicious chorus in his support. It is because, given our cultural history, no such *general* prejudices strike against the dominant groups that there exist no comparable terms of universally understood hatred and contempt applicable to whites, males and heterosexuals as such. ("Civil Rights vs. Civil Liberties: The Case of Discriminatory Verbal Harassment," *Social Philosophy and Policy* 8, no. 2 [Spring 1991]: 81–107)

The asymmetry—someone would say the unfairness—of this is obvious; but symmetry would require us to pretend that epithets hurled at whites and heterosexuals have as much capacity to inflict psychological and material harm as epithets hurled at blacks and gays, and that is simply not so. As Anna Quindlen put it in the opening sentence of a column in the *New York Times* (June 28, 1992), we must "begin . . . with the fact that being called a honky is not in the same league as being called a nigger." Those who blink this fact and argue that insults directed at white males should be met with the same disfavor and penalties as insults directed against minorities are working, whether they know it or not, to preserve the lines of power and cultural authority as they have existed in the past. By insisting that *from now on* there shall be no discrimination, they leave in place the effects of the discrimination that had been practiced for generations. You don't redress discrimination simply by stopping it, for its legacy will live on in the form of habits of thought and action now embedded in the fabric of society. Redress requires active intervention, and active intervention will always be discriminatory in some other direction. The choice is never between discrimination and its opposite but between alternative forms of discrimination. You can only fight discrimination with discrimination.

The same analysis will remove the sting from another apparent asymmetry (Arthur Schlesinger, Jr. [Knoxville. TN: Whittle Direct Books, 1991]), in *The Disuniting of America,* complains that while Actors' Equity tried to prevent a British actor from portraying a Eurasian in the play *Miss Saigon,* the organization did not "apply the same principle" to black actors playing Shakespeare at the same time in the same city. Schlesinger assumes that Actors' Equity was working on the principle that roles should only be played by actors whose ethnicity matched that of the character as written, but I would guess that what was involved was not a principle but a historical perception that in the past white actors had routinely played Asians, blacks, Indians (both American and Asian), Egyptians, Aborigines, Eskimos, etc., and that given the distress and outrage produced by that history, the theater world should be wary (at least for a time) of continuing it. The impulse, in short, is not to be color-blind or color coordinated but color sensitive, and the judgment is that in the 1990s being sensitive to the sensibilities of Asians and blacks is a higher priority than being sensitive to the sensibilities of whites, who have been, and continue to be, doing quite well in the theater and everywhere else. The result is asymmetry of a kind repugnant to those who would hew to a policy of perfect neutrality in these matters, but what is usually meant by perfect neutrality is a policy that leaves in place the effects of the discrimination you now officially repudiate. Neutrality thus perpetuates discrimination, rather than reversing it, for you can only fight discrimination with discrimination.

A final example concerns the controversial effort in Baltimore, Detroit, and other cities to establish schools for young black males; the objection to this is the same leveled at colleges that allow and support all-black fraternities and all-black dormitories: isn't this segregation, and wasn't the elimination of segregation what the civil rights movement was all about? Well, not exactly; while it is certainly true that the watchword of the civil rights struggle was integration—the breaking down of the barriers separating the races—the desire was not so much to fulfill a democratic ideal as it was to gain something of which African Americans had been deprived both by law and by custom, a range of opportunities that stretched across housing, employment, education, transportation, and other areas. Those opportunities could not be realized so long as Jim Crow laws were in effect, and therefore the first step in opening things up was the undoing of those laws in all of the relevant areas of everyday life. But the illegalization of *forced* segregation, under which blacks had no choice but to be separate, does not mean that those same blacks must be forced to be integrated and forsake the right of deciding that present difficulties might be ameliorated by a pattern of segregation they themselves have chosen. After all, the point of the movement was not to give blacks the freedom to be just like whites but to give blacks the freedom, already enjoyed by whites, to order their own affairs, even if they chose to do so by reinstituting a practice they had fought when it was one to which they had been compelled. This seems problematic only if one assumes, as an ahistorical moral logician must, that segregation as an empirical fact—separate facilities for different races—is always the same (distasteful) phenomenon no matter what the historical conditions of its emergence. But I would submit that segregation as a policy of oppression is quite different from segregation as a policy (debatable, to be sure) designed to further the progress of a group once oppressed by it.

That difference can be tracked and filled in by the history of constitutional law in this area. For a long time in this country forced segregation was justified by the phrase "separate but equal"; but then, in 1954, the Supreme Court pointed out that the separate facilities imposed by the white majority were anything but equal, and therefore the Court issued a decision that declared, in effect, "if separate, inherently *un*equal." Now, more than thirty-five years later, new conditions suggest to some that, given the present sorry situation with respect to the education of black males, separate might be a step in the direction of equality. Is this a history of contradictions? I would say no; rather, it is a history illustrative of the different values separateness can take on, depending on the circumstances surrounding it. Separateness is not always the same thing, and in 1992, when it represents a voluntary option rather than a proscribed condition, it names an experiment—a strategy that may or may not work—and not an immoral practice.

Each of my examples—speech codes that assign different values to words when they are uttered by different groups, casting patterns that conform to the ethnic sensitivities of minorities but do not recognize (because they do not exist) the ethnic sensitivities of the Caucasian majority, proposals for segregation by the population that formerly resisted it—points in the same direction, away from an abstract realm in which the calculation of good and evil, better and worse, proceeds as if no one ever had a history and toward the realm of real life, in which human finite agents act according to partial and revisable lights and not according to the full light of revelation. In short, and to use the word that has been hovering at the edges of my argument, my examples turn toward *politics* and to the insight that politics is inescapable for everyone and especially for those whose claim it is to have escaped.

These days you cannot mention politics without calling up the specter of "political correctness." "Politically correct" is a dismissive accusation that only makes sense if it is opposed to a superior alternative. Presumably, what is deficient about "*political* correctness" is that its judgments of right and wrong are made from an angle, from a site of interest, from a position colored by partisan desires. Really correct correctness, on the other hand, would proceed from no angle, no interest, no partisan desire, but from the perspective of truth. The trouble with this requirement, however, is that no human being could meet it because no human being sees truth directly, stands to the side of interest, sees by more than partisan lights. There is no really correct correctness, at least not any we can validate by standards that are themselves not political. "Political correctness" is simply a pejorative term for the condition of operating on the basis of a partial vision, and since that is the condition of all of us, we are all politically correct. To be sure, we are not all politically correct in the same way; the products of different histories, we are all committed to truths, but to truths perpetually in dispute. That is what it means to be partial, or, in an older and preferable vocabulary, fallen. It is with that same vocabulary in mind that I would propose an emendation, the substitution for "politically correct" of the more accurate phrase "faithfully correct," correct from the vantage point of the different faiths we involuntarily inhabit. We are all faithfully correct, true to the convictions that now grasp us and open to the possibility that in the fullness of time we may be grasped by better convictions. This is at once our infirmity and our glory. It is our infirmity because it keeps us from eternity, and it is our glory because it sends us in search of eternity and keeps us from premature rest.

6

BAd COMPANY

One of the things that has been left out of the current debates about the canon and multiculturalism is history. Few have paused to inquire into the relationship between the present controversy and controversies that have erupted in the past. The only exception to this silence about history has been the oft-repeated assertion that the proponents of revised and enlarged curricula are leftover sixties revolutionaries who have traded in their beards and beads for tweed jackets and tenured positions in the universities. So we have this rather thin account of where the multiculturalists come from, but we have had no account at all of where the antimulticulturalists come from, and my intention here is to provide one.

Let me begin by rehearsing some of the main tenets of the antimulticulturalist argument, an argument that has been recently given a concise and forceful exposition by Arthur Schlesinger, Jr., in his book *The Disuniting of America: Reflections on a Multicultural Society*. Schlesinger begins by sounding a call of alarm, a kind of 1991 Paul Revere warning that not the British (or the Russians) but the multiculturalists are coming, and that they threaten the American way of life. He finds the threat in what he calls the "ethnic upsurge," an "unprecedented . . . protest against the Anglocentric culture" that "today threatens to become a counterrevolution against the original theory of America as . . . . a common culture, a single nation." Schlesinger deplores the rejection of what he calls "the old American ideal of assimilation"—the ideal that asks immigrants and minorities to "shed their ethnicity" in favor of the Western Anglo-Saxon tradition. That tradition, says Schlesinger, is the "*unique* source" of ideas of "individual lib-

Originally printed in *Transition* 56 (New York: Oxford University Press, 1992), pp. 60–67. Reprinted by permission of the publisher.

erty, political democracy, the rule of law, human rights, and cultural freedom,'' and he contrasts the virtues of Western individualism with the
"collectivist cultures"—the cultures based on group or ethnic or religious
identity—of Africa and Islam, cultures that, he says, "show themselves
incapable of operating a democracy . . . and who in their tyrannies and
massacres . . . have stamped with utmost brutality on human rights." Given
the practices of what he calls "tribalist" cultures, cultures based in his view
on "despotism, superstition, . . . and fanaticism," Schlesinger finds it absurd that people are now arguing that we should accord those cultures dignity and respect, and that "in this regard the Afrocentrists are especially
absurd." "White guilt," he declares, "can be pushed too far," and he
predicts that the multiculturalist ethnic upsurge will be defeated by the fact
that "the American synthesis has an inevitable Anglo-Saxon coloration."

It is clear from these quotations that for Schlesinger the danger of multiculturalism is not confined to the classroom, but extends to the very fabric
of our society. The message is that unless we beat this challenge back and
reclaim the educational process for the mainstream values upon which
America was built, the unity of the American experience will be replaced
by the balkanization and barbarism characteristic of the alien cultures of
Africa and Islam. "The debate about the curriculum," says Schlesinger,
"is a debate about what it means to be an American," and "what is ultimately at stake is the shape of the American future."

Now when Schlesinger utters these statements, he does so with a sense
that the challenge to the American way is a new one; he calls it "unprecedented," a recent phenomenon invented in his view by a group of self-
serving intellectuals who really don't speak for the communities they pretend to champion; and he implies that until this self-authorized group began
its rabble-rousing efforts, the national business of forging and maintaining
an American identity had not been threatened. But in fact that threat—if it
is a threat—is a very old one, and old too are the arguments Schlesinger
uses against the threat. When he declares that at this moment what is at
stake is the shape of the American future, a future that will be lost if we
do not reject policies and agendas that are simply un-American, he echoes,
knowingly or unknowingly, a series of essays and books written between
1870 and 1925. These books and essays had titles like *The Melting Pot
Mistake* (by Henry Pratt Fairchild), *Our Country* (by Josiah Strong), *The
Passing of the Great Race* (by Madison Grant), and *A Study of American
Intelligence* (by Carl Campbell Brigham). Although they have different focuses, they share a set of attitudes and arguments. First, they are all anti-
immigration tracts that identify immigration, especially from countries other
than those that provided the original "American stock," as a great and
imminent danger. Here is a typical quote: "The most significant unity of

the American people is national unity, and the outstanding problem involved in immigration has been the problem of preserving national unity in the face of the influx of hordes of persons of scores of nationalities" *(Melting Pot)*. In the mind of this writer the chief peril comes from Southern Italians who have a "very small proportion of Nordic blood" in their veins and the "Alpine stock" which "appears to be essentially mongoloid in its racial affiliations," thus raising the fearsome prospect "of introducing into the American population considerable strains of Mongoloid germ plasm."

In addition to being anti-immigration, these tracts are often anti-Catholic or, as they put it, anti-Romanist: Catholics, they argue, are loyal not to this country but to the pope of Rome; and not only do Catholics owe their allegiance to a foreign power, but in allying themselves with this power, they reject the ethos of "free inquiry" and substitute for it the ethos of blind obedience to authority. Nor can one take comfort in the fact that many fall away from Catholicism, for not having been nurtured in the right Protestant virtues, lapsed Catholics react to their newfound freedom by falling into "license and excess" *(Our Country)*. It is because license and excess are thought to be the opposite of the Anglo-Saxon virtues of restraint and moderation that these same books are temperance tracts, and indeed in the eyes of these authors, the growth of immigration, the unchecked breeding of Catholics, and the rise of what one author calls "the liquor power" are seen as manifestations of the same threat.

At bottom, that threat is a racial one. Invoking a popular belief that the path of civilization was moving ever westward, these authors identify that path with the emergence of a pure Western type drawn from Anglo-Saxon stock. The fact of this emergence is explained by reference both to God, who is seen as guiding this adventure with his "mighty hand" *(Our Country)*, and to Darwin, who tended to see in the American "race" a strong confirmation of the theory of natural selection. Indeed, at one point in *The Descent of Man*, Darwin quotes approvingly this statement made by a Protestant minister: "All other . . . events . . . only . . . have purpose and value when viewed in connection with, or rather as subsidiary to, the great stream of Anglo-Saxon emigration to the West."

When all of these ingredients—anti-immigration, anti-excess, anti-Catholicism, and out-and-out racism—are mixed into one brew, the result is something like this:

> It seems to me that God . . . is training the Anglo-Saxon race for an hour sure to come in the world's future. . . . Then this race of unequaled energy—the representative . . . of the largest liberty, the purest Christianity, the highest civilization . . . will spread itself over the earth . . . this powerful race will move down upon Mexico, down upon Central and South America, out upon the

islands of the sea, over upon Africa and beyond. And can anyone doubt that the result . . . will be the survival of the fittest? *(Our Country)*

And what happens in this grand vision to the indigenous peoples of these lands? The answer is as inevitable as it is chilling: "Nothing can save the inferior races but a ready and pliant assimilation." Significantly, the triumph of Anglo-Saxon superiority will also be a triumph of language. Here is the Reverend Nathaniel Clark sounding just like those who today defend the traditional canon: "The English language . . . gathering up into itself the best thought of all the ages, is the great agent of Christian civilization throughout the world"; and another minister also speaks with 1991 accents when he prophesies that "the language of Shakespeare would eventually become the language of mankind" *(Our Country)*. One sees again how comprehensive this vision is: it is informed by a sense of history; it declares a social program and argues for its political implementation; it boasts a theory of language and at least one sacred text; and it is shadowed by dangers—often called perils or menaces—that must be withstood so that a glorious past may flower into an even more glorious future.

In linking these late nineteenth- and early twentieth-century tracts with the recent essay by Arthur Schlesinger, I may seem to be practicing guilt-by-association, just as the McCarthyites did in the fifties when they condemned anyone whose writings or speeches could be shown to include phrases that were also used by Communists. The difference, however, is that where the members of the House Un-American Activities Committee seized on isolated sentences taken out of context, I am arguing for a match at every level, from the smallest detail to the deepest assumptions. It is not simply that the books written today bear some similarities to the books that warned earlier generations of the ethnic menace; they are the same books.

Consider, for example, Laurence Auster's *The Path to National Suicide: An Essay on Immigration and Multiculturalism*. Published in 1990, the book warns, in familiar apocalyptic terms, that "we may be witnessing the beginning of the end of Western Civilization as a whole"; the principal villain is the 1965 immigration act, which shifted immigration priorities from those Nordic European peoples who had furnished America with its original stock to Asian and African peoples from Third World countries. The result, if we do not reverse this policy, will be the end of American culture: "Like ancient Greece after the classical Hellenes had dwindled away and the land was repopulated by Slavonic and Turkic peoples, America will become literally a different country." And what evidence does Auster cite for this dire prediction?—multiculturalism in our schools, which he describes as "an attempt . . . to tear down, discredit and destroy the shared story that has made us a people and impose on us a new story." In that story, which is taught to our children, "the contributions of the American Indian, African, Hispanic (and even *Asian!*) cultures are as important to our civilization's heritage as the Anglo-Saxon contribution." What is obscured is "that the United States has always been an Anglo-Saxon civilization" and that previously immigrants, largely European, tended only to "augment the . . . mix of minorities in our predominately white society." It is this white society that must be preserved in the face of the twin evils of immigration and multiculturalism. Anticipating the objection that demographic projections indicate that in some places like California, whites will soon be a minority, Auster responds that this is just the trouble: "If it is the sheer *number* of Non-Europeans . . . that obligates us to abandon 'our' cultural tradition, is it not an inescapable conclusion that the white majority in this country, if it wishes to preserve that tradition, must place a rational limit on the number of immigrants?"

Now it would be easy to dismiss Auster as a fringe voice. His book is published by something called "The American Immigration Control Foundation"—its financing would no doubt make an interesting story—and his vocabulary is much too blunt to be attractive to mainstream audiences. This is less true, however, of another book, published in 1991, Richard Brookhiser's *The Way of the WASP: How It Made America, and How It Can Save It, So to Speak.* Brookhiser is a senior editor of the *National Review,* a contributor to *Time* and *The New Yorker,* and a former speechwriter for then Vice President Bush. He begins his book by identifying the key white Anglo-Saxon Protestant virtues—Conscience, Antisensuality, Industry, Use, Success, and Civic Mindedness—and contrasts these to a set of non-WASP, presumably African or Asian or Mediterranean, characteristics: Self, Creativity, Ambition, Diffidence, Gratification, and Group Mindedness. (Interestingly, this contrast, which elevates Nordic populations at the expense of populations living in temperate or tropical zones, is an inversion of the much-ridiculed distinction between "ice people" and "sun people" put forward by Leonard Jeffries.) "The *WASP* character," says Mr. Brookhiser, "is the American character" and without it, America "would be another country altogether." Without it, America "is sure to lose its way." It follows then that non-WASPS, and especially immigrants, must submerge their native character in the American one, and this is why "the only price even the most exclusive WASPS exacted was that newly arrived others should become exactly like themselves." This is the strong form of assimilation which produces not a mixture but a surrender and it is exactly the message trumpeted by Henry Fairchild in *The Melting Pot Mistake:* "The true member of the American nationality is not called upon to change in the least. The traits of foreign nationality which the immigrant brings with him are not to be mixed or interwoven. They are to be *abandoned.*"

Unfortunately, according to Brookhiser, that message has not been heeded, and today the WASPS, rather than demanding the assimilation of alternative cultures, are deferring to them, and "the cumulative effect of multiple acts of deference—to other standards, cultures, even species—is insecurity, uncertainty, sheepishness." Brookhiser concludes by urging a return to the older model: "The best favor America can do its newcomers is to present them with a clear sense of what America is, and what they should become . . . the people we want aren't permanent immigrants, but future WASP's."

Auster, Brookhiser, Schlesinger—three very different authors—one a xenophobic nationalist, the second a neoconservative well placed in the corridors of power, the third a moderate progressive identified with the policies of the Kennedy administration; and yet all three tell the same story about the formation of the American character, the necessity of preserving it, and the threat it faces from ethnic upsurges: a story that continues in every respect, from words and phrases to large arguments, a tradition of jingoism, racism, and cultural imperialism. Moreover, that tradition has been implanted into the educational process by a mechanism designed to enforce its assumptions. I am referring to the SAT tests, often put forward as the objective measure of the true merit that is being undermined by the erosion of standards. What most people don't know (in addition to the facts that the tests are culturally biased, statistically unreliable, and easily vulnerable to manipulation by expensive crash courses that teach tricks rather than content) is that the test was devised and administered by an out-and-out racist, indeed by one of the authors I alluded to earlier.

Carl Campbell Brigham's *A Study of American Intelligence,* published in 1923, is an amazing document, even in the context of this particular tradition. It advertises itself as the "first really significant contribution to the study of race differences in mental traits." Brigham announces that he begins by accepting the classification of races found in Madison Grant's *The Passing of the Great Race,* an unashamed racist tract that identifies the Nordic as the superior race and classifies the others in a descending order from Alpine, to Mediterranean, to Eastern and Near Eastern, with the Negro race at the bottom. Clearly Brigham expects to find confirmation of Grant's hierarchy in his data and—surprise, surprise—he is not disappointed. First he finds that army officers are more intelligent than enlisted men, and then he finds that within the enlisted ranks whites are clearly superior to blacks. Turning to immigration he uncovers "a gradual deterioration in the class of immigrants . . . who came to this country in each succeeding five year period since 1802," a deterioration that corresponds to the increasing number of non-Nordic immigrants, whose intelligence is found to be more than five times less than that of immigrants from England and Scotland. "In a very definite way," he concludes, "the results which we obtain by analyzing the army data support Mr. Madison Grant's thesis of the superiority of the Nordic type.''

Happily these results not only corroborate ''the marked intellectual inferiority of the negro,'' but also ''tend to disprove the popular belief that the Jew is highly intelligent.'' Conclusions from these ''results'' quickly follow: ''The average negro child cannot advance through an educational curriculum adapted to the Anglo-Saxon child.'' This is a warning against any form of integrated education, and there is inevitably a warning against the mixing of the races by marriage. If the Nordic type makes the mistake of blending with the other types, ''then it is a foregone conclusion that this future blended American will be less intelligent than the present native born American.'' Unfortunately, Brigham reports, there is a tendency in this direction already, but the ''deterioration of American intelligence is not inevitable'' if the appropriate ''legal steps'' are taken. These steps, he assures us, should be dictated by ''science''—that is, by his results—''and not by political expediency.'' In this last paragraph, Brigham looks back at the several places in his book where he considers the possibility that environmental conditions rather than simple native racial ability might account for his statistics, but he insists, without any argument or documentation, that all studies ''agree in attributing more to original endowment than to environment,'' an insistence that has been repeatedly echoed by SAT testers who are today administering the test he devised; for, two years after the publication of *A Survey of American Intelligence,* Brigham became director of the College Board's testing operation and set in motion the procedures to which, only slightly modified, many of us have been subjected. There has not even been an official repudiation of Brigham's racism, merely a declaration that it did not compromise his scientific objectivity; the library at the Educational Testing Service compound still bears his name.

What does it all mean? Does it mean that Auster, Brookhiser, and Schlesinger are racists? Well, if you mean by racist someone who actively seeks the subjugation of groups thought inferior to his own, none of these qualify. If you mean by racist someone whose views about race, if acted upon in a political way, will lead to the disadvantaging of certain groups, then Mr. Auster is a serious candidate; and if you mean by racism the deployment of a vocabulary that avoids racist talk but has the effect of perpetuating racial stereotypes and the institutions that promote them, then Mr. Brookhiser is in the running; and Mr. Schlesinger, with his talk of the inevitable Anglo-Saxon ''coloration'' of the American character and the necessity of sublimating ethnic strains in a true American amalgam, is a shoo-in.

Of course none of these men would consider themselves racists, and they are concerned in their writings to distance themselves from that label, usually by presenting themselves as the champions not of the Anglo-Saxon race as such but of the political and cultural values given to us by that race two hundred years ago. The appeal, in short, is not to prejudice but to national unity in the face of the danger posed by ethnic balkanization. Auster goes so far as to cast a retroactively benign glow over the entire tradition I have examined: "The concern common to all the historical stages of anti-immigrant sentiment was not race as such but the need for a harmonious citizenry, holding to the same values and political principles and having something of the same spirit." Similar statements can be found in the pages written by Schlesinger, C. Vann Woodward, and others, but the distinction begins to blur when one realizes that in order to produce a harmonious citizenry the traditions and values associated with non–Anglo-Saxon races must be ruthlessly sublimated; "the people we want are future WASPS." There is, to be sure, a difference between a sense of racial superiority that takes the form of legal and political oppression and a sense of racial superiority that issues in a noncoercive call to national unity and common values; but in the end the result is the same, even when it is reached by softer means. Schlesinger, Auster, Brookhiser, etc. may not be racists in the manner of their predecessors, but in the absence of the real thing—although we must always remember David Duke—they will do. "Common values," "national unity," "American character," "one people," a "single nation," the idea of "assimilation"—these are now code words and phrases for an agenda that need no longer speak in the accents of the Know Nothing party of the nineteenth century or the Ku Klux Klan of the twentieth.

Of course there are many who resonate to those words and phrases and who would be horrified to think of them as alibis for a racist politics. To them I would give the advice that has often been given to me when I was taken by a piece of rhetoric I later disavowed: consider the source. I know that this advice goes against the assumption, so strongly embedded in liberal thought, that ideas are to be evaluated on their merit and not on the basis of the historical condition of their emergence, but that assumption itself assumes that ideas exist in some eternal realm or depoliticized marketplace of ideas, and that assumption seems to me to be wrong: ideas are only intelligible within the particular circumstances that give rise to them, and it is within those circumstances that ideas are put to purposes and do work. The ideas that are the stock-in-trade of the argument against multiculturalism originated in bad purposes and have traditionally done bad work, and to those who find them congenial today, I say they will make bad company.

I cannot end without saying one more word about Arthur Schlesinger, or rather about a picture of him that appears in his book. The picture is without any apparent relationship to the facing text, but there is a relationship to be teased out of it nevertheless. Schlesinger's figure in half torso fills the foreground of the picture; his back is toward us, but his head is twisted around so that he faces the camera and looks directly at us, seeming almost ready to step out of the frame. The face is a young one—the year is 1958— and unformed; the features are ethnic, Semitic, even a bit negroid; the look is quizzical, as if he were asking, "what will I become?" The answer he hopes for is represented in the picture by a figure in its middle ground— elegantly dressed, erect, self-contained, finished, looking not at us but at the far horizon. It is Arthur Schlesinger, Sr., a man fully and safely framed, obviously at home in the landscape which just happens to be Harvard Yard. Harvard is the national University, the true symbol of Protestant America; the picture suggests that Protestant America has not yet enclosed the younger Schlesinger in its bosom and that it is not yet certain that he will be able to complete the passage from the heritage of his Jewish grandfather to total assimilation. By the evidence of this book, which nowhere makes mention of that heritage, he has made it.

7

Speaking in Code, or, How to Turn Bigotry and Ignorance into Moral Principles

"I have a dream that one day my four little children will live in a nation where they will not be judged by the color of their skin, but by the content of their character." Martin Luther King, Jr., spoke this famous sentence in his most famous speech on August 28, 1963, at the Lincoln Memorial. I wonder how he would feel if he knew that in 1992 those words and others from his various speeches and writings are being employed in the service of the racism he spent his life combating. In saying this, I might seem to impugn the motives of many who would protest that it could hardly be racism to insist on the color-blind application of the law, but of course the proof is in the pudding, and as one piece of proof let me offer the following statement: "What we want in this country is equal opportunity for everyone, not affirmative action for a few." The author of this statement is David Duke, and everyone knows that he doesn't mean it. That is, he isn't really concerned with the rights and opportunities of everyone; he is concerned that people previously denied equal opportunity are now in the process of gaining it, and he speaks from the position of those who fear that the efforts to empower disadvantaged minorities are being made at their expense. What he's really saying is: "Those niggers and kikes and faggots have come far enough; it's time to stop them before they take our jobs, cheat our children out of a place in college, and try to move in next door." He delivers what is known as a "coded" message, but the code is not difficult to crack, and in fact it is so transparent that for all intents and purposes it is *literal*.

It's like what happens when a sports announcer praises the intelligence or work ethic of an athlete; you understand immediately that the athlete is white, just as you would immediately understand the reverse if the announcer had praised a player's natural athletic ability. Code words are all

around us these days. When a politician declares that we have to stop ca-
tering to special interests and pay attention to the middle class, you know
who the special interests are and you know that the color of the middle
class—symbolically, if not empirically—is white. And when another politician attacks welfare mothers who breed children in order to claim larger
benefits, you know that the real message is composed of two racial stereotypes: (1) the sexually promiscuous black too close to nature to have internalized the restraints of civilization, and (2) the lazy and shiftless negro
made familiar to so many Americans by the comedian Stepin Fetchit. Of
course these stereotypes contradict one another, but the illogic is useful
because it allows the speaker to play on a variety of fears—sexual, economic, demographic, among others—at the same time.

More often than not the audience to whom such coded messages are
addressed is complicit in the transaction they enable. Not only is the code
readily understood, its status *as* code—as something that wears a vocabulary like a disguise—is welcomed by both parties. The speaker does not
deceive the audience but tells it what it wants to hear, and, moreover, tells
it in terms that allow its members to give full rein to their prejudices and
yet appear to repudiate them. Confronted with an analysis like mine, the
code traffickers quickly turn into hard-core literalists who say things like
"How can it be racist to be for equality?" or "My objection is simply to
preferential treatment for anyone," or "I just want decisions to be made
on a basis that is fair." Such demurrers invoke a plain meaning philosophy
of language and also claim a transparency of intention: "I just mean what
I say, and what I say is both innocent and upright."

The question is one of motives. On the campus of the University of
Alabama this year [1991–92], there has been a controversy about the funding of a gay/lesbian alliance. Some members of the university Senate sponsored a resolution asking the state attorney general to rule on the legality
of funding the alliance given the fact of an Alabama law that makes sodomy
(along with other acts, including masturbation) a crime. The authors and
supporters of the resolution announced that their chief concern was to assure university compliance with the law and that they were not engaged in
gay bashing—to which another student responded in a letter to the campus
paper: "[Sure,] and monkeys fly out of my butt . . . if you support an
unnecessary, provocative and insulting resolution targeted at gays, lesbians
and bi-sexuals, the . . . term for what you are doing is gay bashing, and
if you hide your true intentions under a bogus reverence for the law, the
term for what you are doing is hypocritical and cowardly gay bashing"
(*Crimson White*, February 12, 1992, 4).

At an earlier time, not so long ago, someone who wanted to bash gays
wouldn't have bothered to engage in such subterfuge; he could have just

had at it in the confidence that approval and applause would greet his statements. The fact that this is no longer the case is a tribute to the effects of what has been called "political correctness," and what I would call just plain correctness; both in the universities and in the larger society bigotry is out, and talk of tolerance and equality is in. The response of former and present bigots to this distressing state of affairs is to figure out a way of appropriating the new vocabulary so that it transmits the same old messages. The favorite strategy is find a word or concept that seems invulnerable to challenge—law, equality, merit, neutrality—and then to give it a definition that generates the desired outcome. David Duke has it down pat. Faced with the apparently difficult task of promoting a white-supremacist agenda in an age when white supremacy is no longer respectable as a public pose, he hits upon the solution of defining equality as a relationship between persons with no history; no matter what disadvantages have previously been imposed upon someone, he or she is to be treated in exactly the same way as those for whom advantage has been an unquestioned entitlement. The result is nicely illuminated by what someone said long ago about the supposed impartiality of French law: "Yes, the law of France is impartial; it forbids the rich as well as the poor from sleeping under bridges." That is, the law that calls itself impartial is written only to deny those already deprived of resources the one resource that remains available to them; it is only the poor, after all, who would think to sleep under bridges in the absence of anywhere else to sleep. The rich who sleep very well are not the target of the law and indeed are its intended *beneficiaries,* since enforcement will protect them from a sight they find inconvenient and even accusatory. In a similar way, the equality David Duke champions is designed to perpetuate the *in*equalities produced by a history of repression and exclusion; by refusing to take into account what has happened in the past, the rhetoric of "equality for everyone" assures that the privileges of a few will be continued into the future, and, best of all, this policy is able to dress itself in the vocabulary of moral purity. It's like alchemy or magic: now you see white supremacy, but, presto chango, it is given a new description, and now you see "equality for everyone" with no change whatsoever in the practice or the outcome.

Let me say again that the key to this strategy is the hypocrisy that goes along with it, a hypocrisy shared by both the producers and consumers of the coded language. When George Bush declared himself opposed to a civil rights bill because it contained "quotas," his real objection was to a law that addressed the discriminatory practices of his largely white, middle-class supporters; the word *quotas* was used to deflect attention away from the wrongs the legislation was intended to remedy and to focus instead on the wrongs that would supposedly be done to innocent white male bystanders. Bush was thus able to assume a posture of moral rectitude, which could in turn be assumed by the constituency he pandered to. When, later, the political winds shifted and it became apparent that more would be gained than lost by supporting the bill, Bush changed his mind and found that quotas were *not* in fact what the bill mandated, even though almost nothing in its language or its mechanisms had changed.

A smaller but similar sequence was recently played out on my own campus. Several members of the faculty declared themselves outraged by what they believed to be the scandalously high salary of an African American colleague. When a list of top faculty salaries was subsequently published, the name of the supposed offender was nowhere to be found and, indeed, the list was entirely made up of white males. None of those previously outraged turned their outrage in the direction of those who were now revealed to be its appropriate target, and no one thought to acknowledge or regret the harm done to the person who had been vilified on the basis of unsupported speculation. The reason is simple. The real crime of the person so vilified was not his salary—which was, I happen to know, quite unremarkable—but the fact that he was black, lived in a large house, and drove an expensive automobile. It was this profile that could not be tolerated even though it was also the profile of many others on campus, including most of those who had raised the issue in the first place. That issue was framed in terms of equality and fairness, but as the sequence made clear, these words were merely stand-ins for prejudicial attitudes that could no longer be openly displayed but could be displayed under cover of a sanitized vocabulary that proclaimed the ideological innocence of its users.

To be sure, no one was fooled, but fooling people is not the agenda of this form of hypocrisy that flourishes when its dirty little secret is out in the open, as it always is because its moralistic disguise is so thin. That secret— known by everyone—is that talk of equality, standards, and level playing fields is nothing more than a smoke screen behind which there lies a familiar set of prejudices rooted in personal interest. The point has been made by an unlikely source, Herbert Stein, former economic adviser to Richard Nixon and fellow of the American Enterprise Institute, writing in an unlikely place, the *Wall Street Journal:* "Demands for a level playing field are almost always demands that the government do more for me" (September 3, 1991); and I would add that the demand is sometimes that whatever has been done for me in the past should be continued and not extended to others.

My point is that the practices of those who have declared themselves against curricular reform, multiculturalism, affirmative action, deconstruction, feminism, gay and lesbian studies, etc., are informed by a massive bad faith. That bad faith takes many forms, including those I have already discussed, but you see it most clearly in the phrase that has become the rallying cry of the assault on the universities, *political correctness*. Political correctness is supposedly the property of left-leaning academics who have conspired to subvert standards by imposing ideological requirements on the content of courses and by penalizing those who prefer to teach traditional materials by traditional methods. If you read the popular press, you get the impression that the effects of this conspiracy are far-reaching and that those engaged in it are organized in almost a paramilitary fashion. The facts, however, are exactly the opposite. First of all, almost all departments of English, history, philosophy, Romance languages, art history, political science, sociology, and anthropology are still very much dominated by white male professors educated in the fifties and sixties who are still writing and teaching in ways that have remained largely unchanged in the past thirty years. On any college campus I know of—including the campus of Duke University—the percentage of courses and materials that can be termed multicultural rarely rises above ten, more or less the same percentage of women faculty members and quite a bit higher than the percentage of minority faculty members. As in the case of protests against affirmative action, the real complaint is not that the minority voices have taken over but that they are represented at all. "This has gone far enough" means, as it always does, "this should never have happened at all and we have got to stop it."

Second, this small minority, far from being organized and equipped with a police force, is internally self-divided and politically ineffective. The truly political organization is in the hands of neoconservative ideologues who enjoy incredible levels of funding from right-wing foundations—$700,000 a year to the National Association of Scholars alone; who are able to call on the pens of a cadre of fellow-traveling journalists with national audiences—George Will, Cal Thomas, John Leo, among many others; and who can rely on the encouragement and support of highly placed government officials—Lynne Cheney, Henry Hyde, William Bennett, Lamar Alexander, and, on occasion, George Bush. This is *real* political power and *real* political correctness, when an agenda has behind it the triple threat of money, media domination, and governmental regulation. The best evidence of this power has been the ability of this highly political agenda to tag its opponents with the label "political" even as it outspends, out propagandizes, and outpoliticks them. As even *Time* magazine has recently acknowledged, no conservative faculty member has been hounded out of university life or forced to alter the content of a course by administrative pressure, whereas on several campuses pressures brought to bear on administrators by forces on the right have led to the suspension of programs and to the chilling effects of which those same forces shamelessly complain.

The tactics of the ideological right on the microlevel are no less shameless and usually involve the ceaseless circulation of information that is at once deceptive and ignorant. Consider, as just one example, the now-obligatory ridicule by those who have never read it of an article entitled "Jane Austen and the Masturbating Girl." No doubt the article's detractors assume salaciously that the author is peddling a *National Enquirer*–type thesis about the personal life of Jane Austen, whom they probably have also not read. They would, however, be disappointed if they ever passed their eyes over the offending piece, which makes the following rather mild points: (1) in the late eighteenth and early nineteenth century there was a considerable medical literature on masturbation and its dangers and a sub-literature about the special dangers to adolescent girls, whose characters were in the process of being formed; (2) a large part of this literature is given over to the listing of "symptoms" by which parents and others concerned with adolescent education can spot trouble and move, as one might say, to nip it in the bud; (3) Jane Austen's novels are concerned with the moral education of just such young women who display in their animated conversations behavior not unlike those symptoms; (4) perhaps we might have better insight into what Austen is doing if we use this material as a perspective on the interactions between her female characters. Now someone might find this thesis uninteresting and wish to look elsewhere for illumination; but it would be hard—except again for those who had not read the essay—to find in it some threat to the republic or to the integrity of art. "Jane Austen and the Masturbating Girl" is no more threatening or unusual than articles with titles like "John Milton and Glaucoma" or "Samuel Taylor Coleridge and Substance Abuse" or "Dickens and Lung Disease." It is just that masturbation, even in this late age, apparently has a shock value that allows ignorant persons simply to repeat it as a way of frightening persons even more ignorant into believing that the very existence of such a piece means that things have gone terribly wrong.

Ignorance, along with something close to simple lying, is also at work in what Gerald Graff has called the "great *Color Purple* episode." In 1988 Christopher Claussen, a professor at Penn State, said in print that he "would be willing to bet that [Alice Walker's] *The Color Purple* is taught in more English courses today than all of Shakespeare's plays combined." Perhaps, as Graff suggests, Professor Claussen was deliberately exaggerating, but within days his remark was picked up and repeated by the *Wall Street Journal,* and then by Secretary of Education William Bennett, and in the months that followed the circuit of repetition was extended by Hilton Kramer's *New Criterion* and by Lynne Cheney, chair of the National Endowment for the Humanities. Many other citations followed, including one dramatically placed as the last sentence of a paragraph at the end of a section of Dinesh D'Souza's *Illiberal Education*. The trouble is that Claussen's speculation—reported by those I have named not as a speculation but as a *fact*—is without foundation and is contrary to the very hard evidence. Wherever the question has been investigated, the preponderance of Shakespeare readers to Alice Walker readers in college classrooms is something on the order of forty to one, a figure that also describes the situation in the nation's high schools.

And yet it is false and specious assertions like Claussen's, libels on both persons and institutions, that make up the whole of the case against the universities. Think of the prime exhibits: Stanford University, which has been accused of gutting its Western Civ requirement in face of demands by rabid multiculturalists. In fact, the change that has been made in a multi-track, multisectioned course prevents no one from teaching the standard list of Great Books, while opening up a space for those who want to leaven the traditional list—still very much intact as a core—with some texts from alternative cultures. The change, in short, is in the nature of a modification that leaves much of what was in place before exactly as it was. Or consider the supposedly notorious English department at Duke University, a department whose members, according to D'Souza, "demonstrate open contempt for the notion of a 'great book' " (189) and teach courses in which "it is no longer necessary to struggle with *Paradise Lost* to try and figure out what the poem means or what Milton tried to convey" (180). This will be a surprise to the many students who have read Milton with me, with Reynolds Price, with Annabel Patterson, with Regina Schwartz, with Robert Gleckner, all of whom honor the poet and are committed to the study of his works in their historical context; and these will be the same students who progress through an English major that *requires* the great books written by Chaucer, Shakespeare, Milton, and Pope and features massively over-enrolled courses in the classic texts of American literature, Melville, Hawthorne, Twain, Thoreau, Whitman, and Dickinson. The true picture of what goes on at Duke emerged in a statistical study made by the dean of the undergraduate college, who reported that 91.5 percent of the courses given in the department dealt with traditional Western materials and accounted for 95.8 percent of the student enrollment.

When confronted by figures like these, conservative ideologues do not recant or reconsider; rather, they have recourse to a fallback position and say, "Well, maybe they're still teaching the same books, but they're not teaching them in the same way." This is the most curious charge of all, the charge that what is wrong with humanistic studies today is that they have changed. Imagine saying to the chair of a physics department or an economics department or a biology department: "Why aren't you teaching the same things you taught in 1960 when I went to school?" No doubt he or she (another change distressing to some) would reply by declaring it a department's obligation always to be reexamining its assumptions and taking into account the new information brought to light by new methodologies. A department that did nothing but mark time for thirty years would have little claim either in the attention of its students or in the support of the institution. In a curious reversal of logic and common academic sense, the disciplines of literature and history are being faulted for doing their jobs and are being urged to betray the responsibility they assume by taking up space in a university rather than in a museum.

Moreover, the demand that humanistic studies stand still is rooted in the false assumption that they have stood still in the past, the assumption that the present scene represents a departure from a stability and equilibrium that lasted until the day before yesterday. And again, the response must be simply to point to the facts and especially to the fact that the supposed golden age from which today's ideologues have sinfully fallen away was itself regarded as a departure from an even earlier golden age, which was, in its turn, vilified as a corruption of a still prior golden age, and on and on as far back as one can go. Thirty years ago the cries of reactionary protest were directed at the introduction into the curriculum of what we now think of as the classic texts of American literature, and the reasons given were the same one hears today—first, that American literature is a political and not a properly aesthetic category; second, that its canon had not yet stood the test of time; and third, that a focus on these new works would mean that older, more established works would be neglected and forgotten: if we start teaching courses in Faulkner and Hemingway, it was said, we shall no longer teach Shakespeare and Milton. Now the dire prediction is, if we start teaching courses in Alice Walker and Toni Morrison, we shall no longer teach Hemingway and Faulkner. Well, Shakespeare and Milton, Hemingway and Faulkner, Morrison and Walker are *all* being taught, sometimes in fruitful conjunction; the texts that informed the curriculum in our childhood continue to flourish alongside and in the company of texts written by our contemporaries or by ancestors whose vitality and power we are just beginning to recognize.

And as for the accusation that alien meanings are being imposed on traditional texts by irresponsible interpreters, let's look at some of the oft-cited offenders and offenses. George Will cites as obviously ridiculous an interpretation of Shakespeare's *The Tempest* that focuses on colonialism; but the topic of colonialism—of the invasion by white Europeans of non-European cultures—is an explicit point of reference in the play itself. Dinesh D'Souza seems scandalized by the fact that some professors now teach Shakespeare with an eye to gender tensions and class conflict; but of course plays like *The Taming of the Shrew, Twelfth Night,* and *Coriolanus* are about little else, and no one has ever thought otherwise. Reacting to the many different readings of *Hamlet,* D'Souza chortles that "according to this logic, *Hamlet* is not one play but many plays" (178). What he doesn't seem to know is that *Hamlet is* many plays, since during Shakespeare's lifetime there was no one text of the play but a number of performance scripts, all different, and none of them uniquely authorized. In fact, in the late sixteenth and early seventeenth century plays were not regarded as texts, and the printing of them in some permanent form would have seemed an odd action. D'Souza's call for fidelity to the one and only *Hamlet* is really a call for fidelity to an object invented by critics and editors in the nineteenth century.

All of the atrocities cited by our modern defenders of objective scholarship are like that; they are the inventions of ignorance, and what the inventors are ignorant of, more often than not, is history, the history of the disciplines they presume to rebuke, the history of the authors over whose meanings they claim a unique authority, the history of education itself, which in their crooked and cockeyed chronicle knows only two periods, the long, tranquil, and rational past and a present moment of unprecedented conflict and political maneuverings. And it is no accident that the *historical* evidence is what is either ignored or misreported; for it is an article of faith among our neoconservative Jeremiahs that the perspectives of history are distorting, are so many veils that must be pierced or discarded in favor of a perspective that transcends history, a perspective to which *they* have a privileged access (how they got it is never explained), a perspective from which the rest of us sinners have unhappily fallen away.

This is perhaps why the story they tell of an educational process corrupted by strange theories and ideological motives that have only recently polluted the pristine world of the academy will not be shaken by something so paltry as *evidence.* Evidence, after all, is derived from history, and it acquires its force from historical contexts. Consequently, it weighs as nothing against the conviction of someone so self-righteous that he will immediately hear challenges to his present opinions as either tainted in their source or irrelevant to what he already knows to be the case. An ideologue of this type regards assertions made by his opponents not as arguments to be considered but as debating points to be dismissed; and since he isn't taking the back and forth of argument seriously, but as an occasion for forensic display—an occasion not for persuading an audience but diverting it (in two senses of the word)—he is no more invested in what *he* says than he is in what is said by those on the other side. In a sense, then, he doesn't mean anything by his words—he doesn't stand by them—for he regards them as merely instrumental to the promotion of a truth he is in possession of before any discussion begins.

Something of the flavor of this programmatic verbal irresponsibility was on public display during the Clarence Thomas/Anita Hill hearings when at various points it became clear that Arlen Spector, Orrin Hatch, and Alan Simpson would say *anything* and not care a bit about its accuracy or cogency so long as it was sufficiently denigrating, defamatory, and ridiculing. A small version of this strategy has been practiced by Dinesh D'Souza, who, when the going gets tough, or threatens to, will quote a remark supposedly made by Michel Foucault about Jacques Derrida, "He's the kind of philosopher who gives bullshit a bad name." The trouble is that Foucault never said it. It was said first by a member of the Princeton philosophy department, and it was said not about Derrida but about another philosopher. Later, a colleague in the same department borrowed the remark to characterize that small portion of Derrida's work he found unpersuasive as opposed to the bulk of Derrida's writing, which he very much admired. Wrong attribution, wrong target, in fact no target at all, since the second philosopher came not to bury Derrida, but to praise him.

This remark, then, joins the Claussen speculation about the demise of Shakespeare at the hands of Alice Walker, and the abandonment of traditional authors and courses by the Duke English department, and the feminist/gay takeover of the entire academy, as a piece of irresponsible fiction. But I suspect that when he is informed of this fact—and he may well already know it—D'Souza will not retire the anecdote; he will merely wait for another occasion when no one is present who knows better or for an audience whose members feel threatened by authors they have never read and indeed feel threatened *because* they have never read them—and then he will trot it out again. When he does, the remark will provoke laughter, and the laughter will be the laughter of relief, the laughter of those who are happy to see the specter of their fears dissipated by a cheap and inaccurate joke.

As I have already said, at moments like this, ignorance joins with bad faith to produce a coded discourse that flatters the anxieties of those who will do anything to push away suggestions they find uncomfortable. The chief suggestion they wish to push away is the suggestion that their motives are not pure, the suggestion that the slogans they trade in—slogans that breathe moral rectitude—are a thin veneer barely covering attitudes that will not bear examination. In this effort to keep from themselves the true nature of the pieties they ritually invoke, no one is more useful than Martin Luther King, Jr. What better demonstration of the nonracism of a position could there be than the support it finds in the words of the hero, indeed the saint, of the civil rights movement? Here again is the oh-so-convenient sentence: "I have a dream that one day my four little children will . . . not be judged by the color of their skin, but by the content of their character." These words were uttered, as I have already noted, on August 28, 1963, in front of the Lincoln Memorial. That is to say, they were part of a *historical* occasion. Nearly a decade after *Brown v. Board of Education,* blacks were still being denied access to the ballot box, restaurants, hotels, neighborhoods, and King was speaking to those who had marched on Washington in order to press for a comprehensive civil rights bill. The target of his sentence was a set of practices by which persons were barred from participation in the economic, political, and educational process simply because of the color of their skin; he wanted those practices ended, and as part of his effort he crafted an utterance whose power derived largely from its rhetorical and stylistic features, the balance of parallel phrases counterpointed by a heavy use of alliteration and assonance, "not the color of their skin, but the content of their character" (the extra syllable of "character" bringing the whole to a nicely cadenced close).

I do not offer this analysis in order to diminish the achievement of the sentence or to question King's sincerity but merely to emphasize the *local* nature of the achievement in relation to an intention that was *locally* sincere. What King was saying was, "Let's get rid of these Jim Crow laws"; what he was *not* saying was, "Once they are removed, the job will be done and business can then proceed as usual." In fact, other sentences from the same speech indicate that in his view the present moment of action is just the beginning of what will be a moral/religious crusade. There are those who will ask, King says in words that have proved prophetic, "When will you be satisfied?" and he answers "never," not "until justice rolls down like waters and righteousness like a mighty stream," not until "the crooked places shall be made straight and the glory of the Lord will be revealed and all flesh will see it together." These are not the accents of a man who is calling for the reign of neutral principle, or for a merely *procedural* justice that detaches itself from history and pretends that righteousness *already* rolls down in a mighty stream. King is committed to nothing less than the transformation of American life, and the transformation he dreams of will be far from accomplished if legal discrimination is ended while the inequities that have been produced by discrimination are left untouched and continue to do their evil work.

The subtitle of the standard anthology of King's work is *Writings and Speeches That Changed the World.* Those who use those speeches against affirmative action by deriving from them an injunction against any race consciousness whatsoever are working to retard the changes that King began. By detaching his words from the history that produced them, they erase the fact of that history from the slate, and they do so, paradoxically, in order to prevent that history from being truly and deeply altered. As one commentator has put it, a "race-neutral anti-discrimination ordinance does not break the causal chain that leads from past systemic discrimination to subsequent grossly disproportionate racial imbalance''; on the contrary, the ordinance will have "the effect of largely freezing the racial imbalance" that the conveniently expunged history has brought about (Michel Rosenfield, *Affirmative Action and Justice* [New Haven, 1991], 212–213). This effect will not be the accidental by-product of a rigorous morality but the desired outcome of those for whom the morality of the civil rights movement is anathema. "Race-neutral" and "color-blind" are just two more of the coded phrases by means of which Martin Luther King's legacy is not honored but betrayed.

I am not saying that Martin Luther King would have wanted his children to be judged by the color of their skin, as if that were in and of itself an entitlement; but neither would he have wanted the color of their skin to be wholly irrelevant to the determination of what they had to offer to society. He says to those he addresses in 1963, "You have been the veterans of creative suffering." That is, the strength you here display by participating in this march is the product of your trials in the face of racism; and it is by virtue of those trials that you have become what you are, veterans of a war whose terms you did not choose. King salutes his followers not because their skins are black but because the blackness of their skins has generated the experience that has tempered them. In short, *the color of their skin has in some measure been the content of their character,* and a policy that respects that fact and takes it into account—not as the only determinant but as one among others—will be faithful to the spirit of King's vision even if, by a pinched and narrow reading of his words, it is in violation of the letter.

Bowing down to the letter at the expense of the spirit is the practice of those who come to us today as wolves in moralists' clothing, the practice of David Duke when he prattles on about equality, the practice of George Bush when he inveighs against nonexistent quotas in the name of fairness, the practice of my colleagues at Duke when they call sanctimoniously for parity in salaries, the practice of those Alabama students whose passion for the law "forces" them to oppose funding for the gay/lesbian alliance, the practice of the Republican members of the Judiciary Committee when they inveigh against slander in the act of slandering. All of these have become adept at encoding a virulent message in the most high sounding terms. And there are more on the way. As I write, the *New York Times* reports the emergence in Colorado of Shawn Slater, a second-generation David Duke whose motto is "Equal rights for all; special privileges for nobody." The head of the state's Ku Klux Klan, Slater "describes himself as a level-headed suburbanite who fears for his nation." His chief accomplishment has been to orchestrate a Klan demonstration on Martin Luther King's birthday; the predictable counterrally produced violence and national television coverage: "Hundreds of anti-Klan demonstrators, many of them black, reacted just the way Slater had hoped. They threw snowballs and bottles at the Klan and assaulted several police officers.'' Get it? Violent black racists attack peaceful Klan and threaten law and order. Another story: militant gays infect innocent heterosexuals with the plague; or another: rabid feminists destroy family life. Still another: incredibly wealthy Jews control the media and spread the myth of the Holocaust. And one more: deconstructionists subvert standards and tear down Western civilization. This is the package being sold to many who are only too eager to receive it because it enables them to hold up clean hands at the very moment they are sharing in dirty work. The only question is, are *you* buying?

8

THERE'S NO SUCH THING AS FREE SPEECH, AND IT'S A GOOD THING, TOO

Nowadays the First Amendment is the First Refuge of Scoundrels.
—S. Johnson and S. Fish

Lately, many on the liberal and progressive left have been disconcerted to find that words, phrases, and concepts thought to be their property and generative of their politics have been appropriated by the forces of neoconservatism. This is particularly true of the concept of free speech, for in recent years First Amendment rhetoric has been used to justify policies and actions the left finds problematical if not abhorrent: pornography, sexist language, campus hate speech. How has this happened? The answer I shall give in this essay is that abstract concepts like free speech do not have any "natural" content but are filled with whatever content and direction one can manage to put into them. "Free speech" is just the name we give to verbal behavior that serves the substantive agendas we wish to advance; and we give our preferred verbal behaviors *that* name when we can, when we have the power to do so, because in the rhetoric of American life, the label "free speech" is the one you want your favorites to wear. Free speech, in short, is not an independent value but a political prize, and if that prize has been captured by a politics opposed to yours, it can no longer be invoked in ways that further your purposes, for it is now an obstacle to those purposes. This is something that the liberal left has yet to understand, and what follows is an attempt to pry its members loose from a vocabulary that may now be a disservice to them.

Not far from the end of his *Areopagitica,* and after having celebrated

"There's No Such Thing As Free Speech and It's a Good Thing Too" first appeared in *Boston Review*.

the virtues of toleration and unregulated publication in passages that find their way into every discussion of free speech and the First Amendment, John Milton catches himself up short and says, of course I didn't mean Catholics, them we exterminate:

> I mean not tolerated popery, and open superstition, which as it extirpates all religious and civil supremacies, so itself should be extirpate . . . that also which is impious or evil absolutely against faith or manners no law can possibly permit that intends not to unlaw itself.

Notice that Milton is not simply stipulating a single exception to a rule generally in place; the kinds of utterance that might be regulated and even prohibited on pain of trial and punishment constitute an open set; popery is named only as a particularly perspicuous instance of the advocacy that cannot be tolerated. No doubt there are other forms of speech and action that might be categorized as "open superstitions" or as subversive of piety, faith, and manners, and presumably these too would be candidates for "extirpation." Nor would Milton think himself culpable for having failed to provide a list of unprotected utterances. The list will fill itself out as utterances are put to the test implied by his formulation: would this form of speech or advocacy, if permitted to flourish, tend to undermine the very purposes for which our society is constituted? One cannot answer this question with respect to a particular utterance in advance of its emergence on the world's stage; rather, one must wait and ask the question in the full context of its production and (possible) dissemination. It might appear that the result would be ad hoc and unprincipled, but for Milton the principle inheres in the core values in whose name individuals of like mind came together in the first place. Those values, which include the search for truth and the promotion of virtue, are capacious enough to accommodate a diversity of views. But at some point—again impossible of advance specification—capaciousness will threaten to become shapelessness, and at that point fidelity to the original values will demand acts of extirpation.

I want to say that all affirmations of freedom of expression are like Milton's, dependent for their force on an exception that literally carves out the space in which expression can then emerge. I do not mean that expression (saying something) is a realm whose integrity is sometimes compromised by certain restrictions but that restriction, in the form of an underlying articulation of the world that necessarily (if silently) negates alternatively possible articulations, is constitutive of expression. Without restriction, without an inbuilt sense of what it would be meaningless to say or wrong to say, there could be no assertion and no reason for asserting it. The exception to unregulated expression is not a negative restriction but a positive hollowing out of value—we are for *this*, which means we are against *that*—in relation to which meaningful assertion can then occur. It is in reference to that value—constituted as all values are by an act of exclusion—that some forms of speech will be heard as (quite literally) intolerable. Speech, in short, is never a value in and of itself but is always produced within the precincts of some assumed conception of the good to which it must yield in the event of conflict. When the pinch comes (and sooner or later it will always come) and the institution (be it church, state, or university) is confronted by behavior subversive of its core rationale, it will respond by declaring "of course we mean not tolerated ———, that we extirpate," not because an exception to a general freedom has suddenly and contradictorily been announced, but because the freedom has never been general and has always been understood against the background of an originary exclusion that gives it meaning.

This is a large thesis, but before tackling it directly I want to buttress my case with another example, taken not from the seventeenth century but from the charter and case law of Canada. Canadian thinking about freedom of expression departs from the line usually taken in the United States in ways that bring that country very close to the *Areopagitica* as I have expounded it. The differences are fully on display in a recent landmark case, *R. v. Keegstra*. James Keegstra was a high school teacher in Alberta who, it was established by evidence, "systematically denigrated Jews and Judaism in his classes." He described Jews as treacherous, subversive, sadistic, money loving, power hungry, and child killers. He declared them "responsible for depressions, anarchy, chaos, wars and revolution" and required his students "to regurgitate these notions in essays and examinations." Keegstra was indicted under Section 319(2) of the Criminal Code and convicted. The Court of Appeal reversed, and the Crown appealed to the Supreme Court, which reinstated the lower court's verdict.

Section 319(2) reads in part, "Every one who, by communicating statements other than in private conversation, willfully promotes hatred against any identifiable group is guilty of . . . an indictable offense and is liable to imprisonment for a term not exceeding two years." In the United States, this provision of the code would almost certainly be struck down because, under the First Amendment, restrictions on speech are apparently prohibited without qualification. To be sure, the Canadian charter has its own version of the First Amendment, in Section 2(b): "Everyone has the following fundamental freedoms . . . (b) freedom of thought, belief, opinion, and expression, including freedom of the press and other media of communication." But Section 2(b), like every other section of the charter, is qualified by Section 1: "The Canadian Charter of Rights and Freedoms guarantees the rights and freedoms set out in it subject only to such reasonable limits prescribed by law as can be demonstrably justified in a free and democratic society.'' Or in other words, every right and freedom herein granted can be trumped if its exercise is found to be in conflict with the principles that underwrite the society.

This is what happens in *Keegstra* as the majority finds that Section 319(2) of the Criminal Code does in fact violate the right of freedom of expression guaranteed by the charter but is nevertheless a *permissible* restriction because it accords with the principles proclaimed in Section 1. There is, of course, a dissent that reaches the conclusion that would have been reached by most, if not all, U.S. courts; but even in dissent the minority is faithful to Canadian ways of reasoning. ''The question,'' it declares, ''is always one of balance,'' and thus even when a particular infringement of the charter's Section 2(b) has been declared unconstitutional, as it would have been by the minority, the question remains open with respect to the next case. In the United States the question is presumed closed and can only be pried open by special tools. In our legal culture as it is now constituted, if one yells ''free speech'' in a crowded courtroom and makes it stick, the case is over.

Of course, it is not that simple. Despite the apparent absoluteness of the First Amendment, there are any number of ways of getting around it, ways that are known to every student of the law. In general, the preferred strategy is to manipulate the distinction, essential to First Amendment jurisprudence, between speech and action. The distinction is essential because no one would think to frame a First Amendment that began ''Congress shall make no law abridging freedom of action,'' for that would amount to saying ''Congress shall make no law,'' which would amount to saying ''There shall be no law,'' only actions uninhibited and unregulated. If the First Amendment is to make any sense, have any bite, speech must be declared not to be a species of action, or to be a special form of action lacking the aspects of action that cause it to be the object of regulation. The latter strategy is the favored one and usually involves the separation of speech from consequences. This is what Archibald Cox does when he assigns to the First Amendment the job of protecting ''expressions separable from conduct harmful to other individuals and the community.'' The difficulty of managing this segregation is well known: speech always seems to be crossing the line into action, where it becomes, at least potentially, consequential. In the face of this categorical instability, First Amendment theorists and jurists fashion a distinction within the speech/action distinction: some forms of speech are not really speech because their purpose is to incite violence or because they are, as the court declares in *Chaplinsky v. New Hampshire* (1942), ''fighting words,'' words ''likely to provoke the average person to retaliation, and thereby cause a breach of the peace.''

The trouble with this definition is that it distinguishes not between fighting words and words that remain safely and merely expressive but between words that are provocative to one group (the group that falls under the rubric "average person") and words that might be provocative to other groups, groups of persons not now considered average. And if you ask what words are likely to be provocative to those nonaverage groups, what are likely to be *their* fighting words, the answer is anything and everything, for as Justice Holmes said long ago (in *Gitlow v. New York*), every idea is an incitement to somebody, and since ideas come packaged in sentences, in words, every sentence is potentially, in some situation that might occur tomorrow, a fighting word and therefore a candidate for regulation.

This insight cuts two ways. One could conclude from it that the fighting words exception is a bad idea because there is no way to prevent clever and unscrupulous advocates from shoveling so many forms of speech into the excepted category that the zone of constitutionally protected speech shrinks to nothing and is finally without inhabitants. Or, alternatively, one could conclude that there was never anything in the zone in the first place and that the difficulty of limiting the fighting words exception is merely a particular instance of the general difficulty of separating speech from action. And if one opts for this second conclusion, as I do, then a further conclusion is inescapable: insofar as the point of the First Amendment is to identify speech separable from conduct and from the consequences that come in conduct's wake, there is no such speech and therefore nothing for the First Amendment to protect. Or, to make the point from the other direction, when a court invalidates legislation because it infringes on protected speech, it is not because the speech in question is without consequences but because the consequences have been discounted in relation to a good that is judged to outweigh them. Despite what they say, courts are never in the business of protecting speech per se, "mere" speech (a nonexistent animal); rather, they are in the business of classifying speech (as protected or regulatable) in relation to a value—the health of the republic, the vigor of the economy, the maintenance of the status quo, the undoing of the status quo—that is the true, if unacknowledged, object of their protection.

But if this is the case, a First Amendment purist might reply, why not drop the charade along with the malleable distinctions that make it possible, and declare up front that total freedom of speech is our primary value and trumps anything else, no matter what? The answer is that freedom of expression would only be a primary value if it didn't matter what was said, didn't matter in the sense that no one gave a damn but just liked to hear talk. There are contexts like that, a Hyde Park corner or a call-in talk show where people get to sound off for the sheer fun of it. These, however, are special contexts, artificially bounded spaces designed to assure that talking is not taken seriously. In ordinary contexts, talk is produced with the goal of trying to move the world in one direction rather than another. In these contexts—the contexts of everyday life—you go to the trouble of asserting that X is Y only because you suspect that some people are wrongly asserting that X is Z or that X doesn't exist. You assert, in short, because you give a damn, not about assertion—as if it were a value in and of itself—but about what your assertion is about. It may seem paradoxical, but free expression could only be a primary value if what you are valuing is the right to make noise; but if you are engaged in some purposive activity in the course of which speech happens to be produced, sooner or later you will come to a point when you decide that some forms of speech do not further but endanger that purpose.

Take the case of universities and colleges. Could it be the purpose of such places to encourage free expression? If the answer were "yes," it would be hard to say why there would be any need for classes, or examinations, or departments, or disciplines, or libraries, since freedom of expression requires nothing but a soapbox or an open telephone line. The very fact of the university's machinery—of the events, rituals, and procedures that fill its calendar—argues for some other, more substantive purpose. In relation to that purpose (which will be realized differently in different kinds of institutions), the flourishing of free expression will in almost all circumstances be an obvious good; but in some circumstances, freedom of expression may pose a threat to that purpose, and at that point it may be necessary to discipline or regulate speech, lest, to paraphrase Milton, the institution sacrifice itself to one of its *accidental* features.

Interestingly enough, the same conclusion is reached (inadvertently) by Congressman Henry Hyde, who is addressing these very issues in a recently offered amendment to Title VI of the Civil Rights Act. The first section of the amendment states its purpose, to protect "the free speech rights of college students" by prohibiting private as well as public educational institutions from "subjecting any student to disciplinary sanctions solely on the basis of conduct that is speech." The second section enumerates the remedies available to students whose speech rights may have been abridged; and the third, which is to my mind the nub of the matter, declares as an exception to the amendment's jurisdiction any "educational institution that is controlled by a religious organization," on the reasoning that the application of the amendment to such institutions "would not be consistent with the religious tenets of such organizations." In effect, what Congressman Hyde is saying is that at the heart of these colleges and universities is a set of beliefs, and it would be wrong to require them to tolerate behavior, including speech behavior, inimical to those beliefs. But insofar as this logic is persuasive, it applies across the board, for all educational institutions rest on some set of beliefs—no institution is "just there" independent of any purpose—and it is hard to see why the rights of an institution to protect and preserve its basic "tenets" should be restricted only to those that are religiously controlled. Read strongly, the third section of the amendment undoes sections one and two—the exception becomes, as it always was, the rule—and points us to a balancing test very much like that employed in Canadian law: given that any college or university is informed by a core rationale, an administrator faced with complaints about offensive speech should ask whether damage to the core would be greater if the speech were tolerated or regulated.

The objection to this line of reasoning is well known and has recently been reformulated by Benno Schmidt, former president of Yale University. According to Schmidt, speech codes on campuses constitute "well intentioned but misguided efforts to give values of community and harmony a higher place than freedom" (*Wall Street Journal,* May 6, 1991). "When the goals of harmony collide with freedom of expression," he continues, "freedom must be the paramount obligation of an academic community." The flaw in this logic is on display in the phrase "academic community," for the phrase recognizes what Schmidt would deny, that expression only occurs in communities—if not in an academic community, then in a shopping mall community or a dinner party community or an airplane ride community or an office community. In these communities and in any others that could be imagined (with the possible exception of a community of major league baseball fans), limitations on speech in relation to a defining and deeply assumed purpose are inseparable from community membership.

Indeed, "limitations" is the wrong word because it suggests that expression, as an activity and a value, has a pure form that is always in danger of being compromised by the urgings of special interest communities; but independently of a community context informed by interest (that is, purpose), expression would be at once inconceivable and unintelligible. Rather than being a value that is threatened by limitations and constraints, expression, in any form worth worrying about, is a *product* of limitations and constraints, of the already-in-place presuppositions that give assertions their very particular point. Indeed, the very act of thinking of something to say (whether or not it is subsequently regulated) is already constrained—rendered impure, and because impure, communicable—by the background context within which the thought takes its shape. (The analysis holds too for "freedom," which in Schmidt's vision is an entirely empty concept referring to an urge without direction. But like expression, freedom is a coherent notion only in relation to a goal or good that limits and, by limiting, shapes its exercise.)

Arguments like Schmidt's only get their purchase by first imagining speech as occurring in no context whatsoever, and then stripping particular speech acts of the properties conferred on them by contexts. The trick is nicely illustrated when Schmidt urges protection for speech "no matter how obnoxious in content." "Obnoxious" at once acknowledges the reality of speech-related harms and trivializes them by suggesting that they are *surface* injuries that any large-minded ("liberated and humane") person should be able to bear. The possibility that speech-related injuries may be grievous and *deeply* wounding is carefully kept out of sight, and because it is kept out of sight, the fiction of a world of weightless verbal exchange can be maintained, at least within the confines of Schmidt's carefully denatured discourse.

To this Schmidt would no doubt reply, as he does in his essay, that harmful speech should be answered not by regulation but by more speech; but that would make sense only if the effects of speech could be canceled out by additional speech, only if the pain and humiliation caused by racial or religious epithets could be ameliorated by saying something like "So's your old man." What Schmidt fails to realize at every level of his argument is that expression is more than a matter of proffering and receiving propositions, that words do work in the world of a kind that cannot be confined to a purely cognitive realm of "mere" ideas.

It could be said, however, that I myself mistake the nature of the work done by freely tolerated speech because I am too focused on short-run outcomes and fail to understand that the good effects of speech will be realized, not in the present, but in a future whose emergence regulation could only inhibit. This line of reasoning would also weaken one of my key points, that speech in and of itself cannot be a value and is only worth worrying about if it is in the service of something with which it cannot be identical. My mistake, one could argue, is to equate the something in whose service speech is with some locally espoused value (e.g., the end of racism, the empowerment of disadvantaged minorities), whereas in fact we should think of that something as a now-inchoate shape that will be given firm lines only by time's pencil. That is why the shape now receives such indeterminate characterizations (e.g., true self-fulfillment, a more perfect polity, a more capable citizenry, a less partial truth); we cannot now know it, and therefore we must not prematurely fix it in ways that will bind successive generations to error.

This forward-looking view of what the First Amendment protects has a great appeal, in part because it continues in a secular form the Puritan celebration of millenarian hopes, but it imposes a requirement so severe that one would except more justification for it than is usually provided. The requirement is that we endure whatever pain racist and hate speech inflicts for the sake of a future whose emergence we can only take on faith. In a specifically religious vision like Milton's, this makes perfect sense (it is indeed the whole of Christianity), but in the context of a politics that puts its trust in the world and not in the Holy Spirit, it raises more questions than it answers and could be seen as the second of two strategies designed to delegitimize the complaints of victimized groups. The first strategy, as I have noted, is to define speech in such a way as to render it inconsequential (on the model of "sticks and stones will break my bones, but . . ."); the second strategy is to acknowledge the (often grievous) consequences of speech but declare that we must suffer them in the name of something that cannot be named. The two strategies are denials from slightly different directions of the *present* effects of racist speech; one confines those effects to a closed and safe realm of pure mental activity; the other imagines the effects of speech spilling over into the world but only in an ever-receding future for whose sake we must forever defer taking action.

I find both strategies unpersuasive, but my own skepticism concerning them is less important than the fact that in general they seem to have worked; in the parlance of the marketplace (a parlance First Amendment commentators love), many in the society seemed to have bought them. Why? The answer, I think, is that people cling to First Amendment pieties because they do not wish to face what they correctly take to be the alternative. That alternative is *politics*, the realization (at which I have already hinted) that decisions about what is and is not protected in the realm of expression will rest not on principle or firm doctrine but on the ability of some persons to interpret—recharacterize or rewrite—principle and doctrine in ways that lead to the protection of speech they want heard and the regulation of speech they want heard and the regulation of speech they want silenced. (That is how George Bush can argue *for* flag-burning statutes and *against* campus hate-speech codes.) When the First Amendment is successfully invoked, the result is not a victory for free speech in the face of a challenge from politics but a *political victory* won by the party that has managed to wrap its agenda in the mantle of free speech.

It is from just such a conclusion—a conclusion that would put politics *inside* the First Amendment—that commentators recoil, saying things like "This could render the First Amendment a dead letter," or "This would leave us with no normative guidance in determining when and what speech to protect," or "This effaces the distinction between speech and action," or "This is incompatible with any viable notion of freedom of expression." To these statements (culled more or less at random from recent law review pieces) I would reply that the First Amendment has always been a dead letter if one understood its "liveness" to depend on the identification and protection of a realm of "mere" expression distinct from the realm of regulatable conduct; the distinction between speech and action has always been effaced in principle, although in practice it can take whatever form the prevailing political conditions mandate; we have never had any normative guidance for marking off protected from unprotected speech; rather, the guidance we have has been fashioned (and refashioned) in the very political struggles over which it then (for a time) presides. In short, the name of the game has always been politics, even when (indeed, especially when) it is played by stigmatizing politics as the area to be avoided.

In saying this, I would not be heard as arguing either for or against regulation and speech codes as a matter of general principle. Instead my argument turns away from general principle to the pragmatic (anti)principle of considering each situation as it emerges. The question of whether or not to regulate will always be a local one, and we can not rely on abstractions that are either empty of content or filled with the content of some partisan agenda to generate a "principled" answer. Instead we must consider in every case what is at stake and what are the risks and gains of alternative courses of action. In the course of this consideration many things will be of help, but among them will not be phrases like "freedom of speech" or "the right of individual expression," because, as they are used now, these phrases tend to obscure rather than clarify our dilemmas. Once they are deprived of their talismanic force, once it is no longer strategically effective simply to invoke them in the act of walking away from a problem, the conversation could continue in directions that are now blocked by a First Amendment absolutism that has only been honored in the breach anyway. To the student reporter who complains that in the wake of the promulgation of a speech code at the University of Wisconsin there is now something in the back of his mind as he writes, one could reply, "There was always something in the back of your mind, and perhaps it might be better to have this code in the back of your mind than whatever was in there before." And when someone warns about the slippery slope and predicts mournfully that if you restrict one form of speech, you never know what will be restricted next, one could reply, "Some form of speech is always being restricted, else there could be no meaningful assertion; we have always and already slid down the slippery slope; someone is always going to be restricted next, and it is your job to make sure that the someone is not you." And when someone observes, as someone surely will, that antiharassment codes chill speech, one could reply that since speech only becomes intelligible against the background of what isn't being said, the background of what has already been silenced, the only question is the political one of which speech is going to be chilled, and, all things considered, it seems a good thing to chill speech like "nigger," "cunt," "kike," and "faggot." And if someone then says, "But what happened to free-speech principles?" one could say what I have now said a dozen times, free-speech principles don't exist except as a component in a bad argument in which such principles are invoked to mask motives that would not withstand close scrutiny.

An example of a wolf wrapped in First Amendment clothing is an advertisement that ran recently in the Duke University student newspaper, the *Chronicle*. Signed by Bradley R. Smith, well known as a purveyor of anti-Semitic neo-Nazi propaganda, the ad is packaged as a scholarly treatise: four densely packed columns complete with "learned" references, undocumented statistics, and an array of so-called authorities. The message of the ad is that the Holocaust never occurred and that the German state never "had a policy to exterminate the Jewish people (or anyone else) by putting them to death in gas chambers." In a spectacular instance of the increasingly popular "blame the victim" strategy, the Holocaust "story" or "myth" is said to have been fabricated in order "to drum up world sympathy for Jewish causes." The "evidence" supporting these assertions is a slick blend of supposedly probative facts—"not a single autopsied body has been shown to be gassed"—and sly insinuations of a kind familiar to readers of *Mein Kampf* and *The Protocols of the Elders of Zion*. The slickest thing of all, however, is the presentation of the argument as an exercise in free speech—the ad is subtitled "The Case for Open Debate"—that could be objected to only by "thought police" and censors. This strategy bore immediate fruit in the decision of the newspaper staff to accept the ad despite a long-standing (and historically honored) policy of refusing materials that contain ethnic and racial slurs or are otherwise offensive. The reasoning of the staff (explained by the editor in a special column) was that under the First Amendment advertisers have the "right" to be published. "American newspapers are built on the principles of free speech and free press, so how can a newspaper deny these rights to anyone?" The answer to this question is that an advertiser is not denied his rights simply because a single media organ declines his copy so long as other avenues of publication are available and there has been no state suppression of his views. This is not to say that there could not be a case for printing the ad, only that the case cannot rest on a supposed First Amendment obligation. One might argue, for example, that printing the ad would foster healthy debate, or that lies are more likely to be shown up for what they are if they are brought to the light of day, but these are precisely the arguments the editor *disclaims* in her eagerness to take a "principled" free-speech stand.

What I find most distressing about this incident is not that the ad was printed but that it was printed by persons who believed it to be a lie and a distortion. If the editor and her staff were in agreement with Smith's views or harbored serious doubts about the reality of the Holocaust, I would still have a quarrel with them, but it would be a different quarrel; it would be a

quarrel about evidence, credibility, documentation. But since on these matters the editors and I are in agreement, my quarrel is with the reasoning
that led them to act in opposition to what they believed to be true. That
reasoning, as I understand it, goes as follows: although we ourselves are
certain that the Holocaust was a fact, facts are notoriously interpretable and
disputable; therefore nothing is ever really settled, and we have no right to
reject something just because we regard it as pernicious and false. But the
fact—if I can use that word—that settled truths can always be upset, at
least theoretically, does not mean that we cannot affirm and rely on truths
that according to our present lights seem indisputable; rather, it means exactly the opposite: in the absence of absolute certainty of the kind that can
only be provided by revelation (something I do not rule out but have not
yet experienced), we must act on the basis of the certainty we have so far
achieved. Truth may, as Milton said, always be in the course of emerging,
and we must always be on guard against being so beguiled by its present
shape that we ignore contrary evidence; but, by the same token, when it
happens that the present shape of truth is compelling beyond a reasonable
doubt, it is our moral obligation to act on it and not defer action in the
name of an interpretative future that may never arrive. By running the First
Amendment up the nearest flagpole and rushing to salute it, the student
editors defaulted on that obligation and gave over their responsibility to a
so-called principle that was not even to the point.

Let me be clear. I am not saying that First Amendment principles are
inherently bad (they are *inherently* nothing), only that they are not always
the appropriate reference point for situations involving the production of
speech, and that even when they are the appropriate reference point, they
do not constitute a politics-free perspective because the shape in which they
are invoked will always be political, will always, that is, be the result of
having drawn the relevant line (between speech and action, or between
high-value speech and low-value speech, or between words essential to the
expression of ideas and fighting words) in a way that is favorable to some
interests and indifferent or hostile to others. This having been said, the
moral is not that First Amendment talk should be abandoned, for even if
the standard First Amendment formulas do not and could not perform the
function expected of them (the elimination of political considerations in
decisions about speech), they still serve a function that is not at all negli-
gible: they slow down outcomes in an area in which the fear of overhasty
outcomes is justified by a long record of abuses of power. It is often said
that history shows (itself a formula) that even a minimal restriction on the
right of expression too easily leads to ever-larger restrictions; and to the
extent that this is an empirical fact (and it is a question one could debate),

there is some comfort and protection to be found in a procedure that requires you to jump through hoops—do a lot of argumentative work—before a speech regulation will be allowed to stand.

I would not be misunderstood as offering the notion of "jumping through hoops" as a new version of the First Amendment claim to neutrality. A hoop must have a shape—in this case the shape of whatever binary distinction is representing First Amendment "interests"—and the shape of the hoop one is asked to jump through will in part determine what kinds of jumps can be regularly made. Even if they are only mechanisms for slowing down outcomes, First Amendment formulas by virtue of their substantive content (and it is impossible that they be without content) will slow down some outcomes more easily than others, and that means that the form they happen to have at the present moment will favor some interests more than others. Therefore, even with a reduced sense of the effectivity of First Amendment rhetoric (it can not assure any particular result), the counsel with which I began remains relevant: so long as so-called free-speech principles have been fashioned by your enemy (so long as it is *his* hoops you have to jump through), contest their relevance to the issue at hand; but if you manage to refashion them in line with your purposes, urge them with a vengeance.

It is a counsel that follows from the thesis that there is no such thing as free speech, which is not, after all, a thesis as startling or corrosive as may first have seemed. It merely says that there is no class of utterances separable from the world of conduct and that therefore the identification of some utterances as members of that nonexistent class will always be evidence that a political line has been drawn rather than a line that denies politics entry into the forum of public discourse. It is the job of the First Amendment to mark out an area in which competing views can be considered without state interference; but if the very marking out of that area is itself an interference (as it always will be), First Amendment jurisprudence is inevitably self-defeating and subversive of its own aspirations. That's the bad news. The good news is that precisely *because* speech is never "free" in the two senses required—free of consequences and free from state pressure—speech always matters, is always doing work; because everything we say impinges on the world in ways indistinguishable from the effects of physical action, we must take responsibility for our verbal performances—*all* of them—and not assume that they are being taken cares of by a clause in the Constitution. Of course, with responsibility comes risks, but they have always been our risks, and no doctrine of free speech has ever insulated us from them. They are the risks, respectively, of permitting speech that does obvious harm and of shutting off speech in ways that might deny us the benefit of Joyce's *Ulysses* or Lawrence's *Lady Chatterly's Lover* or Titian's paintings. Nothing, I repeat, can insulate us from those risks. (If there is no normative guidance in determining when and what speech to protect, there is no normative guidance in determining what is art—like free speech a category that includes everything and nothing—and what is obscenity.) Moreover, nothing can provide us with a principle for deciding which risk in the long run is the best to take. I am persuaded that at the present moment, right now, the risk of not attending to hate speech is greater than the risk that by regulating it we will deprive ourselves of valuable voices and insights or slide down the slippery slope toward tyranny. This is a judgment for which I can offer reasons but no guarantees. All I am saying is that the judgments of those who would come down on the other side carry no guarantees either. They urge us to put our faith in apolitical abstractions, but the abstractions they invoke—the marketplace of ideas, speech alone, speech itself—only come in political guises, and therefore in trusting to them we fall (unwittingly) under the sway of the very forces we wish to keep at bay. It is not that there are no choices to make or means of making them; it is just that the choices as well as the means are inextricable from the din and confusion of partisan struggle. There is no safe place.

Postscript

When a shorter version of this essay was first published, it drew a number of indignant letters from readers who took me to be making a *recommendation:* let's abandon principles, or let's dispense with an open mind. But, in fact, I am not making a recommendation but declaring what I take to be an unavoidable truth. That truth is not that freedom of speech should be abridged but that freedom of speech is a conceptual impossibility because the condition of speech's being free in the first place is unrealizable. That condition corresponds to the hope, represented by the often-invoked "marketplace of ideas,'' that we can fashion a forum in which ideas can be considered independently of political and ideological constraint. My point, not engaged by the letters, is that constraint of an ideological kind is *generative* of speech and that therefore the very intelligibility of speech (as assertion rather than noise) is radically dependent on what free-speech ideologues would push away. Absent some already-in-place and (for the time being) unquestioned ideological vision, the act of speaking would make no sense, because it would not be resonating against any background understanding of the possible courses of physical or verbal actions and their possible consequences. Nor is that background accessible to the speaker it constrains; it is not an object of his or her critical self-consciousness; rather, it constitutes the field in which consciousness occurs, and therefore the productions of consciousness, and specifically speech, will always be political (that is, angled) in ways the speaker cannot know.

In response to this, someone might say (although the letters here discussed do not rise to this level) that even if speech is inescapably political in my somewhat rarified sense, it is still possible and desirable to provide a cleared space in which irremediably political utterances can compete for the public's approval without any one of them being favored or stigmatized in advance. But what the history of First Amendment jurisprudence shows is that the decisions as to what should or should not enjoy that space's protection and the determination of how exactly (under what rules) that space will first be demarcated and then administered are continually matters of dispute; moreover, the positions taken in the dispute are, each of them, intelligible and compelling only from the vantage point of a deeply assumed ideology, which, like the ideology of speech in general, dare not, and indeed cannot, speak its name. The structure that is supposed to permit ideological/political agendas to fight it out fairly—on a level playing field that has not been rigged—is itself always ideologically and politically constructed. This is exactly the conclusion reached reluctantly by Robert Post in a piece infinitely more nuanced than the letter he now writes. At the end of a long and rigorous analysis, Post finds before him "the startling proposition that the boundaries of public discourse cannot be fixed in a neutral fashion" ("The Constitutional Concept of Public Discourse: Outrageous Opinion, Democratic Deliberation, and *Hustler v. Falwell*," *Harvard Law Review* 103, no. 3 [January 1990]: 683). "The ultimate fact of ideological regulation," he adds, "cannot be blinked." Indeed not, since the ultimate fact is also the *root* fact in the sense that one cannot get behind it or around it, and that is why the next strategy—the strategy of saying, "Well, we can't get beyond or around ideology, but at least we can make a good faith try"—won't work either. In what cleared and ideology-free space will the "try" be made? one must ask, and if the answer is (and it must be by Post's own conclusion) that there is no such cleared space, the notion of "trying" can have no real content. (On a more leisurely occasion I would expand this point into an argument for the emptiness of any gesture that invokes a regulative ideal.)

No such thing as free (nonideologically constrained) speech; no such thing as a public forum purged of ideological pressures or exclusions. That's my thesis, and waiting at the end (really at the beginning) of it is, as my respondents have said, politics. Not, however, politics as the dirty word it becomes in most First Amendment discussions, but politics as the attempt to implement some partisan vision. I place the word "vision" after "partisan" so as to forestall the usual reading of partisan as "unprincipled," the reading Post attributes to me when he finds me "writing on the assumption that there is some implicit and mutually exclusive dichotomy between politics and principle.'' In fact, my argument is exactly the reverse: since it is only from within a commitment to some particular (not abstract) agenda that one feels the deep urgency we identify as "principled," politics is the *source* of principle, not its opposite. When two agendas square off, the contest is never between politics and principle but between two forms of politics, or, if you prefer, two forms of principle. The assumption of an antagonism between them is not mine, but Post's, and it is an assumption he doubles when he warns of the danger of "unprincipled self-assertion." This is to imagine selves as possibly motivated by "mere" preference, but (and this is the same point I have already made) preference is never "mere" in the sense of being without a moral or philosophical rationale; preference is the precipitate of some defensible (and, of course, challengeable) agenda, and selves who assert it, rather than being unprincipled, are at that moment extensions of principle. Again, it is Post, not me, who entertains a picture of human beings "as merely a collection of Hobbesian appetites." I see human beings in the grip of deep (if debatable) commitments, commitments so constitutive of their thoughts and actions that they cannot help being sincere. Franklin Haiman and Cushing Strout (two other correspondents) could not be more off the mark when they brand me cynical and opportunistic. They assume I am counseling readers to set aside principle in favor of motives that are merely political, whereas in fact I am challenging that distinction and counseling readers (the counsel is superfluous) to act on what they believe to be true and important, and not to be stymied by a doctrine that is at once incoherent and (because incoherent) a vehicle for covert politics.

In general, the letter writers ignore my challenge to the binaries on which their arguments depend, and take to chiding me for failing to respect distinctions whose lack of cogency has been a large part of my point. Thus, Professor Haiman solemnly informs me that an open mind is not the same as an empty one; but, in my analysis—which Professor Haiman is of course not obliged to accept but is surely obliged to note—they *are* the same. An open mind is presumably a mind not unduly committed to its present contents, but a mind so structured, or, rather, *un*structured, would lack a framework or in-place background in relation to which the world (both of action and speech) would be intelligible. A mind so open that it was anchored by no assumptions, no convictions of the kind that order and stabilize perception, would be a mind without gestalt and therefore without the capacity of keeping anything *in*. A consciousness not shored up at one end by a belief (not always the same one) whose negation it could not think would be a sieve. In short, it would be empty.

Professor Strout ventures into the same (incoherent) territory when he takes me to task for "confusing toleration with endorsing" and "justifying" with "putting up with." The idea is that a policy of allowing hate speech does not constitute approval of hate speech but shifts the responsibility for approving or disapproving to the free choice of free individuals. But this is to assume that the machinery of deliberation in individuals is purely formal and is unaffected by what is or is not in the cultural air. Such an assumption is absolutely necessary to the liberal epistemology shared by my respondents, but it is one that I reject because, as I have argued elsewhere, the context of deliberation is cultural (rather than formal or genetic), and because it is cultural, the outcome of deliberation cannot help being influenced by whatever notions are current in the culture. (Minds are not free, as the liberal epistemology implies, for the same reason that they cannot be open.) The fact that David Duke was rudely and provocatively questioned by reporters on "Sixty Minutes" or "Meet the Press" was less important than the fact that he was on "Sixty Minutes" and "Meet the Press" in the first place, for these appearances legitimized him and put his views into national circulation in a way that made them an unavoidable component of the nation's thinking. Tolerating may be different from endorsing from the point of view of the tolerator, who can then disclaim responsibility for the effects of what he has not endorsed, but, if the effects are real and consequential, as I argue they are, the difference may be cold comfort.

It is, of course, *effects* that the liberal epistemology, as represented by a strong free-speech position, cannot take into account, or can take into account only at the outer limits of public safety ("clear and imminent danger," "incitement to violence"). It is, therefore, perfectly apt for Professor Haiman to cite Holmes's dissent in *Abrams,* for that famous opinion at once concisely states the modern First Amendment position and illustrates what I consider to be its difficulties, if not its contradictions. Holmes begins by acknowledging the truth basic to my argument: it makes perfect sense to desire the silencing of beliefs inimical to yours, because if you did not so desire, it would be an indication that you did not believe in your beliefs. But then Holmes takes note of the fact that one's beliefs are subject to change, and comes to the skeptical conclusion that since the course of change is unpredictable, it would be unwise to institutionalize beliefs we may not hold at a later date; instead, we should leave the winnowing process to the marketplace of ideas unregulated by transient political pressures.

This sounds fine (even patriotic), but it runs afoul of problems at both ends. The "entry" problem is the one I have already identified in my reply to Professor Post: the marketplace of ideas—the protected forum of public discourse—will be structured by the same political considerations it was designed to hold at bay; and therefore, the workings of the marketplace will not be free in the sense required, that is, be uninflected by governmental action (the government is given the task of managing the marketplace and therefore the opportunity to determine its contours). Things are even worse at the other end, the exit or no-exit end. If our commitment to freedom of speech is so strong that it obliges us, as Holmes declares, to tolerate "opinions . . . we . . . believe to be fraught with death" (a characterization that recognizes the awful consequentiality of speech and implicitly undercuts any speech/action distinction), then we are being asked to court our own destruction for the sake of an abstraction that may doom us rather than save us. There are really only three alternatives: either Holmes does not mean it, as is suggested by his instant qualification ("unless . . . an immediate check is required to save the country"), or he means it but doesn't think that opinions fraught with death could ever triumph in a free market (in which case he commits himself to a progressivism he neither analyzes nor declares), or he means it and thinks deadly opinions could, in fact, triumph, but is saying something like *"qué será, será,"* (as it would appear he is in a later dissent, *Gitlow v. New York*). Each of these readings of what Holmes is telling us in *Abrams* and *Gitlow* is problematic, and it is the problems in the position born out of these two dissents that have been explored in my essay. The replies to that essay, as far as I can see, do not address those problems but continue simply to rehearse the pieties my analysis troubles. Keep those cards and letters coming.

9

Jerry Falwell's Mother, or, What's the Harm?

In the November 1983 issue of *Hustler,* publisher Larry Flynt printed what represented itself as an interview with Jerry Falwell in which the then leader of the Moral Majority recalled his first experience of sexual intercourse. According to the interview, entitled "Jerry Falwell Talks about His First Time" and labeled in small print at the bottom of the page "ad parody," the encounter took place in a very smelly outhouse when Falwell was dead drunk. His partner on that occasion, also drunk, was said to be his mother, who was portrayed both as sexually aggressive and extremely unattractive. In a deposition Flynt was asked whether he intended to convey to the magazine's readers that Falwell was a liar and a hypocrite, and he replied, "Yeah." He was then asked if it was his objective to destroy Falwell's integrity, and he responded, I wanted "to assassinate it."

So what we have here is a false and malicious depiction of two people with the avowed intention of wounding one so that his ability to function as a minister of the gospel would be destroyed. The Supreme Court held that Flynt's action was a constitutionally protected instance of free speech and that Falwell could not recover damages either for libel, invasion of privacy, or intention to inflict emotional harm. The Court gave as its reason for so holding, first, that as a public figure Falwell had willingly entered an arena in which he could expect to be the target of "robust" criticism and, second, that the parody interview was a part "of the free flow of ideas . . . on matters of public interest" that is so essential to the health of a democracy.

Now you may wonder what "idea" was here conveyed except the false one that Falwell and his mother were committing incest, and you may wonder, too, what contribution to public discussion could be made by the spreading of this false idea. The Court's answer is that while the interview parody was certainly outrageous, malicious, gross, repugnant, and responsible for the infliction of harm, it is a cousin or near-relation to political cartoons and caricatures, which are often not "reasoned or even-handed, but slashing and one-sided" and "calculated to injure the feelings of the subject of the portrayal." The Court goes on to list several examples, including an early cartoon portraying George Washington as an ass and later portrayals of Lincoln's "gangling posture, Teddy Roosevelt's glasses and teeth and Franklin D. Roosevelt's jutting jaw and cigarette holder," and concludes that because "our political discourse would have been considerably poorer without them," we must tolerate the parody of Falwell and his mother.

But think about those examples: what they do is *exaggerate* traits and qualities the persons in question actually *had*—Lincoln *was* gangly, Teddy Roosevelt *did* have prominent teeth, and his cousin Franklin's jaw *did* jut—but Falwell and his mother did not engage in the actions attributed to them in the parody interview. There seems to be a slip in the logic here, but before we talk about what it is and analyze what produces it, let me offer you a second, equally slippery example, a 1985 Seventh Circuit decision, *American Booksellers Association, Inc. v. Hudnut*. The case turns on an Indianapolis ordinance that defined pornography as "the graphic sexually explicit subordination of women" and therefore productive of a harm for which there should be legal remedies. Note that pornography is not said to *depict* subordination but to *be* subordination on the reasoning that its very circulation puts into the world images of women that encourage violent actions against them. Pornography, the ordinance drafters declared, "is an aspect of dominance. It does not persuade people so much as change them . . . pornography is not an idea; pornography is the injury." After rehearsing these arguments the court declares itself to be persuaded by them: "we accept the premises of this legislation"; and then, in an act of apparent generosity, the court elaborates those "premises" by analogizing the harms generated by pornography to the effects of "Nazi propaganda," which, we are reminded, "led to the deaths of millions." And yet, after having said this and more, the court declines to reach the conclusion mandated by its own logic because that logic "simply demonstrates the power of pornography as speech"; and while it may be true that "pornography is what pornography does, so is other speech." That is, the court is finally not swayed by the mere fact that pornography is productive of harmful effects because the same thing can be said of any form of speech.

This, however, is a curious line of reasoning to find in a First Amendment case, since there would not even be a First Amendment were it not for the assumption that speech is distinguishable from action on the basis of their respective effects, one being a form of behavior that is safely confined to the realms of cognition and expression, the other a form of behavior whose consequences spill out into the world, where they can threaten a person's bodily comfort or livelihood or quality of life. The importance of this distinction becomes apparent when we imagine rewriting the First Amendment so that it reads not "Congress shall make no law abridging freedom of speech" but "Congress shall make no law abridging freedom of *action*." The oddness of the formulation would strike one immediately, since the regulating and therefore the abridging of action is the whole of what governments do, and a constitution that forswore the regulation of action would be declaring itself *and* the government out of business. In order for the First Amendment, in its strong form, to make sense, speech must occupy a special category that makes its freedom from regulation compatible with the operation of normal governmental functions. And yet in reaching its decision the *American Booksellers* court explicitly *denies* that speech is special, when, in order to avoid approving the Indianapolis ordinance, it assimilates pornography to a definition of speech—it is what it does—that renders it indistinguishable from action.

To be sure, the point I am making has been made before, and in response to it First Amendment purists acknowledge that while speech may, on occasion, have harmful real-world effects, the proper response is not regulation of speech but the production of *more* speech in an effort to counteract the speech that may have wounded you. But the *American Booksellers* court rejects this argument too, noting that under present law "racial bigotry, anti-semitism," and other verbal assaults that "influence the culture" are protected even though "none is . . . answerable by more speech." But if the effects of injurious speech cannot be combated by additional speech, it would seem that the marketplace of ideas will not function as it is meant to, as a means of winnowing the true from the false in the course of a free and open competition. This is traditionally the strongest of the rationales usually given for the First Amendment: it allows the whole truth to emerge by providing a breathing space in which partial truths can fight it out. But the *American Booksellers* court rejects that too, declaring that the eventual "domination of truth" is not "a necessary condition for the application of the amendment." While that may at times be an "outcome" of free speech, it is not, the court insists, the reason for free speech. And were we to ask, "What then *is* the reason for free speech?" we would find that the court never tells us. What it does tell us is that words can cause harm, that the harms may be irremediable, that the truth may be damaged, but that, nevertheless, words are speech and speech cannot be abridged. But even at this level, the position is incoherent, because by the court's own admission, pornography *isn't* speech, at least as speech is defined for First Amendment purposes.

Now some would say that there needn't *be* a reason given for the First Amendment's flat prohibition of the abridging of speech. My colleague William Van Alstyne, for example, declares roundly that "the First Amendment does not link the protection it provides with any particular objective and may . . . be deemed to operate without regard to anyone's view of how well the speech it protects may or may not serve such an objective." But it is one thing to say that the First Amendment, as written, does not include a rationale for its existence, and another to say that therefore there is not, nor need be, and such rationale. There is always the category of "it goes without saying," as in "it goes without saying that underlying the laws against theft is a theory of property and property rights, even if the law does not itself come accompanied with an account of the theory that underlies it." Most discussion of the doctrine of free speech, including the discussions by John Milton and J.S. Mill, begin by specifying the *values* or *goals* that a free-speech regime will presumably protect or encourage; and traditionally those values and goals have been thought to be (1) the emergence of truth as the product of public discussion, (2) the self-fulfillment of individuals who are best served if they have access to as many views and arguments as possible, and (3) the maintenance and furtherance of democratic process, of the serious business of self-government by an informed population. Once this list, or some other similarly drawn, is in place, it becomes possible and indeed obligatory to ask of any instance of speech that has been the cause of distress or harm to some citizens: "Does this speech contribute to the healthy flourishing of the relevant values, or is it positively dangerous to their continued existence?" If you don't ask this question, or some version of it, but just say that speech is speech and that's it, you are *mystifying*—presenting as an arbitrary and untheorized fiat—a policy that will seem whimsical or worse to those whose interests it harms or dismisses. Free-speech purists will respond that once you start regulating speech because it has been deemed unrelated to the protection of a set of core values, you never know where the process will stop; you never know what will be regulated *next*. To this I would respond, but if you don't provide a rationale for the toleration of a particular form of speech, but simply declare that the Constitution made me do it, you will have characterized the Constitution as an irrational document. The choice is clear; either acknowledge that, like other items in the Constitution, the First Amendment has a *purpose* and that in the light of that purpose some acts of toleration make sense and some don't; or acknowledge that the free-speech clause has no purpose beyond itself, and face the conclusion that there is no compelling—that is, serious—reason for adhering to it. It may sound paradoxical, but the First Amendment has a positive claim on us only if we understand it to be self-qualifying: you will not abridge speech that is supportive of the values in the name of which we have joined together. Regulation of other forms of speech—speech either irrelevant to the maintenance of those values or subversive of them—should not be regarded as an exception to the amendment but as a fulfillment of its mandate.

Hard-line First Amendment advocates will vigorously protest this account of their sacred text, but the protest is belied by their own activities, for they typically play the regulation game behind their own backs. They insist up front that they read the text without exceptions ("shall make *no* law") but then smuggle in the exceptions by declaring them not really to be speech or to be speech "bridged with action." The more successful these maneuvers, the *less* likely are those who perform them to confront the hard questions that a more forthright analysis—an analysis that acknowledged the difficulties of its own vocabulary—would bring to the surface. The result are opinions that are superficially cogent but deeply incoherent: the *Hustler/ Falwell* opinion, in which eight grown men and one grown woman tell us with a collective straight face that they can't tell the difference between exaggerating someone's already big teeth and falsely portraying a man and his mother in the act of incest; or the *American Booksellers* opinion, in which the court eliminates one by one the traditional reasons for reaching the decision it nevertheless reaches. It is hard when reading these opinions not to feel that the entire enterprise has gone off the rails and that you are in the hands either of charlatans or idiots.

In fact, you are in the hands of persons who are captive to a faulty theory, a theory that has produced among other philosophical curiosities this oft-repeated dictum: "Under the first amendment there is no such thing as a false idea" (*Gertz v. Robert Welch,* 1974). What this means (it couldn't possibly mean what it says, because what it says is obviously false) is that for First Amendment purposes a court will suspend its judgments as to truth and falsity or right and wrong in order to give expression the widest possible scope. In adopting this stance, the court falls in with the deep skepticism first expressed by Oliver Wendell Holmes in *Abrams v. the United States* (1919). Holmes begins by observing that belief and conviction are notoriously changeable: the truth for which we would die (and perhaps kill) today may become tomorrow's scorned error, and, mutatis mutandis, the outlandish opinion we would today rule out of court may become tomorrow's cherished orthodoxy. In short, you never know for certain, and because you never know for certain, it is the better part of both prudence and wisdom (is this, by the way, a privileged truth, a preferred idea?) to refrain from judging speech on the basis of its ideational content. Better instead to leave the task of winnowing the wheat from the chaff to time and the marketplace of ideas.

But even as it develops this skeptical jurisprudence in the decades since

Abrams, the court puts alongside it a jurisprudence based on the insight that some forms of expression seem either so worthless (i.e., obscenity) or dangerous (i.e., "fighting words") that they simply cannot be tolerated. It doesn't take a rocket scientist to see that a jurisprudence that first indemnifies all forms of speech and then sets about distinguishing good speech from bad will be slightly schizophrenic, and it is no surprise to find that many First Amendment decisions, including the two I have analyzed, run contradictory arguments simultaneously. To be sure, the tradition is to some extent aware that what its right hand gives (total freedom of speech) its left hand takes away, but in the face of criticism it invokes a succession of distinctions designed to save its coherence. Basically the strategy is to declare that the forms of speech found unworthy or intolerable are not really speech and that therefore we do not compromise our free-speech principles by regulating them. Moreover, the strategy is reversible when a court comes upon a form of physical action whose consequences it is willing to tolerate; for then it can say that while the behavior in question appears to be action, it is really speech—expressive of some idea—and therefore outside the scope of judicial attention. It's a wonderful game, and those who play it are limited only by their own considerable ingenuity. See, it looks like speech, but it's really action; or, it looks like action, but it's really speech; or, it looks like intimidation, harassment, libel, and group vilification, but it's really the expression of an idea. Armed with this marvelously flexible instrument, a court can have its First Amendment and make you eat it too, can tell you, for example, that the malicious depiction of a man having incest with his mother is an idea; can tell you that the repeated portrayal of women as the appropriate objects of degradation is an idea (presumably the idea that women are inferior and can be exploited at will); can tell you that the genocidal message of Nazis who wanted to march through Skokie is an idea (presumably the idea that Jews are vermin and should be removed from the face of the earth).

This is where the idea that there is no such thing as a false idea (and therefore no such thing as a true idea, like the idea that women are full-fledged human beings or the idea that Jews shouldn't be killed) gets you; it prevents you, as a matter of principle, from inquiring into the real world of consequences of allowing certain forms of so-called speech to flourish. Behind the principle (that there is no such thing as a false idea) lies a vision of human life as something lived largely in the head. There is an entire book to be written about the stigmatization and devaluation of the *body* in First Amendment jurisprudence, but for the moment I will point out that First Amendment jurisprudence works only if you assume that mental activities, even when they emerge into speech, remain safely quarantined in the cortex and do not spill over into the real world, where they can inflict harm.

This assumption is crazy, and the frantic and sometimes comical manipulation of the speech/action distinction by courts is an involuntary and unwitting acknowledgment of just how crazy it is.

"Crazy" is perhaps too flip a word for a practice that would be more politely described as "categorical analysis." Categorical analysis proceeds by asking essentially taxonomic questions such as "Is it speech?" and refraining from further inquiry once an answer is forthcoming. In a "rights regime," a regime whose chief concern is to protect the autonomy of individuals, categorical analysis turns an indifferent and dismissive eye to the effects produced by the exercised rights. At times those effects are noted in passing, especially when, as in *Hustler* and *American Booksellers,* they are acknowledged to be harmful to specific persons or groups; but they are noted only so that they can be declared beside the analytical point because they are the unintended (and therefore accidental) consequences of adhering to a prior principle. When the harms seem particularly grievous, as in the case of the Holocaust survivors who were told that they must endure a parade of Nazis marching through their neighborhood with the intent of disseminating anti-Semitic propaganda, the court will typically announce the regret with which it refuses a judicial remedy, and then solemnly declare that this is the price we must pay (one wonders exactly who the "we" are here) for living in a democracy.

The alternative to categorical analysis is "balancing." Balancing does not exclude the asking of taxonomic questions but does not consider its work done once the answers have been given. Rather, it opens its inquiry to include the rights and interests of the persons or groups affected by the exercise of some enumerated right. The balancer, as Kathleen Sullivan has observed, is more like a grocer than a zoologist. The job of a balancing court is "to place competing rights and interests on a scale and weigh them against each other" so that "the outcome is not determined at the outset, but depends on the relative strengths of a multitude of factors" ("Postliberal Judging: The Roles of Categorization and Balancing," *University of Colorado Law Review* 63 [1992]: 293–294). In the context of First Amendment issues, a balancing court will consider *more* and will not assume that the only relevant sphere of action is the head and larynx of the individual speaker.

The arguments against balancing begin and end with the assertion that as procedure it is ad hoc rather than principled. The fact that a balancing court will take into account "a multitude of factors" weighs against balancing because, given the variety and volatility of the relevant factors, each case will turn out to be unique, and we will end up with a jurisprudence of particulars, a jurisprudence endlessly sensitive to local and transient pressures. The answer to the objection is that this is the jurisprudence we already have. A so-called principled analysis is, as we have seen, ad hoc behind its back; it is continually engaged in saving its own appearances by inventing (and then reinventing as needed) distinctions that hide from itself what it is really doing. Balancing is always going on, even when—no, especially when—it is categorically renounced. As Sullivan puts it, "the use of taxonomies obscures" the balancing maneuvers, but it is always the case that the announced rule has been "precipitated . . . from an implicit prior weighing of substantive values" (311).

What would balancing look like if it were applied to free-expression cases (as it is in Canada, a country that at last look has not gone down the totalitarian tubes)? It would look very much like the procedure suggested by Learned Hand many years ago. If you are faced with a conflict between the rights of speakers and the rights and interests of those who are hurt or intimidated by the speech in question, "ask whether the gravity of the 'evil' discounted by its improbability justifies such invasion of free speech as is necessary to avoid the danger" (*U.S.* v. *Dennis,* 183 F.2d 212). Ask, that is, whether the harms supposedly caused by the offending speech are indeed likely to materialize, and if the answer is "yes," weigh those harms against the harms that may be produced by regulation. In short, you substitute for the categorical question "Is it speech?" (and for the assumption that an affirmative answer is the end of the matter) the three-part question "Given that it is speech what does it do, do we want it to be done, and is more to be gained or lost by moving to curtail it?" Note that a test like this one will not, in every case, lead to the regulation of injurious speech; in many instances the calculation the test directs will lead to the conclusion that in *this* instance regulating injurious speech will be more costly than tolerating it.

Often the outcome will depend strongly on the social or institutional context in which the offending speech occurs. You would expect, for example, that the free-speech interest would be comparatively small in a military context, in which the maintenance of lines of authority is a priority; and the same would hold for a medical context, in which the value of patient care would take precedence over the value of free expression. But in a college or university context, in which the presumption in favor of speech interests would usually be very strong, the threshold of toleration would be very high, although not so high as to preclude regulation no matter what the circumstances. For even within the category of educational institutions there are differences that should be taken into account. First, there are differences between different locations on the same campus. In a free-speech zone, an area set aside for soapbox oratory, the scope of toleration is very broad indeed, although it would not extend to speech that was assaultive, intimidating, or harassing. In semipublic areas—dormitory corridors, common rooms, cafeterias—freedom of speech would be constrained by the purposes for which such areas were established. In the classroom, the established purpose—instruction as mandated by college and department requirements—would place obligations on both teachers and students that would restrict both the subjects to be discussed and the manner of their discussion. And, second, there are differences between campuses differently constituted. Recently at a Catholic college the staff of the student newspaper was dismissed because, in the president's judgment, a series of articles mocking the church's position on contraception was subversive of the values on which the institution rested. Presumably the president of a non–church-related school, a school whose primary value was free inquiry, would have responded to the situation differently or would have felt no obligation to respond to it at all. I am not saying that in my judgment the president of the Catholic school made the right decision, only that, given the nature of the "shop" over which he presided, he had to decide one way or the other, just as his secular counterpart would have to decide whether or not he had to decide.

This way of reasoning, in which different kinds of institutions will have different responsibilities in relation to the same speech, will be anathema to those who hold that it is the very essence of a college or university to foster free expression and the uninhibited circulation of ideas, however offensive they might be to some members of the community. This makes sense if we think of colleges and universities primarily as "open forums" whose purpose is to encourage unregulated investigation of any matter under the sun. But there is another way to think about the campus scene, not as a free-speech forum but as a workplace where people have contractual obligations, assigned duties, pedagogical and administrative responsibilities, and so on, and if we accept *that* analogy rather than the analogy of a protected zone of unimpaired expression, the stigmatization of intimidating or assaultive speech will be a straightforward extension of existing law under Title VII of the 1964 Civil Rights Act. It was under Title VII that a court in Florida recently held the operators of a shipyard culpable for permitting so-called girlie calendars to remain on the walls of the shop even after receiving the detailed complaints of many women employees (*Robinson v. Jacksonville Shipyards, Inc.*, U.S. District Court, Middle District of Florida, Jacksonville Division, No. 86-927-J-12 January 18, 1991).

Predictably, the shipyard managers raised a "free-speech" defense, but the court rejected it. The workplace, the court stated, is for working, and therefore an "employer may lawfully withhold its consent for employees to engage in expressive activities" if those activities produce "special harms" in the form of a work atmosphere destructive of the abilities of some employees to perform their duties with a sense of dignity and safety. If the campus is a workplace, as it surely is by any definition of workplace that could be imagined, then it would seem appropriate and even obligatory for an administration to take judicial note of activities—including verbal activities—that make the workplace intolerable for some members of the community. As the Florida court points out, this does not mean that the protection afforded disagreeable speech "in the world at large" is being withdrawn; a workplace is *not* the world at large (I would wonder whether anything is); and even if "the speech at issue" is elsewhere "fully protected," the court "must balance" against this presumed protection "the governmental interest in cleaning the workplace of impediments to the equality of women," and I would add, to the equality of African Americans or gays or Jews or any other group entitled to move freely and without undue anxiety within the precincts of a public, commercial, or educational institution.

The Florida case allows us to see exactly what balancing means. It means that the value of the "free-speech" interest, which is a real interest, will vary with the underlying purpose for which some social space has been organized. It is only in the most peculiar and eccentric of social spaces, like a Hyde Park corner, where the production of speech has no purpose other than itself that absolute toleration will make sense, and it is one of the oddities of "official" First Amendment rhetoric that such peculiar spaces are put forward as the norm. That is, First Amendment rhetoric presupposes the ordinary situation as one in which expression is wholly unconstrained and then imagines situations of constraint as special. But the truth is exactly the reverse: the special and almost-never-to-be-encountered situation is one in which you can say what you like with impunity. The ordinary situation is one in which what you can say is limited by the decorums you are required to internalize before entering. Regulation of free speech is a defining feature of everyday life, not because the landscape is polluted by censors, but because the very condition of purposeful activity (as opposed to activity that is random and inconsequential) is that some actions (both physical and verbal) be excluded so that some others can go forward. In the very few contexts in which the idea of "free expression" is really taken seriously— call-in talk shows and major league baseball games (and even these have their limits)—it is understood that the speech freely produced is tolerated because it doesn't matter. When speech does matter, when it is produced in the service of some truth or preferred agenda, the sign of its mattering is the fact that only some forms of it will be welcome.

If we keep in mind that regulation of speech is constitutive of meaningful discourse, we will not regard proposals for regulation as anomalous but receive them as suggested modifications of a condition—the condition of productive constraint—that has always obtained. In this light the question "Should hate speech be regulated?" will lose much of its sexiness, for it will no longer be heard as an extraordinary question provoked by extraordinary circumstances, as a limit case that tests the resources of philosophy; rather, it will be heard as a perfectly ordinary question, no more or less difficult than the question of whether spectators at a trial can applaud or boo the statements of opposing counsels. And how would I answer the question? I would say, "It depends," an answer that may be philosophically unsatisfying but one that is responsive to the messy contingency of a world that defies the neatness of philosophical formulations. In such a world rules are hard to come by, and one must make do with rules of thumb. My rule of thumb is "Don't regulate unless you have to," because the machinery of regulation is liable to cause more problems than it solves; but on some occasions the balance may tip in the other direction, and then, however reluctantly and cautiously, you may have to consider regulatory action.

To this some will object that it's fine to talk about caution and reluctance, but once you begin to regulate and discipline, there is no natural place to stop; and what begins as a small and limited restriction may in time flower into full-fledged tyranny. This is known in the trade as the "slippery slope" argument, and it says that, given the danger of going down the regulatory road, it is safer never to begin. But the slippery slope argument is another one of those exercises in *abstract* reasoning that imagines a worse-case scenario every time because nothing fills up its landscape but its own assumptions. That is, the slippery slope argument assumes that there is nothing in place, no underbrush, to stop the slide; but in any complexly organized society there will always be countervalues to invoke and invested persons to invoke them. Slippery slope trajectories are inevitable only in the *head,* where you can slide from A to B to Z with nothing to retard the acceleration of the logic. In the real world, however, the step even from A to B will always meet with resistance of all kinds from persons differently positioned, and, as a matter of fact, the chances of ever getting to Z are next to nothing. Somewhere along the route some asserted interest will stop the slide, and a line will be drawn beyond which regulators will be prevented from going, at least for a time, until new pressures and new resistances provoke a new round of debates, at the end of which still another line will be provisionally drawn.

But that means, someone will say, that you will always be drawing the line in an ad hoc and inconsistent way, and the line should be drawn in a *principled* way. And I would reply that the so-called principled drawing of the line by distinctions like that between speech and action, or between content and time-place-manner regulations, or between high- and low-value speech, or between fighting words and words that are merely expressive, will be equally ad hoc and context sensitive; there is no other way to draw a line except in the context of an act of judgment that rests on disputable definitions and stipulations of value. What is a fighting word today may not be one tomorrow and may not be everyone's fighting word even today; what is low-value speech under one set of conditions may become high under another. Line drawing, in short, will always be a political and contestable action and therefore inseparable from the biases and blindnesses inherent in politics.

Wouldn't it be better, then, to refrain from drawing any lines at all? It might be better if it were possible, but you can only refrain if you know exactly where the line should *not* be drawn; and you cannot know *that* without already having drawn the line beyond which you will refuse to draw any others; and you can only do *that* with the help of definitions and distinctions that will involve you in drawing just the kind of line you have pledged to avoid. If you resolve, for example, to draw the line at speech, you first must define speech; but since any definition of what speech is will be controversial—will seem to some to be too narrowly exclusive, ruling out symbolic expression, and will seem to others to be overly inclusive, running the risk that *everything* will be labeled speech—the line you then draw with the intention of not going beyond it to draw any other lines will already be in violation of that intention.

When the difficulty of drawing lines is discussed, the problem is usually said to be that one never knows where to stop; but in fact the problem is that one never knows where to begin, and that is because one has always already begun. When you go to draw a line, you are confronted not with a blank slate but with a slate covered by the line drawings of those who have gone before you. Nor would it be possible to wipe the slate clean and begin afresh; for were that to be done (and it is hard even to imagine), you would find yourself in a wasteland without guidelines or directions. Deliberation on alternative courses of action can only occur against a background of actions previously considered and taken. Were that background to be removed or declared "off limits" as a resource for decision making (two actions that are in fact impossible), the result would not be a clearer vision of the task but *no* vision of the task.

The content of what I am calling the background is everything that goes without saying; and it is my thesis that unless there is a category of what goes without saying, filled with assumptions so deeply in place that challenges to them are literally unimaginable, nothing could be said. The history of past sayings is not an obstacle to be swept away but the very ground of our present intelligibility. Modern First Amendment doctrine wishes to rise above that ground and ascend to an intelligibility that is hostage to no past whatsoever. It wishes, that is, to justify its actions from scratch, without reference to the views or interests of anyone who has ever lived. This is the impossible dream of liberalism, and its impossibility is well illus-

trated by the apparently intractable difficulty of defining obscenity. Walter
Berns has observed that the "Founders took for granted" that obscene speech
was "beyond the pale" and did not feel obliged to give reasons for its
nonprotection; but in 1971, "when the Court suddenly was challenged to
state the reasons why the word 'fuck' ought not to be publicly uttered, it
found it was unable to supply a reason" (*The First Amendment and the
Future of American Democracy* [New York, 1976], 196). The reason for
this inability to supply a reason is that by 1971 what had come to be taken
for granted was that nothing could be taken for granted. Therefore the only
truly good reason would be one that could be pursued down to its very
roots and shown to rest on nothing other than itself. A reason whose force
was a function of particular circumstances would not be good enough because it could always be challenged by imagining alternative circumstances.
Define obscenity with reference to "the average man," and you will be
confronted by the above-average men like Joyce and Lawrence who have
written "obscene" masterpieces. Define obscenity by invoking community
standards, and you will be told that there are many communities each of
which has a standard of its own. Define obscenity as that which engages
prurient interests, and you will be asked whose sensibility will determine
what is and is not prurient? After all, "one man's vulgarity is another's
lyric." In this famous pronouncement by Justice Harlan in *Cohen v. California,* one hears the same skepticism voiced by Justice Powell when he
declares that there is no such thing as a false idea; and the result is the
same, a self-imposed incapacity to make distinctions that would seem perfectly obvious to any well-informed teenager. When Harlan observes that
"no general principle exists" for determining what is and is not obscene,
he means that there is no principle that would hold for any and all situations
that could be imagined by a determined and resourceful intelligence; but by
that standard there are no principles to be found anywhere, and we might
as well throw up our hands and say that anything goes because we cannot
think of a way to insulate our judgments against all possible future objections.

It is a counsel of despair generated by a refusal to be satisfied with reasons that share the infirmities of an imperfect and contingent world. And
finally it is the world that free-speech purists spurn in favor of a cognitive
utopia where invariant principles form the basis of unshakable deductions.
No accidents, no politics, no bodies, just abstract concepts that speak to
one another without interference from base and fleshly appetites. It is a
Cartesian fantasy that finds a perfect literary instantiation in Ray Bradbury's
Fahrenheit 451. In that poor man's *1984,* civilization has fallen to the book
burners, one of whom is gradually awakened from his television-induced
lethargy and in the end is persuaded to join the resistance. What he finds is

a community so dedicated to the preservation of what the censors would destroy that its members make themselves into living repositories of books and parts of books. Each man or woman is responsible for memorizing a renowned text, and the accomplishment of the task is marked by the absorption of the living person into the text of which he or she is now the vehicle. "We're nothing more than dust jackets for books," one of them says. They are, quite literally, "walking heads," and their disdain for the material dross of their bodies is reflected in a casual willingness to undergo plastic surgery and the alteration of fingerprints. It is the ultimate Great Books seminar populated by readers who have purged themselves of affectional desires and desire only to serve the Harvard Library of Classics. They, of course, think of themselves as the remnant, as the bearers of the only thing that makes life worth living. They think that nothing could be worse than living a life without Plato or Milton or Boswell's *Life of Johnson*. But lots of things are worse, and some of them are things that the First Amendment, as it is now interpreted, allows and, by allowing, encourages. If she were alive, you could ask Jerry Falwell's mother.

10

LIBERALISM DOESN'T EXIST

(This brief essay was written in response to Stephen Carter's "Evolutionism, Creationism, and Treating Religion as a Hobby" [Duke Law Journal, *December 1987, 977–996]. Professor Carter explores the tension between liberal rationalism and the claims of religious faith. He observes that although liberalism presents itself as the protector of religious freedoms, it is "really derogating religious belief in favor of other, more 'rational' methods of understanding the world." "Liberalism," he says, is "curiously intolerant." It is my thesis that there is nothing curious about liberalism's intolerance, since like any other ideology it must reject as beyond the pale anything that does not conform to its constitutive principle. The fact that liberals like to think of themselves as the apostles of tolerance only testifies to the strength of the blindness with which they pursue their (non)program.)*

I find Stephen Carter's argument compelling and incisive, and my only quarrel is with its conclusion when he urges a "softened liberal politics" that would "acknowledge and genuinely cherish the religious beliefs that for many Americans provide their fundamental worldview."[1] He confesses that he has "not yet worked out the details"[2] of such a politics, and it is my contention that he never could for reasons he himself enumerates. The chief reason is that liberalism is informed by a faith (a word deliberately chosen) in reason as a faculty that operates independently of any particular world view. It is therefore committed at once to allowing competing world views equal access to its deliberative arena, and to disallowing the claims of any one of them to be supreme, unless of course it is demonstrated to

Originally printed in *Duke Law Journal* (Durham, N.C., 1987), p. 997. Reprinted with permission of the publisher.

be at all points compatible with the principles of reason. It follows then that liberalism can only "cherish" religion as something under its protection; to take it seriously would be to regard it as it demands to be regarded, as a claimant to the adjudicative authority already deeded in liberal thought to reason. This liberalism cannot do because, as Carter points out, if you take away the "primacy of reason"[3] liberal thought loses its integrity, has nothing at its center, becomes just one more competing ideology rather than a procedure (and it is in procedure or process that liberalism puts its faith) that outflanks or transcends ideology. The one thing liberalism cannot do is put reason *inside* the battle where it would have to contend with other adjudicative principles and where it could not succeed merely by invoking itself because its own status would be what was at issue.

Indeed, liberalism depends on not inquiring into the status of reason, depends, that is, on the assumption that reason's status is obvious: it is that which enables us to assess the claims of competing perspectives and beliefs. Once this assumption is in place, it produces an opposition between reason and belief, and that opposition is already a hierarchy in which every belief is required to pass muster at the bar of reason. But what if reason or rationality itself rests on belief? Then it would be the case that the opposition between reason and belief was a false one, and that every situation of contest should be recharacterized as a quarrel between two sets of belief with no possibility of recourse to a mode of deliberation that was not itself an extension of belief. This is in fact my view of the matter and I would defend it by asking a question that the ideology of reason must repress: where do reasons come from? The liberal answer must be that reasons come from nowhere, that they reflect the structure of the universe or at least of the human brain; but in fact reasons always come from somewhere, and the somewhere they come from is precisely the realm to which they are (rhetorically) opposed, the realm of particular (angled, partisan, biased) assumptions and agendas. What this means is that not all reasons (or reasonable trains of thought) are reasons for everyone. If (to take a humble literary example) I am given as a reason for preferring one interpretation of a poem to another the fact that it accords with the poet's theological views, I will only hear it as a reason (as a piece of weighty evidence) if it is *already* my conviction that a poet's aesthetic performance could be influenced by his theology; if, on the other hand, I see poetry and theology as independent and even antagonistic forms of life (as did many of those new critics for whom the autonomy of the aesthetic was an article of faith) this fact will not be a reason at all, but something obviously beside the (literary) point. Similarly, a lawyer may give as a reason for acquitting his client the fact that his action was not intentional, but both the fact and the reason it becomes will be perspicuous only because the boundaries between the intentional and the unintentional have been drawn in ways that could themselves be contested, even if at the moment they are not being contested but assumed. It is not that reasons can never be given or that they are, when given, incapable of settling disputes, but that the force they exert and their status *as* reasons depends on the already-in-place institution of distinctions that themselves rest on a basis no firmer (no less subject to dispute) than the particulars they presently order. In short, what is and is not a reason will always be a matter of faith, that is, of the assumptions that are bedrock within a discursive system which because it rests upon them cannot (without self-destructing) call them into question. (Nor can one avoid this conclusion by invoking supposedly abstract—i.e. contentless—logical operations like the "law of contradiction"; for just what is and is not a contradiction will vary depending on the distinctions already in place; a contradiction must be a contradiction between something and something else and the shape of those somethings will always be the product of an interpretive rather than a formal determination.)

It follows then that persons embedded within *different* discursive systems will not be able to hear the other's reasons *as* reasons, but only as errors or even delusions. This, I think, is Carter's point when he observes that "to the devout fundamentalist . . . evolutionary theory is not simply contrary to religious teachings; . . . it is *demonstrably* false."[4] I take the stress on the word *demonstrably* to mean that Carter understands fully that the clash between liberals and fundamentalists is a clash between two faiths, or if you prefer (and it is my thesis that these two formulations are interchangeable) between two ways of thinking undergirded by incompatible first principles, empirical verification and biblical inerrancy. Given this incompatibility it would not be possible for either party to "cherish" or "take" seriously the commitments and conclusions of the other, for to do so would be to abandon the foundation on which it rests. The fundamentalist cannot measure the statements in *Genesis* against a standard of scientific fact because for him the proper direction of measurement is the other way around; he knows (in the only sense that knowledge can possibly have) that whatever does not accord with the Word of God cannot be true; and he knows further that untruths are dangerous and should not be allowed to flourish. And in the eyes of the liberal, the pronouncements of fundamentalists are no less dangerous and for the same reason: they flow from ignorance and bigotry, and if they go unchecked they may succeed in turning the nation away from reason. Accordingly the liberal feels obliged to quarantine religious pronouncements, to confine them to contexts (the home, the Church) that present the least risk of general infection. He cannot allow them to enter into the general political conversation because he does not regard them, and *could* not regard them, as issuing from a respectable point of view on a par with the points of view, for example, of libertarians or utilitarians. And by his lights he is right: the debate between those who would maximize individual freedom and those who would achieve the greatest good for the greatest number is conducted according to principles (of argument and evidence) to which all parties subscribe; not only do fundamentalists not subscribe to these principles, they stigmatize them as the diabolical tools of godless humanism; obviously they cannot be given a place in the arena for they refuse to play by the rules. Of course, a humane society (another key notion in liberal thought) does not kill or imprison people just because they believe foolish and unprofitable things; indeed it is the disinclination to punish those with whom you disagree that distinguishes the liberal from his "fanatic" opposite; those who believe obviously false things must be protected. Nevertheless, this does not require that the fundamentalist be taken seriously, for according to liberal assumptions, he gave up his claim to serious consideration when he abandoned the rule of reason.

All of this is implicit (and sometimes explicit) in Carter's argument. Why then does he cling to the hope of "softening the tension inherent in the liberal principle of neutrality toward religion"?[5] The answer, I think, is that he mistakes the essence of liberalism when he characterizes it as "steeped . . . in skepticism, rationalism and tolerance."[6] "Tolerance" may be what liberalism claims for itself in contradistinction to other, supposedly more authoritarian, views; but liberalism is tolerant only *within* the space demarcated by the operations of reason; any one who steps outside that space will not be tolerated, will not be regarded as a fully enfranchised participant in the marketplace (of ideas) over which reason presides. In this liberalism does not differ from fundamentalism or from any other system of thought; for any ideology—and an ideology is what liberalism is—must be founded on some basic conception of what the world is like (it is the creation of God; it is a collection of atoms), and while the conception may admit of differences within its boundaries (and thus be, relatively, tolerant) it cannot legitimize differences that would blur its boundaries, for that would be to delegitimize itself. A liberalism that did not "insist on reason as the only legitimate path to knowledge about the world"[7] would not be liberalism; the principle of a rationality that is above the partisan fray (and therefore can assure its "fairness") is not incidental to liberal thought; it *is* liberal thought, and if it is "softened" by denying reason its priority and rendering it just one among many legitimate paths, liberalism would have no content. Of course it is my contention (and Carter's too I think) that liberalism doesn't have the content it believes it has. That is, it does not have at its center an adjudicative mechanism that stands apart from any particular moral and political agenda. Rather it is a very particular moral agenda (privileging the individual over the community, the cognitive over the affective, the abstract over the particular) that has managed, by the very partisan means it claims to transcend, to grab the moral high ground, and to grab it from a discourse—the discourse of religion—that had held it for centuries. This victory certainly sets liberalism apart from the ideologies it has vanquished, but because the victory is political, liberalism cannot finally claim to be different from its competitors. Liberalism, however, defines itself by that difference—by its not being the program of any particular group or party—and therefore in the absence of that difference one can only conclude, and conclude nonparadoxically, that liberalism doesn't exist.

PART II

11

THE LAW WISHES TO HAVE A FORMAL EXISTENCE

Achieving Plain and Clear Meanings

The law wishes to have a formal existence. That means, first of all, that the law does not wish to be absorbed by, or declared subordinate to, some other—nonlegal—structure of concern; the law wishes, in a word, to be distinct, not something else. And second, the law wishes in its distinctness to be perspicuous; that is, it desires that the components of its autonomous existence be self-declaring and not be in need of piecing out by some supplementary discourse; for were it necessary for the law to have recourse to a supplementary discourse at crucial points, that discourse would be in the business of specifying what the law is, and, consequently, its autonomy would have been compromised indirectly. It matters little whether one simply announces that the principles and mechanisms of the law exist ready-made in the articulations of another system or allows those principles and mechanisms to be determined by something they do not contain; in either case, the law as something independent and self-identifying will have disappeared.

In its long history, the law has perceived many threats to its autonomy, but two seem perennial: morality and interpretation. The dangers these two pose are, at least at first glance, different. Morality is something to which the law wishes to be related, but not too closely; a legal system whose conclusions clashed with our moral intuitions at every point so that the categories *legally valid* and *morally right* never (or almost never) coincided would immediately be suspect; but a legal system whose judgments perfectly meshed with our moral intuitions would be thereby rendered superfluous. The point is made concisely by the Supreme Court of Utah in a case where it was argued that the gratuitous payment by one party of the other party's mortgage legally obligated the beneficiary to repay. The court rejected the argument, saying "that if a mere moral, as distinguished from a legal, obligation were recognized as valid consideration for a contract, that would practically erode to the vanishing point the necessity for finding a consideration."[1] That is to say, if one can infer directly from one's moral obligation in a situation to one's legal obligation, there is no work for the legal system to do; the system of morality has already done it. Although it might seem (as it does to many natural law theorists) that such a collapsing of categories recommends itself if only on the basis of efficiency (why have two systems when you can make do with one?), the defender of a distinctly legal realm will quickly answer that since moral intuitions are notoriously various and contested, the identification of law with morality would leave every individual his or her own judge; in place of a single abiding standard to which disputing parties might have recourse, we would have many standards with no way of adjudicating between them. In short, many moralities would make many laws, and the law would lack its most saliently desirable properties, generality and stability.

It is here that the danger posed by morality to law, or, more precisely, to the rule (in two senses) of law intersects with the danger posed by interpretation. The link is to be found in the desire to identify a perspective larger and more stable than the perspective of local and individual concerns. Morality frustrates that desire because, in a world of more than one church, recourse to morality will always be recourse to someone's or some group's challengeable moral vision. Interpretation frustrates that desire because, in the pejorative sense it usually bears in these discussions, interpretation is the name for what happens when the meanings embedded in an object or text are set aside in favor of the meanings demanded by some angled, partisan object. Interpretation, in this view, is the effort of a morality, of a particular, interested agenda, to extend itself into the world by inscribing its message on every available space. It follows then that, in order to check the imperial ambitions of particular moralities, some point of resistance to interpretation must be found, and that is why the doctrine of formalism has proved so attractive. Formalism is the thesis that it is possible to put down marks so self-sufficiently perspicuous that they repel interpretation; it is the thesis that one can write sentences of such precision and simplicity that their meanings leap off the page in a way no one—no matter what his or her situation or point of view—can ignore; it is the thesis that one can devise procedures that are self-executing in the sense that their unfolding is independent of the differences between the agents who might set them in motion. In the presence (in the strong Derridean sense) of such a mark or sentence or procedure, the interpretive will is stopped short and is obliged to press its claims within the constraints provided by that which it cannot override. It must take the marks into account; it must respect the self-declaring reasons; it must follow the route laid down by the implacable procedures, and if it then wins it will have done so fairly, with justice, with reason.

Obviously then, formalism's appeal is a function of the number of problems it solves, or at least appears to solve: it provides the law with a palpable manifestation of its basic claim to be perdurable and general; that is, not shifting and changing, but standing as a point of reference in relation to which change can be assessed and controlled; it enables the law to hold contending substantive agendas at bay by establishing threshold requirements of procedure that force those agendas to assume a shape the system will recognize. The idea is that once a question has been posed as a *legal* question—has been put into the proper *form*—the answer to it will be generated by relations of entailment between that form and other forms in the system. As Hans Kelsen put it in a book aptly named *The Pure Theory of Law*,

> The law is an order, and therefore all legal problems must be set and solved as order problems. In this way legal theory becomes an exact structural analysis of positive law, free of all ethical-political value judgments.[2]

Kelsen's last clause says it all: the realms of the ethical, the political, and of value in general are the threats to the law's integrity. They are what must be kept out if the law is to be something more than a misnomer for the local (and illegitimate) triumph of some particular point of view.

There are at least two strong responses to this conception of law. The first, which we might call the "humanistic" response, objects that a legal system so conceived is impoverished, and that once you have severed procedures from value, it will prove enormously difficult, if not impossible, to relink them in particular cases. Since the answers generated by a purely formal system will be empty of content (that, after all, is the formalist claim), the reintroduction of content will always be arbitrary. The second response, which we might call "radical" or "critical," would simply declare that a purely formal system is not a possibility, and that any system pretending to that status is already informed by that which it purports to exclude. Value, of both an ethical and political kind, is already inside the gate, and the adherents of the system are either ignorant of its sources or are engaged in a political effort to obscure them in the course of laying claim to a spurious purity. In what follows, I shall be elaborating a version of the second response, and arguing that however much the law wishes to have a formal existence, it cannot succeed in doing so, because—at any level from the most highly abstract to the most particular and detailed—any specification of what the law is will already be infected by interpretation and will therefore be challengeable. Nevertheless, my conclusion will not be that the law fails to have a formal existence but that, in a sense I shall explain, it always succeeds, although the nature of that success—it is a political/rhetorical achievement—renders it bitter to the formalist taste.

We may see what is at stake in disputes about formalism by turning to a recent (July, 1988) opinion delivered by Judge Alex Kozinski of the United States Court of Appeals for the Ninth Circuit.[3] The case involved the desire of a construction partnership called Trident Center to refinance a loan at rates more favorable than those originally secured. Unfortunately (or so it seemed), language in the original agreement expressly blocked such an action, to wit that the " '[m]aker shall not have the right to prepay the principal amount hereof in whole or in part' for the first 12 years."[4]

Trident's attorneys, however, pointed to another place in the writing where it is stipulated that "[i]n the event of a prepayment resulting from a default . . . prior to January 10, 1996 the prepayment fee will be ten percent"[5] and argued that this clause gives Trident the option of prepaying the loan provided that it is willing to incur the penalty as stated. Kozinski is singularly unimpressed by this reasoning, and, as he himself says, dismisses it "out of hand,"[6] citing as his justification the clear and unambiguous language of the contract. Referring to Trident's contention that it is entitled to precipitate a default by tendering the balance plus the ten percent fee, Kozinski declares that "the contract language, cited above, leaves no room for this construction,"[7] a judgment belied by the fact that Trident's lawyers managed to make room for just that construction in their arguments. It is a feature of cases like this that turn on the issue of what is and is not "expressly" said that the proclamation of an undisputed meaning always occurs in the midst of a dispute about it. Given Kozinski's rhetorical stance, the mere citation (his word, and a very dangerous one for his position) of the contract language should be sufficient to end all argument, but what he himself immediately proceeds to do is argue, offering a succession of analyses designed to buttress his contention that "it is difficult to imagine language that more clearly or unambiguously expresses the idea that Trident may not unilaterally [more is given away by this word than Kozinski acknowledges] prepay the loan during its first 12 years."[8] If this were in fact so, it would be difficult to imagine why Kozinski should feel compelled to elaborate his opinion again and again. I shall not take up his points except to say that, in general, they are not particularly persuasive and usually function to open up just the kind of interpretive room he declares unavailable. Thus, for example, he reasons that Trident's interpretation "would result in a contradiction between two clauses of the contract" whereas the "normal rule of construction . . . is that courts must interpret contracts, if possible, so as to avoid internal conflict."[9] But it is no trick at all (or at least not a hard one) to treat the two clauses so that they refer to different anticipated situations and are not contradictory (indeed that is what Trident's lawyers do): in the ordinary course of things, as defined by the rate and schedule of payments set down in the contract, Trident will not have the option of prepaying; but in the extraordinary event of a default, the prepayment penalty clause will then kick in. To be sure, Kozinski is ready with objections to this line of argument, but those objections themselves trace out a line of argument and operate (no less than the interpretations he rejects) to fill out the language whose self-sufficiency he repeatedly invokes.

In short (and this is a point I shall make often), Kozinski's assertion of ready-made, formal constraints is belied by his efforts to stabilize what he supposedly relies on, the plain meaning of absolutely clear language. The act of construction for which he says there is no room is one he is continually performing. Moreover, he performs it in a way no different from the performance he castigates. Trident, he complains, is attempting "to obtain judicial sterilization of its intended default,"[10] and the reading its lawyers propose is an extension of that attempt rather than a faithful rendering of what the document says. The implication is that *his* reading is the extension of nothing, proceeds from no purpose except the purpose to be scrupulously literal. But his very next words reveal another, less disinterested purpose: "But defaults are messy things and they are supposed to be. . . . Fear of these repercussions is strong medicine that keeps debtors from shirking their obligations. . . ."[11] And he is, of course, now administering that strong medicine through his reading, a reading that is produced not by the agreement, but by his antecedent determination to enforce contracts whenever he can. The contrast then is not (as he attempts to draw it) between a respect for what "the contract clearly does . . . provide"[12] and the bending of the words to an antecedently held purpose, but between two bendings, one of which by virtue of its institutional positioning—Kozinski is after all the judge—wins the day.

Except that it doesn't. In the second half of the opinion there is a surprise turn, one that alerts us to the larger issue Kozinski sees in the case and explains the vehemence (often close to anger) of his language. The turn is that Kozinski rules for Trident, setting aside the district court's declaration that the clear and ambiguous nature of the document leaves Trident with no cause of action and setting aside, too, the same court's sanction of Trident for the filing of a frivolous lawsuit. In so doing Kozinski is responding to Trident's second argument, which is that "even if the language of the contract appears to be unambiguous, the deal the parties actually struck is in fact quite different" and that "extrinsic evidence" shows "that the parties had agreed Trident could prepay at any time within the first 12 years by tendering the full amount plus a 10 percent prepayment fee."[13] Kozinski makes it clear that he would like to reject this argument and rely on the traditional contract principle of the parol evidence rule, the rule (not of evidence but of law) by which "extrinsic evidence is inadmissible to interpret, vary or add to the terms of an unambiguous integrated written instrument."[14] He concedes, however, that this rule has not been followed in California since *Pacific Gas & Electric Co. v. G. W. Thomas Drayage & Rigging Co.*,[15] a case in which the state supreme court famously declared that there is no such thing as a clear and unambiguous document because it is not "feasible to determine the meaning the parties gave to the words from the instrument alone."[16] In other words (mine, not the court's), an instrument that seems clear and unambiguous on its face seems so because "extrinsic evidence"—information about the conditions of its production including the situation and state of mind of the contracting parties, etc.—is already in place and assumed as a background; that which the parol evidence rule is designed to exclude is already, and necessarily, invoked the moment writing becomes intelligible. In a bravura gesture, Kozinski first expresses his horror at this doctrine ("it . . . chips away at the foundation of our legal system")[17] and then flaunts it by complying with it.

> While we have our doubts about the wisdom of *Pacific Gas*, we have no difficulty understanding its meaning, even without extrinsic evidence to guide us . . . we must reverse and remand to the district court in order to give plaintiff an opportunity to present extrinsic evidence as to the intentions of the parties.[18]

That is, "you say that words cannot have clear and constant meanings and that, therefore, extrinsic evidence cannot be barred; I think you are wrong and I hereby refute you by adhering strictly to the rule your words have laid down."

But of course he hasn't. The entire history of the parol evidence rule—the purposes it supposedly serves, the fears to which it is a response, the hopes of which it is a repository—constitutes the extrinsic evidence within whose assumption the text of the case makes the sense Kozinski labels "literal." When he prefaces his final gesture (the judicial equivalent of "up yours") by saying "As we read the rule," he acknowledges that it is *reading* and not simply receiving that he is doing.[19] And to acknowledge as much is to acknowledge that *Pacific Gas* could be read differently. Nevertheless, the challenge Kozinski issues to the Traynor court is pertinent; for what he is saying is that the question of whether or not it is possible to produce " 'a perfect verbal expression' "[20]—an expression that will serve as a "meaningful constraint on public and private conduct"[21]—will not be settled by the pronouncement of a court. Either it is or it isn't; either a court or a legislature or a constitutional convention can order words in such a way as to constrain what interpreters can then do with them or it cannot. The proof will be in the pudding, in what happens to texts or parts of texts that are the repository of that (formalist) hope. The parol evidence rule will not have the desired effect if no one could possibly follow it.

That this is, in fact, the case is indicated by the very attempt to formulate the rule. Consider, for example, the formulation found in section 2–202 of the Uniform Commercial Code.

> Terms with respect to which the confirmatory memoranda of the parties agree or which are otherwise set forth in a writing intended by the parties as a final expression of their agreement with respect to such terms as are included therein may not be contradicted by evidence of any prior agreement or of a contemporaneous oral agreement but may be explained or supplemented
>> (a) by course of dealing or usage of trade (Section 1–205) or by course of performance (Section 2–208); and
>> (b) by evidence of consistent additional terms unless the court finds the writing to have been intended also as a complete and exclusive statement of the terms of the agreement.[22]

One could pause at almost any place to bring the troubles lying in wait for would-be users of this section to the surface, beginning perhaps with the juxtaposition of "writing" and "intended," which reproduces the conflict supposedly being adjudicated. (Is the writing to pronounce on its own meaning and completeness or are we to look beyond it to the intentions of the parties?) Let me focus, however, on the distinction between explaining or supplementing and contradicting or varying. The question is how can you tell whether a disputed piece of evidence is one or the other? And the answer is that you could only tell if the document in relation to which the evidence was to be labeled one or the other declared its own meaning; for only then could you look at "it" and then at the evidence and proclaim the evidence either explanatory or contradictory. But if the meaning and completeness of the document were self-evident (a wonderfully accurate phrase), explanatory evidence would be superfluous and the issue would never arise. And on the other hand, if the document's significance and state of integration are not self-evident—if "it" is not complete but must be pieced out in order to become what "it" is—then the relation to "it" of a piece of so-called extrinsic evidence can only be determined after the evidence has been admitted and is no longer extrinsic. Either there is no problem or it can only be solved by recourse to that which is in dispute.

Exactly the same fate awaits the distinction between "consistent additional terms" and additional terms that are inconsistent. "Consistent in relation to what?" is the question; the answer is "consistent in relation to the writing." But if the writing were clear enough to establish its own terms, additional terms would not be needed to explain it (subsection *[b]*, you will remember, is an explanation of "explained or supplemented"), and if additional terms are needed there is not yet anything for them to be consistent or inconsistent with. The underlying point here has to do with the distinction—assumed but never examined in these contexts—between inside and outside, between what the document contains and what is external to it. What becomes clear is that the determination of what is "inside" will always be a function of whatever "outside" has already been assumed. (I use quotation marks to indicate that the distinction is interpretive, not absolute.) As one commentary puts it, "questions concerning the admissibility of parol evidence cannot be resolved without considering the nature and scope of the evidence which is being offered," and "thus the court must go beyond the writing to determine whether the writing should be held to be a final expression of the parties'. . . agreement."[23]

Nowhere is this more obvious than in the matter of *trade usage,* the first body of knowledge authorized as properly explanatory by the code. Trade usage refers to conventions of meaning routinely employed by members of a trade or industry, and is contrasted to *ordinary usage,* that is, to the meanings words ordinarily have by virtue of their place in the structure of English. The willingness of courts to regard trade usage as legitimately explanatory of contract language seems only a minor concession to the desire of the law to find a public—i.e., objective—linguistic basis, but in fact it is fatal, for it opens up a door that cannot be (and never has been) closed. In a typical trade usage case, one party is given the opportunity to "prove" that the words of an agreement don't mean what they seem to mean because they emerged from a special context, a context defined by the parties' expectations. Thus, for example, in one case it was held that, by virtue of trade usage, the shipment term "June-Aug." in an agreement was to be read as excluding delivery in August;[24] and in another case the introduction of trade usage led the court to hold that an order for thirty-six-inch steel was satisfied by the delivery of steel measuring thirty-seven inches.[25] But if "June-Aug." can, in certain persuasively established circumstances, be understood to exclude August, and "thirty-six" can be understood as meaning thirty-seven, then anything, once a sufficiently elaborated argument is in place, can mean anything: "thirty-six" could mean seventy-five, or, in relation to a code so firmly established that it governed the expectations of the parties, "thirty-six" could mean detonate the atomic bomb.

If this line of reasoning seems to slide down the slippery slope too precipitously, consider the oft cited case of *Columbia Nitrogen Corporation v. Royster Company*.[26] The two firms had negotiated a contract by which Columbia would purchase from Royster 31,000 tons of phosphate each year for three years, with an option to extend the term. The agreement was marked by "detailed provisions regarding the base price, escalation, minimum tonnage and delivery schedules,"[27] but when phosphate prices fell, Columbia ordered and accepted only one-tenth of what was specified. Understandably, Royster sued for breach of contract, and was awarded a judgment of $750,000 in district court. Columbia appealed, contending that, in the fertilizer industry,

> because of uncertain crop and weather conditions, farming practices, and government agricultural programs, express price and quantity terms in contracts . . . are mere projections to be adjusted according to market forces.[28]

One would think that this argument would fail because it would amount to saying that the contract was not worth the paper it was printed on. If emerging circumstances could always be invoked as controlling, even in the face of carefully negotiated terms, why bother to negotiate? Royster does not make this point directly, but attempts to go the (apparently) narrower route of section 202. After all, even trade usage is inadmissible according to that section if it contradicts, rather than explains, the terms of the agreement, and as one authority observes, "it is hard to imagine a . . . 'trade usage' that contradicts a stated contractual term more directly than did the usage in *Columbia Nitrogen Corporation*."[29] The court, however, doesn't see it that way. Although the opinion claims to reaffirm "the well established rule that evidence of usage of trade . . . should be excluded whenever it cannot be reasonably construed as consistent with the terms of the contract,"[30] the reaffirmation undoes itself; for by making the threshold of admissibility the production of a "reasonable construal" rather than an obvious inconsistency (as in 31,000 is inconsistent with 3,100), the court more or less admits that what is required to satisfy the section is not a demonstration of formal congruity but an exercise of rhetorical skill. As long as one party can tell a story sufficiently overarching so as to allow the terms of the contract and the evidence of trade usage to fit comfortably within its frame, that evidence will be found consistent rather than contradictory. What is and is not a "reasonable construal" will be a function of the persuasiveness of the construer and not of any formal fact that is perspicuous before some act of persuasion has been performed.

The extent to which this court is willing to give scope to the exercise of rhetorical ingenuity is indicated by its final dismissal of the contention by Royster that there is nothing in the contract about adjusting its terms to reflect a declining market. "Just so," says the court, there is nothing in the contract about this and that is why its introduction is not a contradiction or inconsistency. Since "the contract is silent about adjusting prices and quantities . . . it neither permits or prohibits adjustment, and this neutrality provides a fitting occasion for recourse to usage of trade . . . to supplement the contract and explain its terms."[31] Needless to say, as an interpretive strategy this could work to authorize almost anything, and it is itself authorized by the first of the official comments on section 202 (and why a section designed supposedly to establish the priority of completely integrated writings is itself in need of commentary is a question almost too obvious to ask): "This section definitely rejects (a) any assumption that because a writing has been worked out which is final on some matters, it is to be taken as including all the matters agreed upon."[32] Or in other words, just because a writing says something doesn't mean that it says everything relevant to the matter; it may be silent on some things, and in relation to those things parol evidence is admissible. But of course the number of things on which a document (however interpreted) is silent is infinite, and consequently there is no end to the information that can be introduced if it can be linked narratively to a document that now becomes a mere component (albeit a significant one) in a larger contractual context.

One way of doing this is exemplified by the majority opinion in *Masterson v. Sine*,[33] a case in which the attempt of a bankruptcy trustee to exercise an option to purchase a particular piece of property (on the grounds that the right of option belongs to the estate) was challenged by parol evidence tending to show that it was the intention of the drafting parties to keep the property in the family (Mr. Masterson and Mrs. Sine were brother and sister) and "that the option was therefore personal to the grantors and could not be exercised by the trustee."[34] The trial court excluded the evidence, ruling that the written contract was a complete and final embodiment of the terms of the agreement and said nothing about the assignability of the option. The court, in the person of Chief Justice Traynor (the same Traynor who in Kozinski's eyes commits the villainy of *Pacific Gas*), responds by declaring that, yes, "the deed is silent on the question of assignability,"[35] but that this very silence was a reason for admitting the evidence, not as a gloss on the agreement as written, but as proof of a collateral agreement—an agreement made on a related, but adjoining matter—that was entered into orally. The beauty of this recharacterization of the situation is that it manages at once to save the integrity of the integrated agreement and to create another agreement whose honoring has the effect of setting aside what the integrated agreement seems to say. This is all managed by telling another story about the negotiations. The parties conducted not one, but two negotiations; in one, the question of the conveying of the land and the option of the conveyers to repurchase was settled; in another (orally conducted), the question of reserving the option to members of the family was settled. The demands of formalism are at once met and evaded, a result that led two dissenting justices to complain that the parol evidence rule had been eviscerated, that the decision rendered all instruments of conveyance, no matter how full and complete, suspect, and that the reliance one might previously have placed upon written agreements had been materially undermined.

This conclusion might seem to be the one I, myself, was moving toward in the course of presenting these examples, for surely the moral of *Columbia Nitrogen, Warren's Kiddie Shoppe, Dekker Steel, Pacific Gas*, and *Masterson v. Sine* (and countless others that could be adduced) is that the parol evidence rule is wholly ineffective as a stay against interpretive assaults on the express language of contracts and statutes. But the moral I wish to draw goes in quite another direction, one that reaffirms (although not in a way formalists will find comforting) the power both of the parol evidence rule and of the language whose "rights" it would protect, to "provide a meaningful constraint on public and private conduct."[36] It is certainly the case that *Masterson v. Sine*, like *Columbia Nitrogen* and the others, indicates that no matter how carefully a contract is drafted it cannot resist incorporation into a persuasively told story in the course of whose unfolding its significance may be altered from what it had seemed to be. But the same cases also indicate that the story so told cannot be any old story; it must be one that fashions its coherence out of materials that it is required to take into account. The important fact about *Masterson* is not that in it the court succeeds in getting around the parol evidence rule, but that it is the parol evidence rule—and not the first chapter of Genesis or the first law of thermodynamics—that it feels obliged to get around. That is, given the constraints of the institutional setting—constraints that help shape the issue being adjudicated—the court could not proceed on its way without raising and dealing with the parol evidence rule (and this would be true even if the rule had not been invoked by the eager trustee); consequently, the *path* to the result it finally reaches is constrained, in part, by the very doctrine that result will fail to honor.

One sees this clearly in the route the court takes to the discovery that there are not one, but two agreements. It is not enough, the court acknowledges, to observe that if an agreement is silent on a matter, information pertaining to it is admissible, for the official comment to section 2–202 adds that "if the additional terms are such that, if agreed upon, they would certainly have been included in the document in the view of the court, then evidence of their alleged making must be kept from the trier of fact."[37] In other words, the court must determine whether or not the additional terms that would make up a collateral agreement are such that persons contemplating the original agreement would certainly have considered including them; for if they were such and were not included, their exclusion was intentional and the original writing must be regarded as complete. In *Masterson,* the court reasons that the inexperience of the parties in land transactions made it unlikely that they would have been aware "of the disadvantages of failing to put the whole agreement in the deed" and rules that therefore "the case is not one in which the parties 'would certainly' have included the collateral agreement"[38] had they meant to enter into it. Again the point is not so much the persuasiveness of such reasoning (in another landmark case a New York court found the same reasoning unpersuasive),[39] but the fact that it must be produced, and this requirement would have held even if the reasoning had been rejected. It was open to the court (as a note to the case indicates) to find that in a particular instance what the parties would naturally have done was, in fact, not done and that "the unnatural actually happened";[40] but had the court so found, the official comment would have been honored even as it was declared to be inapposite, for that finding (or some other in the same line of country) would have been rendered obligatory by the existence of the comment. It is always possible to "get around" the comment as it is always possible to get around the parol evidence rule—neither presents an absolute bar to reaching a particular result; there is always work that can be done—but the fact that it is the comment you are getting around renders it constraining even if it is not, in the strict sense, a constraint.

In short, the parol evidence rule is of more service to the law's wish to have a formal existence than one might think from these examples. The service it provides, however, is not (as is sometimes claimed) the service of safeguarding a formalism already in place, but the weaker (although more exacting) service of laying down the route by which a formalism can be fashioned. I am aware, of course, that this notion of the formal will seem strange to those for whom a formalism is what is "given" as opposed to something that is made. But, in fact, efficacious formalisms—marks and sounds that declare meanings to which all relevant parties attest—are always the product of the forces—desire, will, intentions, circumstances, interpretation—they are meant to hold in check. No one has seen this more clearly than Arthur Corbin who, noting that "sometimes it is said that 'the courts will not disregard the plain language of a contract or interpolate something not contained in it,' "[41] offers for that dictum this substitute.

If, after a careful consideration of the words of a contract, in the light of all the relevant circumstances, and of all the tentative rules of interpretation based upon the experience of courts and linguists, a plain and definite meaning is achieved by the court, a meaning actually given by one party as the other party had reason to know, it will not disregard this plain and definite meaning and substitute another that is less convincing.[42]

There are many words and phrases one might want to pause over in this remarkable sentence *(relevant, tentative, experience, actually)*, but for our purposes the most significant word is *achieved* and, after that, *convincing*. *Achieved* is a surprise because, in most of the literature, a plain meaning is something that constrains or even precludes interpretation, while in Corbin's statement it is something that interpretation helps fashion; once it is fashioned, the parol evidence rule can then be invoked with genuine force: you must not disregard this meaning—that is, the meaning that has been established in the course of the interpretive process—for one that has not been so established. *Convincing* names the required (indeed the only) mode of establishing, the mode of persuasion, and what one is persuaded *to* is an account (story) of the circumstances ("relevant" not before, but as a result of, the account) in relation to which the words of the agreement could only mean one thing. Of course, if an alternative account were to become more rather than less convincing—perhaps in the course of appeal—then the meanings that followed from *its* establishment would be protected by the rule from the claims of meanings to which the court had not been persuaded. As Corbin puts it in another passage, "when a court says that it will enforce a contract in accordance with the 'plain and clear' meaning of its words . . . the losing party has merely urged the drawing of inferences . . . that the court is unwilling to draw."[43] That is, the losing party has told an unpersuasive story, and consequently the meanings it urges—i.e., the inferences it would draw—strike the court as strained and obscure rather than plain.

There are, then, two stages to the work done by the parol evidence rule: in the first its presence on the "interpretative scene" works to constrain the path interpreters must take on their way to telling a persuasive story (an account of all the "relevant" circumstances); then, once the story has been persuasively told, the rule is invoked to protect the meanings that flow from that story. The phrase that remains to be filled in is *persuasive story*. What is one and how is it, in Corbin's word, *achieved?* The persuasiveness of a story is not the product merely of the arguments it explicitly presents, but of the relationship between those arguments, and other, more tacit, arguments—tantamount to already-in-place beliefs—that are not so much being urged as they are being traded on. It is this second, recessed, tier of arguments—of beliefs so much a part of the background that they are partly determinative of what will be heard as an argument—that does much of the work of fashioning a persuasive story and, therefore, does much of the work of filling in the category of "plain and clear" meaning. What kinds of arguments or (deep) assumptions are these? It is difficult to generalize (and dangerous, since generalization would hold out the false promise of a *formal* account of persuasion), but one could say first of all that they will include, among other things, beliefs one might want to call "moral"—dispositions as to the way things are or should be as encoded in maxims and slogans like "order must be preserved" or "freedom of expression" or "the American way" or "the Judeo-Christian tradition" or "we must draw the line somewhere." It follows, then, that whenever there is a dispute about the plain meaning of a contract, at some level the dispute is between two (or more) visions of what life is or should be like.

Consider, for example, still another famous case in legal interpretation, *In Re Soper's Estate*.[44] The facts are the stuff of soap opera. After ten years of marriage, Ira Soper faked his suicide and resurfaced in another state under the name of John Young. There he married a widow who died three years later, whereupon he married another widow with whom he lived for five years when he again committed suicide, but this time for real. The litigation turns on an agreement Soper-Young made with a business partner according to which, upon the death of either, the proceeds of the insurance on the life of the deceased would be delivered to his wife. The question of course is who is the wife, Mrs. Young or the long-since deserted Mrs. Soper, who, to the surprise of everyone, appeared to claim her rightful inheritance.

The majority rules for the second wife, Gertrude, while a strong dissent is registered on behalf of the abandoned Adeline. There is a tendency in both opinions to present the case as if it were a textbook illustration of a classic conflict in contract law between the view (usually associated with Williston) that determinations of the degree of integration and of the meaning of an agreement are to be made by looking to the agreement itself and the view (later to be associated with Corbin) that such issues can only be decided by ascertaining the intentions of the drafting parties, that is, by going outside the agreement to something not in it but in the light of which it is to be read. But, in fact, what the case illustrates is the impossibility of this very distinction. It is the minority that raises the banner of literalism, arguing that since a man can have only one lawful wife and since Adeline was, at the time of the agreement, "the only wife of Soper then living,"[45] the word *wife* must refer to her. The majority replies that by the same standard of literalism, the agreement contains no mention of a Mr. Soper, referring only to a Mr. Young whose only possible wife was a lady named Gertrude; and indeed, observes Justice Olson, the document can only be read as referring to Adeline by bringing in the same kind of oral evidence her lawyers (and the dissent) now wish to exclude. An inquirer merely looking at the document might well conclude "that two different men are involved," for after all,

> in what manner may either establish relationship to the decedent as his "wife" except by means of oral testimony. . . . . Adeline, to establish her relationship, was necessarily required to and did furnish proof, principally oral, that her husband, Ira Collins Soper, was in fact the same individual as John W. Young, [and] Gertrude by similar means sought to establish her claim.[46]

The moral is clear even though the court does not quite draw it: rather than a dispute between a reading confining itself to the document and one that goes outside it to the circumstances from which it emerged, this is a dispute between two opposing accounts of the circumstances. Depending on which of these accounts is the more persuasive—that is, on which of the two stories about the world, responsibility, and wives is firmly in place— the document will acquire one of the two literal meanings being proposed for it. The majority finds itself persuaded by Gertrude's story and thus can quite sincerely declare that the "agreement points to no one else than Gertrude as Young's 'wife,' "[47] and Justice Olsen with an *e* (I resist the temptation to inquire into the *différance* of this difference) can declare with equal sincerity that "I am unable to construe this word to mean anyone else than the only wife of Soper then living."[48] Indeed he *is* unable, since as someone who subscribes to the moral vision that underlies Adeline's claims— a vision in which responsibilities once entered into cannot be weakened by obligations subsequently incurred—*wife* can only refer to the first in what is, for him, a nonseries; just as, for Justice Olson with an *o*—in whose morality obligations in force at the time of agreement take precedence over obligations recognized by a legal formalism—*wife* can only refer to the person "all friends and acquaintances . . . recognized . . . as his wife."[49]

The majority says that the "question is not just what words mean literally but how they are intended to operate practically on the subject matter,"[50] but its own arguments show that words *never* mean literally except in the context of the intention they are presumed to be effecting, and that rather than being determined by the meanings the words already (by right) have, the intentional context—established when one or the other party succeeds in being convincing—determines the meaning of the words. In short, the issue is not nor could ever be the supposed choice between literal and contextual reading, but the relative persuasiveness of alternative contexts as they are set out in ideologically charged narratives. That is why it is no surprise to find Justice Olsen's confidence declaration of linguistic clarity ("I am unable to construe") preceded by a rehearsal of the moralizing story that produces that clarity:

> Much is said in the opinion as to the wrong done to the innocent woman whom
> he purported to marry. Nothing is said about the wrong done to the lawful wife.
> To have her husband abandon her and then purport to marry another, and live in
> cohabitation with such other, was about as great a wrong as any man could inflict
> upon his wife.[51]

The majority thinks that something else (setting aside as of no account the
relationship between the deceased and Gertrude) is the greater wrong and
therefore thinks, like the minority, that the meaning of the agreement is
obvious and inescapable. Again, my point is not to discredit the reasoning
of either party nor to dismiss the claim of each to have pointed to a formal
linguistic fact; rather, I wish only to observe once again that such formal
facts are always "achieved" and that they are achieved by the very means—
the partisan urging of some ideological vision—to which they are then rhetorically (and not unreasonably) opposed.

Contract's Two Stories

That is to say—and in so saying I rehearse the essence of my argument—
the law is continually creating and recreating itself out of the very materials
and forces it is obliged, by the very desire to *be* law, to push away. The
result is a spectacle that could be described (as the members of the critical
legal studies movement tend to do) as farce, but I would describe it differently, as a signal example of the way in which human beings are able to
construct the roadway on which they are traveling, even to the extent of
"demonstrating" in the course of building it that it was there all the while.
The failure of both legal positivists and natural law theorists to find the set
of neutral procedures or basic moral principles underlying the law should
not be taken to mean that the law is a failure, but rather that it is an amazing kind of success. The history of legal doctrine and its applications is a
history neither of rationalistic purity nor of incoherence and bad faith, but
an almost Ovidian history of transformation under the pressure of enormously complicated social, political, and economic urgencies, a history in
which victory—in the shape of *keeping going*—is always being wrested
from what looks like certain defeat, and wrested by means of stratagems
that are all the more remarkable because, rather than being hidden, they are
almost always fully on display. Not only does the law forge its identity out
of the stuff it disdains, it does so in public.

If this is true of the law's relation to interpretation, it is equally true of
its relation to morality, as one can readily see by inquiring into the doctrine
of consideration. Consideration is a term of art and refers to the "bargain"

element in a transaction, what X did or promised to do in return for what Y did or promised to do. It is an article of faith in modern contract law that only agreements displaying consideration—a mutuality of bargained-for exchange—are legally enforceable. The intention of this requirement is to separate the realm of legal obligation from the larger and putatively more subjective realm of moral obligation, and the separation is accomplished (or so it is claimed) by providing *formal* (as opposed to value-laden) criteria of the intention to be legally bound. There are all kinds of reasons why one might make promises or perform actions conferring benefits on another and all kinds of after-the-fact analyses of what the promise signified or what the action contemplated. But if there is something tangible offered in return for something tangible requested, the transaction is a legal one and the machinery of legal obligation kicks in.

> The existence of consideration helps to provide *objective evidence* that the parties intended to make a binding agreement. It helps courts distinguish those agreements that were intended by the parties to be legally enforceable from . . . promises of gifts which neither party expected to be enforceable in court.[52]

By demarcating an area of legal obligation that is distinct from and independent of the larger matrix of obligations that make up our social existence, consideration plays a role similar to the role played (if it could be played) by the parol evidence rule. Just as Judge Kozinski says (or wants to say) to *Trident*, "you may not like the agreement you made, you may wish you could have it to do all over again, or that you had employed another negotiating team, but, nevertheless, this is the agreement you signed and this is what it says, and that's all there is to it," so might he or another judge say, "you may repent of your bargain because you know that someone will pay more for the automobile, but you promised to hand it over if he gave you $500; he has given you $500 and you must hand it over." Like absolutely express language (if there could be such a thing), consideration is a device for severing the moment of legal transaction from its surrounding circumstances and reducing it to a form that will stand out over the circumstances in which it is later examined by a court.

Consideration is, thus, a part of the law's general effort to disengage itself from history and assume (in two senses) a shape that time cannot alter. Consideration can be said to be such a shape because it has no content or, rather (it amounts to the same thing), can have any content whatsoever provided that there is an exchange and it is voluntary. That is why courts frequently declare that they will not inquire into the adequacy of the consideration,[53] will not, that is, inquire into whether or not the two parties were equally informed or received equivalent benefits from the exchange or

were equally powerful actors in the market. To do so would be to reintro-
duce the very issues—of equity, of the distribution of resources, of fairness,
of relative capacity, of *morality*—that consideration is designed to bracket.
(All of those issues will return by routes students of contract law know very
well, but the fact that such indirection is required is crucial.) With these
issues bracketed, the act of contracting (or so it is claimed) becomes purely
rational as the parties play out their formal roles in response to a mechanical
requirement, the requirement of consideration. This, in turn, requires the
court to be similarly mechanical (formal) lest it substitute for the bargain
two free agents rationally made a bargain it would have preferred them to
make.

> Where a party contracts for the performance of an act . . . his estimate of the
> value should be left undisturbed. . . . There is . . . absolutely no rule by which
> the courts can be guided, if once they depart from the value fixed by the prom-
> isor. If they attempt to fix some standard, it must necessarily be an arbitrary one,
> and . . . then the result is that the court substitutes its own judgment for that of
> the promisor, and, in doing this, makes a new contract.[54]

That is, presumably the parties had reasons for bargaining as they did and
the court should not set aside those reasons for reasons that seem to it to
be more compelling. The court's responsibility is to "judge" the contract
only in a weak sense, to determine whether or not it displays the requisite
shape of consideration; were it to exercise judgment in a stronger sense by
inquiring into the conditions of the contract, the court would pass from
being an instrument of *the* law into an instrument of *a* morality. The result,
should this occur, would be exactly the same result that attends the failure
to constrain interpretation, the making of a new contract in place of the
contract it is the job of the law to enforce. The conclusion is one toward
which I have been pointing since the beginning of this essay: interpretation
and morality are not simply twin threats to the autonomy and integrity of
the law; they are the *same* threat. Interpretation is the name for the activity
by which a particular moral vision makes its hegemonic way into places
from which it has been formally barred.

If interpretation and morality pose the same threats to the law's self-
identity, what is true of one is likely to be true of the other, and, as we
have seen, what is true of interpretation is that it is already inside the precincts that would exclude it. The parol evidence rule does not—cannot—
work because the integrated agreement it is designed to protect can only
come into being—achieve integration—by the very (interpretive) means it
stigmatizes. Similarly, the distinction between legal and moral obligation
will not work because any specification of a legal obligation is itself already

linked with a morality. The large point here is one that I cannot pause to argue (although it has been argued elsewhere by me and by others): that the requirement of procedures that are neutral between contending moral agendas cannot be met because, in order even to take form, procedures must promote some rationales for action and turn a blind eye to others. This is spectacularly true of the procedures built around the doctrine of consideration, a doctrine that finally makes sense not as an alternative to morality, but as the very embodiment of the morality of the market, a morality of arm's length dealing between agents without histories, gender, or class affiliation. Whatever one thinks of this conception of transaction and agency, it is hardly one that has bracketed moral questions; rather it has decided them in a particular way, and, moreover, in a way that is neither necessary nor inevitable. As E. Allan Farnsworth points out, the principle of "direct bilateral exchanges" is not the "only possible basis for an economic system." Other societies have "distributed their resources by sharing, based on notions of generosity, rather than by bargaining, based on notions of self-interest."[55] Historically, the morality of self-interest and with it the requirement of consideration triumphed after a determined effort by Lord Mansfield to make contractual and moral obligations one and the same. The fact that he failed does not mean that morality had been eliminated as an issue in contract law, but that one morality—the morality of discrete, one-shot transactions—became so firmly established that it won the right to call itself "mere procedure" and was able to set up a watchdog—called "consideration"—whose job it was to keep the other moralities at bay.

In a way, this outcome is inevitable given what we have already noted about the law. In order to be law, it must define itself *against* particular moral traditions. It follows then that the first thing a moral tradition must do after having captured the law (or some portion of its territory) is present itself as being beyond or below (it doesn't really matter) morality. This, in turn, dictates the strategy by which any alternative morality will have to make its way: it will have materially to alter the law while maintaining all the while that it is preserving what it alters. Just as the winning interpretation of a contract must persuade the court that it is not an interpretation at all but a plain and clear meaning, so the winning morality must persuade the court (or direct the court in the ways of persuading itself) that it is not a morality at all but a perspicuous instance of fidelity to the law's form.

Contract law performs this feat of legerdemain by implanting within consideration theory a number of subversive concepts, chief among which is the concept of "contract implied in law." Nominally, a contract implied in law is a category in the taxonomy of contracts, but in fact it exceeds the taxonomy and threatens to render it incoherent. The category can be best grasped by contrasting it with its two neighbors, the express contract and the contract implied in fact. An express contract is a written or oral agreement whose terms explicitly state the basis for consideration: I will do or promise to do this if you will do or refrain from doing or promise to do or promise to refrain from doing that. Now, it may be that the parties have entered into no formal agreement but comport themselves in relation to one another in ways that could only be explained by the existence of the requisite contractual intentions. If you bring a broken item to a repair shop and leave it, your action and the action of the shop's agent are intelligible only within the assumption that, in return for his professional skill, you have obligated yourself to the payment of a reasonable fee. In short, you and the shop have entered into a contract even though it has not been expressly stated; it is a contract implied in fact, implied, that is, by the behavior of the contracting parties.

Express contracts and contracts implied in fact are thus different ways of signifying an intention to be party to a bargained-for exchange. Contracts implied in law, on the other hand, are not attempts to ascertain and enforce the parties' intentions with respect to a contemplated transaction, but are imposed by a court on persons irrespective of the intentions they had or the actions they performed. Indeed, a contract implied in law is a judgment by a court that a party *ought* to have had a certain intention or performed in a certain way and for the purposes of justice and equity that intention or performance will now be imputed to him along with the obligations that follow. The notion of a contract implied in law springs all the bolts that consideration is designed to secure and provides the means for a court to do what, under contract law, a court is not supposed to do, make a new contract in accordance with its conception of morality. Contract implied in law is a wild-card category inserted into the heart of contract doctrine, and, moreover, courts that employ it often admit as much.

[A] contract implied in law is not a contract at all, but an obligation imposed by law for the purpose of bringing about justice and equity without reference to the intent or agreement of the parties and, in some cases, in spite of an agreement between the parties.

. . . It is a non-contractual obligation that is to be treated procedurally as if it were a contract. . . .[56]

In other words, the notion of implied in law contract does not belong in contract law, for as Stanley Henderson puts it, "the substantive right to recover benefits conferred upon another does not respect legal categories, particularly that of contract."[57]

What, then, is it doing there? The question implies that contract law would be better off were it not there or were its presence accounted for in ways that could better support a claim to consistency. Consistency, however, is not a feature of contract law but is its (always precarious) achievement, and it is an achievement whose possibility depends on *not* resolving the conflicts contract doctrine displays. It is *because* it is a world made up of materials that pull in diverse directions that contract law can succeed in its endless project of making itself into a formal whole. Rather than being an embarrassment, the presence in contract doctrine of contradictory versions of the enterprise is an opportunity. It is in the spaces opened by the juxtaposition of apparently irreconcilable impulses—to be purely formal and intuitively moral—that the law is able to exercise its resourcefulness.

These spaces continue to be opened up even in documents supposedly designed to close them. Here, for example, is section 86 of *Restatement Second,* a section that promises to adjudicate the tension between legal and moral notions of obligation, but ends up reproducing it.

86. Promise for Benefit Received
   (1) A promise made in recognition of a benefit previously received by the promisor from the promisee is binding to the extent necessary to prevent injustice.
   (2) A promise is not binding under Subsection (1)
      (a) if the promisee conferred the benefit as a gift or for other reasons the promisor has not been unjustly enriched; or
      (b) to the extent that its value is disproportionate to the benefit.[58]

Although the form of the section is straightforward and declarative, its content, as Grant Gilmore has observed, is hesitant and even schizophrenic.[59] By recognizing promises for benefits received, subsection (1) seems to send us firmly in the direction of an expansive notion of legal obligations (in "classic doctrine" such promises are not enforceable because the benefit was unsought and is not part of a bilateral exchange); but wait, "what Subsection (1) giveth, Subsection (2) largely taketh away";[60] such a promise will not be binding if the benefit came to the promisor in the form of a gift, that is, in a form not involved in a mutual transaction, that is, in a form without consideration. As Henderson observes, the requirement of subsection (2) "means that events are to be screened by consideration tests in order to determine which promises are to be included within the category of promises binding without consideration"; and he concludes that the restriction as stated "erodes the policy of growth manifested" in subsection (1).[61] But this is to regard the section as if it were (or aspired to be) a logical statement when, in fact, it is a set of directions for accomplishing a particularly difficult (but essential) task, the task of maintaining the formal basis of contract law while at the same time making room for the substantive concerns formalism desires to exclude. By first opening the door to moral obligation (that is, to promises for benefits received) and then allowing it to enter only if it can be provided with an origin in a bargained-for exchange, section 86 delivers a message, and delivers it not despite but because of its schizophrenia: "Relax the requirement of consideration, but do so in the guise of honoring it." The message is repeated and given a rationale almost charming in its transparency in official comment (a):

> Enforcement of promises to pay for benefit received has sometimes been said to rest on "past consideration" or on the "moral obligation" of the promisor, and there are statutes in such terms in a few states. Those terms are not used here: "past consideration" is inconsistent with the meaning of consideration stated in [section] 71, and there seems to be no consensus as to what constitutes a "moral obligation."[62]

Or in other words, "look, if we want to revise contract doctrine so as to bring it more into line with our moral intuitions, we would be well advised to do so in terms that contract doctrine, as presently formulated, will find acceptable. Neither past consideration nor moral obligation are such terms. The first fails because it is too obviously ex post facto with respect to the moment of exchange that constitutes a contract; it too nakedly asks us simply to declare that what had not been bargained for in the past—an unasked-for benefit—was in fact the basis of a bargain. The second fails because it provides us with no standard—there seems to be no consensus as to what constitutes a 'moral obligation'—and too nakedly acknowledges that the basis of law is variable. It is true that in some precincts these terms have been used, but the liabilities they present are greater than the advantage of employing them. If we want our moral intuitions to be incorporated into the law we will have to make sure that they look like what the law wants them to look like."

That is to say, they must be worked into a form that matches the picture of consideration, the picture of a freely chosen giving up of something in return for something just as freely proffered. The fact (as I take it to be) that the notion of choosing one's obligations independently of historical pressures is a fiction, that the "inside" of the isolated and "free" transaction is always determined by an "outside" it does not acknowledge, may be philosophically compelling; but the morality of consideration—of exchanges uninfluenced by anything except the opportunities offered by the moment of transaction—is too firmly embedded in contract doctrine to be embarrassed by any analysis of it. So firmly is it embedded that anticonsideration impulses can be harbored and even nurtured in contract doctrine where, rather than undermining the orthodox view, they provide it with the flexibility it needs. Henderson notes that "the judicial process, in spite of a consistent adherence to the test of bargain, exhibits a recurring tendency to appeal to a code of moral duties in order to justify enforcement of a particular promise."[63] Henderson calls this a "fundamental inconsistency" that has been "built into contract analysis,"[64] but like most "buildings" this structure is neither accidental nor unproductive. In order to be what it claims to be—something, rather than everything or nothing—contract law must uphold a view of transaction in which its features are purely formal; but in order to be what it wants to be—sensitive to our always changing intuitions about how people ought to behave—contract law must continually smuggle in everything it claims to exclude. I must emphasize again that the so-called formal view of legal obligation was never really formal at all, but was the extension of a social vision from which it was detached at the moment of that vision's triumph. The tension between consideration doctrine with its privileging of the autonomous and selfish agent and the doctrine of moral obligation with its acknowledgment of responsibilities always and already in place is a tension between two contestable conceptions of life; it is just that one of them has won the right to occupy the pole marked *formal* (i.e., unattached to any particular agenda) in a powerful (because constitutive of an institutional space) opposition. Given that victory, the fact that the claim of consideration doctrine to be merely formal cannot finally be upheld is of no practical consequence; it is upheld by the rhetorical structure it has generated, and in order to alter that structure you must appear to be upholding it too. As I have already said, you can only get around consideration doctrine by elaborately honoring it.

And how do you do that? In many ways, including several we have already observed. You develop a taxonomy of contractual kinds, one of which violates the principles of the taxonomy; you produce a document (the *Restatement*) that, in the guise of clarifying the state of the law, presents its contradictions in a form that further institutionalizes them; you announce as a "principle of law" that the improver of another's land has no right to relief, and then in the next sentence you declare this principle "merely . . . technical" (i.e., legal) and dismiss it in favor of "an equitable remedy";[65] you develop and expand notions like promissory estoppel, duress, incapacity, unconscionability, and unjust enrichment and then expand them to the point where there is no action that cannot be justified in their terms; you invoke the distinction between public and private, even as you allow public pressures to determine the distinction's boundaries. In short, you tell two stories at the same time, one in which the freedom of contracting parties is proclaimed and protected and another in which that freedom is denied as a possibility and undermined by almost everything courts do. But in order to make them come out right, you tell the two stories as if they were one, as if, rather than eroding the supposedly formal basis of contract law, the second story merely refines it at the edges and leaves its primary assertions (which are also assertions of the law's stability) intact.

Illustrations of this process are everywhere in the law; indeed, as I have been arguing, the process *is* the law. For a conveniently concise and naked instance of the process at work, we can turn to a classic case of promise for benefit received, *Webb v. McGowin.*[66] Webb was an employee of a lumber company and one of his duties was to clear the upper floor of a mill by dropping pine blocks weighing 75 pounds to the ground below. "While so engaged" he saw McGowin directly below him, in the line of drop as it were, and rather than allowing the block to fall, he diverted its course by falling with it, "thus preventing injuries to McGowin," but causing himself to suffer "serious bodily injuries" as a result of which "he was badly crippled for life and rendered unable to do physical or mental labor." A grateful McGowin, "in consideration . . . [for] having prevented him from sustaining death or serious bodily harm" promised to pay Webb $15 every two weeks for the rest of his life.[67] That sum was paid until McGowin's death, whereupon the payments were discontinued and Webb sued to recover the unpaid installments.

This recital of the facts is taken, the court acknowledges, from the appellant's brief, and therefore it is obvious from the beginning where the court's (moral) sympathies lie. It is also obvious what obstacles stand in the way of transforming those sympathies into a legal remedy. Webb's action seems, on its face, to be a spontaneous and gratuitous expression of fellow feeling and, as such, ineligible for a remedy that requires evidence of a transaction of a pecuniary nature. Late in the opinion, however, the court declares that

> [t]he case at bar is clearly distinguishable from that class of cases where the consideration is a mere moral obligation or conscientious duty unconnected with receipt by promisor of benefits of a material or pecuniary nature.[68]

But as the facts are initially encountered, the category of "mere moral obligation or conscientious duty" seem to be the one they clearly instantiate. The clarity to which the court refers is not a clarity it finds, but a clarity it *achieves,* and, accordingly, the story the opinion really tells is of that achievement.

The work is already being done as the facts are rehearsed. Presumably the action that initiates the entire affair took place in a split second, but in the court's presentation of it (or rather, the presentation adopted wholesale from the appellant's brief), that action occurs in slow motion: from the first report that the appellant was "in the act of dropping the block" (a locution that already extends the duration of something that must have occurred in the blink of an eye) to his decision (if that is the word) to drop with it, four long sentences intervene. What these sentences do is transform an instantaneous and instinctive response to an unanticipated situation of crisis into a deliberative and considered act. The first sentence does this by beginning the sequence all over again and further dilating it: "As he started to turn the block loose"—"in the act" now has stages, starting and whatever is the next stage after starting. (Of course starting could itself be further subdivided; there is no end to the process of drawing out process.) This stop action technique then leads to the revelation of what is stopping the action: "he saw J. Greeley McGowin, testator of the defendants, on the ground below and directly under where the block would have fallen." The recital has now become a drama of the "Perils of Pauline" variety. Will the block fall? What will Webb do? The "would have fallen" tells us that it didn't, in fact, fall and provokes in us the desire to know how Webb prevented it from falling; but before that desire is fulfilled two additional sentences inform us, first of what "would have" happened if the block would have fallen ("Had he turned it loose it would have struck McGowin with such force as to have caused him serious bodily harm or death"), and, second, how (by what deliberative route) Webb arrived at the course of action he finally (by now this is a full-fledged melodrama) took: "The only safe and reasonable way to prevent this was for appellant to hold to the block and divert its direction in falling from the place where McGowin was standing and the only safe way to divert it so as to prevent its coming into contact with McGowin was for appellant to fall with it to the ground below." A teacher of freshman composition might be moved to criticize this sentence and the entire passage for repetition ("to the ground below" or some variant appears five times in almost as few sentences) and prolixity ("so as to prevent its coming into contact"), but such criticism would miss the effect and the intention behind it, to stretch a punctual moment into a sequence long enough to allow the playing out of freely chosen alternatives and consequences, so that when, in the next sentence, we are told (with a brevity whose force is in direct proportion to the prolixity of what comes before it) "Appellant did this," our sense of what he did is complicated enough, has a sufficient number of stages and spaces, so that we can regard it as the action of a rational and free agent.

So far, so good, but the court is only halfway home; conferring rationality and choice on Webb is quite an achievement, but in order to be truly efficacious it has to be matched by another: Webb's rationality and choice must be shown to exist in a reciprocal relation with the rationality and choice of McGowin. After all, on the facts, his promise to compensate occurs *after* the benefit has been received and the element of bargain for a consideration is conspicuously absent. The court deals with this difficulty in two simultaneously performed moves: it quantifies the benefit and it transforms the quantified benefit into one the promisor (McGowin) had requested. The benefit is quantified not by a legal or even an economic argument but by a blatantly moral one (remember that the point of these moves is to disengage the case from the realm of "mere moral obligation"). After all, the court reasons, the preservation of life is something for which physicians charge, and patients willingly pay a price; therefore, "[l]ife and preservation of the body have material, pecuniary values, measurable in dollars and cents." [69] That is to say, since if Webb were a physician and McGowin his patient what passed between them would be regarded as a transaction in relation to services rendered, let us so regard it. At the same time that an analogy is thus extended into a legal fact, the present case is related to an earlier one in which "a promise . . . to pay for the past keeping of a bull which had escaped from defendant's premises and had been cared for by plaintiff was valid," on the reasoning that in the circumstances the promise was "equivalent to a previous request." [70] The court conveniently ignores the fact about the case that would distinguish it from the one before it (since the bull is a fungible good, both its value to the owner and the value of the service rendered by the person who cares for it can be monetarized) and seizes the opportunity to exclaim that the service Webb rendered McGowin was "far more material than caring for his bull." [71] In this sentence, the two contradictory strains of the court's argument mesh perfectly: the moral argument that nothing could be more material than the saving of life becomes (through an equivocation on the notion of material) a reason for finding that the present instance of saving a life becomes (through an equivocation on the notion of material) a reason for finding that the present instance of saving life is both more and less than moral, that is, legal and quantifiable. The moral urgency of the court's desire produces the sleight of hand by which the case is disengaged from moral consideration. It is then that the court can triumphantly declare that "[t]he case at bar is clearly distinguishable from that class of cases where the consideration is a mere moral obligation." [72]

It remains only for the court to consolidate its gains in a final breathtaking move. Since it has now been shown that "the promisor received a material benefit constituting a valid consideration for his promise," we can regard the promise as "an affirmance or ratification of the services rendered carrying with it the presumption that a previous request for the service was made," and the court proceeds immediately to so regard it:

McGowin's express promise to pay appellant for the services rendered was an affirmance or ratification of what appellant had done raising the presumption that the services had been rendered at McGowin's request.[73]

And when was this request made? Why in that infinitely extended moment when all the alternative courses of action and all the attendant consequences passed through Webb's mind, and now we are told implicitly, passed through McGowin's mind also. The commercial transaction of voluntary and free agents that seemed so obviously lacking in this case is now supplied and given a location in the minds of two parties who never spoke a word to one another.

The conclusion is as inevitable as it is fantastic: "the services rendered by the appellant were not gratuitous." That is, they were bargained for. And how do we know that? Because of "[t]he agreement of McGowin to pay and the acceptance of payment by appellant." In other words, the meaning and shape of what McGowin did on August 3, 1925, becomes clear in the light of what the two parties did later. In "real-life" terms, the reasoning is familiar and uncontroversial; often we only know what we did when subsequent actions provide us with a retrospective understanding of our actions. But the world of contract is not "real life"; it is (or is supposed to be) formal, and in *that* world, events are discrete and discernible in terms of punctual intentions and foreseen (not retroactively constructed) consequences. The court is dangerously close, here, to falling into the language of "past consideration," but the danger is avoided because, in the context of the opinion's story (of which this is the conclusion), what Webb and McGowin did after the event (make a promise and accept a payment) becomes proof of what they actually (not as a matter imposed after the fact) did *in* the event, enter into a transaction. The fact that they never spoke or in any real sense ever met and the fact that the transaction they are said to have entered into is bizarre ("if you will risk being crippled for life I will pay you thirty dollars a month") might give one pause, but pausing is not what the opinion encourages (except in those earlier moments when the space in which this "bargain" will be inserted is being opened up) and the court quickly moves to a brisk exit: "Reversed and remanded."

In a brief but revealing concurring opinion, Judge William H. Samford further pulls back a curtain that had never really been closed. He admits that the opinion he now joins is "not free from doubt" and acknowledges that according to "the strict letter of the rule" Webb's recovery would be barred, but then he simply declares the "principle" the court has been following, a principle whose articulation he attributes to Chief Justice Marshall when he said "I do not think that law ought to be separated from justice."[74] The effect is complex but swift; what Justice Marshall says amounts to a denial of the law's independence, but the fact that he, the most respected jurist in U.S. history, said it makes his pronouncement a *legal* one and therefore one that can be invoked as a legal justification for departing from the rule of law. Once again, and on several levels of constructions supporting constructions (Justice Samford legitimizes the creative work of his brethren by linking it in a narrative to the equally creative work of a now authoritative predecessor), the legal establishment reaffirms its commitment to a formal process it is in the act of setting aside. Once again, the two stories have been told and then made into the single story that assures the continuity of the tradition.

The Amazing Trick

An unsympathetic reader of the previous paragraph, or indeed of this entire essay, might say that what I have shown is that what works in the law is what you can get away with, precisely the observation made by some members of the critical legal studies movement in essays that point, as I have, to the contradictions that fissure legal doctrine. The difference between those essays and this one lies in the conclusions that follow (or are said to follow) from the analysis. The conclusion often (but not always) reached by critical legal studies proponents is that the inability of legal doctrine to generate logically consistent outcomes from rules and distinctions that have a clear formal basis means that the entire process is at once empty and insidious. The process is empty because its results are entirely ad hoc—lacking firm definitions or borders, the concepts of doctrine can be manipulated at will and in any direction one pleases—and the process is insidious because these wholly ad hoc determinations are presented to us as if they had been produced by an abstract and godly machine. Here is a representative statement from a well-known essay by Clare Dalton that anticipates many of my own arguments.

> . . . we need . . . to understand . . . how doctrinal inconsistency necessarily undermines the force of any conventional legal argument, and how opposing arguments can be made with equal force. We need also to understand how legal argumentation disguises its own inherent indeterminacy and continues to appear a viable way of talking and persuading.[75]

By "doctrinal inconsistency" Dalton means (1) the inability of doctrine to keep itself pure—as she points out, the poles of supposedly firm oppositions are defined in terms of one another and thus cannot do the work they pretend to do—and (2) the presence in doctrine of contradictory justificatory arguments that are deployed by lawyers and jurists in an ad hoc and opportunistic manner. That is why "opposing arguments can be made with equal force": given the play in the logic of justification, the facts of a case can, with equal plausibility, be made to generate any number of outcomes, no one of which is deduced from a firm base of principle. Nevertheless, Dalton's complaint continues, the law's apologists present these outcomes as if they issued from a procedure that was as determinate as it was impersonal.

To this I would reply, first, that doctrinal inconsistency undoes conventional argument only when the arguments are removed from the local occasion of their emergence and then put to the test of fitting with one another independently of any particular circumstances. But since it is only in particular circumstances that arguments weigh or fail to weigh, the inconsistency Dalton is able to document is not fatal and is embarrassing only if the context is not law and its workings, but philosophy and its requirements. Law, however, is not philosophical (except when it borrows philosophy's arguments for its own purposes) but pragmatic, and from the pragmatic standpoint, the inconsistency of doctrine is what enables law to work. Dalton inadvertently says as much when, in the same sentence, she denies force to conventional argument because of its inconsistency, and then complains that conventional argument, again because of its inconsistency, has too much force. This is not so much a contradiction as it is a distinction (not quite spelled out) between two kinds of force, one good and one bad. The good force is the force of determinate procedure, and that is what the law lacks; the bad force is the force of rhetorical virtuosity, and that the law has in shameful abundance. But the rhetorical nature of law is a shameful fact only if one requires that it operate algorithmically, and that is the requirement (of which there are hard and soft versions) of the position Dalton rejects. By stigmatizing the law's rhetorical content, she makes herself indistinguishable from her opponents for, like them, she measures the law by a standard of rational determinacy; it is just that where they give the law high marks, she finds it everywhere failing.

My point is that while Dalton's description of the law is exactly right, it is a description of strengths rather than weaknesses. When Dalton observes that the law's normative statements are so vaguely formulated ("fairness," "what justice requires," "good faith") that the moment of "normative choice" is deferred "until an individual judge is required to make an individual decision,"[76] all she means is that while the law's normative formulations specify the vocabulary and conceptual "neighborhood" of decision making, they set no limits to what a judge can do with that vocabulary on the way to reaching a plausible (in the sense of recognizably legal) result. In the absence of a mechanical decision procedure there is ample room for judicial maneuvering (although, as I have shown, that maneuvering is itself far from free), and if the "individual decision" is strong enough—if the story it tells seems sufficiently seamless—it will have constituted the norm it triumphantly invokes as its justification. That is the trouble, Dalton might respond: the law is at once thoroughly rhetorical and engaged in the effacing of its own rhetoricity. Exactly, I would reply, and isn't it marvelous (a word intended nonevaluatively) to behold. It may be true that "we have no reliable, and therefore no legitimate, basis for allocating responsibility between contracting parties,"[77] but while the legitimacy is not ready-made in the form of some determinate system of rules and distinctions, it is continually being achieved by the very means Dalton rehearses in such detail.

Consider, for example, her discussion of the interplay between the doctrines of consideration and reliance. She has been retelling the story of *Second Restatement* sections 71 and 90 (one defining contract obligation in terms of consideration, the other concerning "Contracts Without Consideration") and notes the avoidance both in the *Restatement* and in the cases that invoke it of "the knotty questions of how their coexistence should be imagined."[78] The mechanisms of avoidance, as she describes them, include the sequential application of the doctrines so that each of them seems to be preserved and a clash between them is forever deferred; the stipulating of different measures of recovery in a way that suggests a distinction that an analysis of the cases does not support; and the elaboration of different vocabularies that cause "reliance rhetoric to sound different from consideration rhetoric," although when the occasion demands, the two vocabularies can draw together and begin "to appear indistinguishable."[79] Impressive as this is, it is only a partial catalog of the mechanisms at the law's disposal, mechanisms that allow a distinction that cannot finally be maintained to be reinforced and, at other times, relaxed and, at still other times, conveniently forgotten. The story is an amazing one, and Dalton accurately characterizes it as "the story of how what appears impossible is made possible."[80]

It is, in short, the story of rhetoric, the art of constructing the (verbal) ground upon which you then confidently walk. Reviewing a case that displays the law's virtuosity at its height, Harry Scheiber exclaims

> One is reminded . . . of a dazzling double-feint, backhand flying lay-up shot by a basketball immortal. Only in slow motion replay does one comprehend the whole move; and only then does one realize that defiance of gravity is an essential component of it![81]

Scheiber calls this the law's "amazing trick," the trick by which the law rebuilds itself in mid-air without ever touching down. It is a trick Dalton and others decry, but it is the trick by which law subsists and it is hard to imagine doing without it. The alternatives would seem to be either the determinate rationality that every critical legal analysis shows to be impossible, or the continual exposure of the sleights of hand by means of which the "amazing trick" is performed. But if the latter alternative were followed, and every legal procedure turned into a debunking analysis of its enabling conditions, decisions would never be reached and the law's primary business would never get done. Perhaps this is the result we want, but somehow I doubt it, and therefore I tend to think that the law's creative rhetoricity will survive every effort to deconstruct it.

It need hardly be said that I am not the first to declare that the operations of law are rhetorical. That is, in fact, Dalton's point, although she makes it as an indictment, and it is the point of others who (in the tradition of Cicero) see the law's rhetoricity more positively. Under the rubric of "rhetorical jurisprudence," Steven J. Burton has elaborated a description of the law that accords on many points with one presented here. His basic thesis is that the "local law of a society represents a possible organization of human relations, and a public commitment to bring it into empirical being"; each application of the law "brings that imagined world into being in some respect."[82] In this argument, the organization of human relations is not something the law follows or replicates, but something the law produces, and produces by means it invents. Rather than proceeding as science does (at least in pre-Kuhnian characterizations) to adjust its presentation in order "better to fit the world," law is a practical discipline, operating to change the world so as "better to fit the representation."[83] That change is brought about by a discourse that creates the authorities it invokes. "A rhetorical understanding concerns the criteria of evidence implicit in a local legal discourse, and thus the effects of the discourse on what the participants will take seriously as law or legal argument, with or without good reasons."[84] A rhetorical jurisprudence does not ask timeless questions; it inquiries into the local conditions of persuasion, into the reasons that *work;* and what it finds interesting about the law's normative claims is not whether or not they can be cashed (in strict terms they cannot), but the leverage one can achieve by invoking them.

> Whether such concepts of law are sound or not, they continue in legal discourses to influence the thoughts and actions of many legal actors. Accordingly, they are an important object for rhetorical study within the effort to understand legal practices.[85]

Not surprisingly, I find this very agreeable. I become nervous only when Burton shifts from describing the law as rhetorical to a claim that this description, if heeded, could have beneficial consequences for legal practice:

"A rhetorical criticism draws attention to features of law that are neglected by a legal discourse as a first step toward possibly improving that discourse as *legal*."[86] The course of this improvement is not spelled out, but it seems to follow from the reasoning that once we know the law to be rhetorical we will be better able to function within it, better able to "listen, deliberate and justify action."[87] But knowing that the law is rhetorical could improve it only if we were thereby insulated against that rhetoricity—in which case Burton would be harboring a desire for the scientific jurisprudence his essay rejects—or if that knowledge made the law *more* rhetorical than it already is—a goal that is incoherent since the condition of being rhetorical, of being tied to the exigencies and pressures of the moment, admits of no degree. Once "the law is understood from the practical point of view as a system of reasons for action,"[88] one does not either gain a distance from those reasons or become more compelled by them because one has achieved that understanding. They will still be *local* reasons, as they were before they were so named by Burton, and they will still occur in the context of local pressures rather than in the context of some overall recognition that they are local. The lesson of the law's rhetoricity—the lesson that reasons are reasons only within the configurations of practice and are not reasons that generate practice from a position above it—must be extended to itself. It can no more serve as a master thesis than the formalist theses it replaces. Formalists at least make their mistake legitimately, since it is *their* position that local practices follow or should follow from master principles; it cannot, without internal contradiction, be a rhetorician's position, even when the master principle is rhetoric itself.

Burton's flirtation with what I have elsewhere called "antifoundationalist theory hope" (the hope that by becoming aware of the rhetoricity of our foundations we gain a [nonrhetorical] perspective on them that we didn't have before) is a fully developed romance in the work of James Boyd White. White defines the law (correctly in my view) as "a set of resources for thought and argument."[89] This set, he argues, is open and includes the concepts thought to be basic to the enterprise. As a result, the law is at every level creative, constructing its principles even as it applies them.

> For in speaking the language of the law the lawyer must always be ready to try to change it: to add or to drop a distinction, to admit a new voice, to claim a new source of authority, and so on. One's performance is in this sense always argumentative, not only about the result one seeks to obtain but also about the version of the legal discourse that one uses—that one creates—in one's speech and writing. That is, the lawyer is always saying not only "here is how this case should be decided," but also "here—in this language—is the way this case and similar cases should be talked about. The language I am speaking is the proper

language of justice in our culture." . . . in this sense legal language is always argumentatively constitutive of the language it employs.[90]

To this point I am more or less with White, but he loses me when his description of the way the law works leads to a demand for a new form of legal practice: "This means that one question constantly before us as lawyers is what kind of culture we shall have."[91] But the question "before us" is always a legal one, couched in terms of legal categories and possible courses of action. The fact that a legal question can always be shown to have a source in presupposed cultural values does not mean that it is the business of a legal inquiry to discover or revise those values. Of course one could always engage in that business, but to do so would not be to practice law as the institution's members now recognize it. White himself makes the point when he observes again and again that the workings of the law are local; "it always starts in a particular place among particular people," and therefore "one cannot idealize" it by saying "here is how it should go in general."[92] But here is White, idealizing it and saying how it should go in general: it should provoke a continuing philosophical discussion of the society's values and goals. But were it to do that, it would not be law but moral philosophy; the irremediably local perspective of the law—its rootedness in particular disputes requiring particular, and timely, solutions—leaves no room for the extended reflections White recommends and indeed brands them as inappropriate. The judge who was always stopping to "put his (or her) fundamental attitudes and methods to the test of sincere engagement with arguments the other way"[93] would not be doing his or her job as a judge, but would be doing something else, something valuable no doubt but, in legal terms, something inept and even irresponsible (unless the exercise were preliminary, strategically, to the announcing of a conclusion, in which case the practice of testing one's attitudes against alternatives would not be engaged in for its own sake—as an effort to expand one's consciousness—but for the sake of a goal to which a limited consciousness was precommitted).

White's mistake is to conflate the perspective from which one might ask questions about the nature of law (is it formal or moral or rhetorical?) with the perspective from which one might ask questions in the hope that the answers will be of use in getting on with a legal job of work. It is from the first perspective (the perspective of metacritical inquiry) that one might decide that the law is a process in which "we, and our resources, are constantly remade by our own collective activities,"[94] but those who are immersed in that process do not characteristically act with the intention of furthering that remaking, but with the intention either of winning or deciding. With respect to that intention, an account of the law's rhetoricity will either be irrelevant (i.e., it does not touch on the issues the case raises) or dangerous (introducing it could weaken the position you are defending) or, in some circumstances, marginally helpful (as when you remind yourself and your fellows that the law is not an exact science and is less severe than science in its demand for proof). But even in this last instance, the thesis of the law's rhetoricity will not have generated a new way of practicing law; it will merely have added one more resource to a practice that will still be shaped, in large measure, by the goals the law will continue to have, the goals of winning an argument or crafting an opinion. These are result-oriented activities, and to engage in them seriously is to have already foreclosed on the openness to alterity that White would have us adopt.

White doesn't see this because he thinks that if openness to alterity characterizes the law (as opposed to the more closed characterizations offered by formalist theorists), then legal actors should themselves be open; since history shows that the law "provides a ground for challenge and change,"[95] it is with the motives of challenge and change that we should act. But while challenge and change are often the by-products of the resourcefulness legal actors display, they are not the motives for which legal action is usually taken. It may be, as White contends, that, as a rhetorical process, the law can never be closed to the interpretive pressures of alternative conversations and displays a "structural openness,"[96] but it does not follow that those who practice the law do so with the intention of being thus open. That is the intention of those who would practice a form of critical self-consciousness. As it turns out, this is precisely the future White envisions for the law: the legal agent will continually "doubt the adequacy of any language, and seek to be aware of the limits of her own forms of thought and understanding"; she will be committed to "many-voicedness" and be "profoundly against monotonal thought and speech, against the single voice, the single aspect of the self or culture dominating the rest."[97] White calls this vision "rhetorical," but it is a strange rhetoric that imagines conflict finally dissolved in the wash of a many-voiced pluralism. The truth is that White's hopes for the law are not rhetorical, but transcendental; he regards the scene of persuasion as only temporary, and, like Habermas, he looks forward to a time when all parties will lay down their forensic arms and join together in the effort to build a new and more rational community. This "commitment to openness,"[98] this determination to "be tentative and poetic,"[99] may be admirable, but it is hard to see what place it could have in a process that *demands* single-voiced judgments, even if that voice can be shown to be plurally constituted. White may begin by acknowledging and celebrating difference, but in the end he cannot tolerate it.

The same can be said of Peter Goodrich, a British-style critical legal scholar who, in books and essays of enormous erudition and sophistication, elaborates "a concept of a rhetoric of legal language" that emphasizes "the rhetorical, sociolinguistic and loosely pragmatic dimensions and contexts of any communicational practice."[100] Goodrich begins by observing that a "defining feature of all formalism" is the "rejection of history,"[101] that is, of the circumstantial background that informs the supposedly self-sufficient and self-declaring rule or doctrine. He then finds the source of the formalist dream in "the distinction between logic and rhetoric,"[102] a distinction that produces the traditional categories of the philosophy of language with a pure semantic kernel at their center and the meanings generated by social and political conditions on the stigmatized periphery. Of course, Goodrich insists, this pure center is a "mythology,"[103] "palpably more rhetorical than actual,"[104] a device for diverting attention away from "the actual 'social facts' or historical and particular 'forms of life' that determine the substantive meaning of legal rules."[105] Once we realize that meaning is "the product or outcome of communication between socially organized individuals and groups," we will see that there could not possibly be any "self-articulated unities of discourse" because every text, no matter how apparently autonomous "implies other meanings, other texts, other discourses, and will constantly exceed the boundaries of any given instance of discourse."[106]

Once again I find myself in agreement both with the analysis and its conclusion: rather than a formal mechanism applying determinate rules to self-declaring fact situations, the law is "preeminently the discourse of power,"[107] that is, a discourse whose categories, distinctions, and revered formulas are extensions of some political program that does not announce itself as such. Goodrich supports and extends this conclusion by analyzing cases in which, as he shows, it is "the persuasive function of particular rhetorical techniques"[108] rather than any independent logic that generates decisions, even though the decision will always be presented as one that reinforces "the distinction between the formal normative character of the legal process and the substantive content . . . of a dispute."[109] The law, in short, is continually engaged in effacing the ideological content of its mechanisms so that it can present itself as "a discourse which is context independent in its claims to universality and reason."[110] In this way it rhetorically establishes its independence from the very social and political values that are its content, exercising a "constant, centripetal, endeavor to maintain itself by differentiation and by exclusion of the discourses and languages which surround it."[111] By "avoiding or excluding the . . . implications of its own institutionalization," its status as a historical and contingent discourse, and "by treating legal problems of syntax or of a lexico-grammatical kind . . . the law manages and controls . . . the hierarchy of social and political relations . . . while apparently doing no more than prohibiting and facilitating certain generic and inherently uncontroversial, legally specific, activities and functions."[112]

Except for the somewhat inflated vocabulary, this is precisely my account of the law as a discourse continually telling two stories, one of which is denying that the other is being told at all. The difference is that for Goodrich this account amounts to a scandal, whereas for me it simply brings to analytical attention the strategy by which the law fashions out of alien materials the autonomy it is obliged to claim. Were the law to deploy its categories and concepts in the company of an analysis of their roots in extralegal discourses, it would not be exercising, but dismantling its authority; in short it would no longer be law. Goodrich quite correctly sees that one can become an adept in the law only by forgoing an inquiry into the sources of the norms you internalize: "the entire process of socialization into the legal institution is a question of learning deference and obedience, a question of explicit and implicit education into the requisite modes of interaction—the forms, procedures and languages—of the different levels, functions and topics of the legal system."[113] But he thinks that we can and should undo this education and bring into the foreground everything it labored to occlude. We must "challenge the hermetic security . . . of substantive jurisprudence";[114] we must make visible the "alternative meanings"[115] that legal meanings ignore:

> In reading the law, it is constantly necessary to remember the compositional, stylistic and semantic mechanisms which allow legal discourse to deny its historical and social genesis. It is necessary to examine the silences, absences and empirical potential of the legal text, and to dwell upon the means by which it appropriates the meaning of other discourses and of social relations themselves, while specifically denying that it is doing so.[116]

Now, it may be that "in reading the law"—that is, in subjecting it to a sociological or deconstructive analysis—one must remember everything forgotten in the course of its self-constitution, but the practice of law requires that forgetting, requires legal discourse to "appropriate the meaning of other discourses . . . while specifically denying that it is doing so." And if you reply that a practice so insulated from a confrontation with the contingency of its foundations is unworthy of respect, I would reply, in turn, that every practice is so insulated and depends for its emergence as a practice—as an activity distinct from other activities—on a certain ignorance of its debts and complicities. As I have put the matter elsewhere,

> "Forgetfulness," in the sense of not keeping everything in mind at once, is a condition of action, and the difference between activities . . . is a difference between differing species of forgetfulness.[117]

This is true even of the practice of remembering what other practices have forgotten, for in order to engage in that practice, Goodrich must himself forget (or at the very least bracket) the empirical conditions that give rise to law and constrain its operation, conditions including the need for procedures to adjudicate disputes, and the pressure for prompt remedies and decisions. I am not criticizing Goodrich, only pointing out that his project is no more free of forgettings than the project he excoriates. His mistake is to think that it could be so free, and he thinks *that* because he believes in the possibility of a general discourse that takes account of everything and excludes nothing. There is no such discourse, only the particular discourses that gain their traction by the very means Goodrich laments. And, indeed, it is more than a little ironic that Goodrich finally scorns the material setting of the law's exercise and seeks to set it, instead, in the leisurely precincts (no less material but differently so) of a philosophy seminar. The law, however, is not philosophy; it is law, although, like everything else it can become the object of philosophical analysis, in which case it becomes something different from what it is in its own terms. To be sure, the phrase "in its own terms" refers to the very construct Goodrich would expose, and exposing it as a construct is a perfectly respectable thing to do. That is not, however, what the law can do and still remain operative as law. It is certainly true, as Goodrich both asserts and demonstrates, that the law is not "best read in its own terms," [118] but that does not mean that the law is best not *practiced* in its own terms, for it is only by deploying its own terms confidently and without metacritical reservation that it can be practiced at all. Goodrich ends his book with a call for the "interdisciplinary study of law . . . aimed . . . at breaking down the closure of legal discourse." [119] This is a worthy project and one that (with his help) is already succeeding; but it is an academic project determined in its shape by norms of academic inquiry (themselves forms of closure): once the seminar is over and the grip of philosophy's norms has been relaxed, legal discourse will once again be closed (although the shape of its closure will be endlessly revisable) and the law will resume the task of simultaneously declaring and fashioning the formal autonomy that constitutes its precarious, powerful being.

I cannot conclude without speaking briefly to three additional points. First of all, my account and defense of the law's rhetoricity—of the strategies by which it generates outcomes from concerns and perspectives it ostentatiously disavows—should not be taken as endorsing those outcomes. Although much of legal theory is an effort to draw a direct line between some description of the law's workings and the rightness (or wrongness) of particular decisions, it has been my (antitheoretical) point that "rightness" is automatically conferred on any decision the system produces, that is to say, any decision that follows from the persuasive marshalling of certain arguments. As soon as an argument has proven to be persuasive to the relevant parties—a court, a jury—we say of it that it is right, by which we mean that it is now the law (nothing succeeds like success), that it is *legally* right. Of course, we are still free to object to the decision on other grounds, to find it "wrong" in moral terms or in terms of the long-range health of the republic. In that event, however, our recourse would not be to an alternative form of the legal process but to alternative arguments that would be successful—that is persuasive—within the same general form. In my view, the legal process is always the same, an open, though bounded, forum where forensic battles are contingently and temporarily won; therefore, preferred outcomes are to be achieved not by changing the game but by playing it more effectively (and what is and is not "more effective" is itself something that cannot be known in advance). In short, even if the cases I discuss were to be decided differently, were to be reversed or overturned, the routes of decision would be as I have described them here.

This brings me to my second point. It might seem that, by saying that the legal process is always the same, I have made the law into an ahistorical abstraction and endowed it with the universality and stability my argument so often denies. However, this objection (which I raise myself because others will certainly raise it) turns on a logical quibble and on the assumption that one cannot at the same time be true to history and contingency and make flat categorical statements about the way things *always* are. But what I am saying is that things always are historical and contingent; that is, I am privileging history by refusing to recognize a check on it—a determinate set of facts, a monumentally self-declaring kind of language—and it is only a philosophical parlor trick that turns this insistence on historicity into something ahistorical. The alternative to my account would be one in which the law's operations were grounded in a reality (be it God or a brute materiality or universal moral principles) independent of historical process, and it seems curious to reason that, because I do not allow for that reality, I am being unhistorical. To be sure, the possibility that such an independent reality may reveal itself to me tomorrow remains an alive one, but it is not a possibility that can weigh on my present understanding of these matters, nor would it be the case that the act of ritually acknowledging it (as Dalton, Burton, White, and Goodrich urge) would be doing anything of consequence. Nevertheless, there is a sense in which the present essay is not historical: it doesn't do historical *work;* that is, it does not chart in any detail any of the differently contingent courses the law has taken in the areas it has marked out for its own. That work, however, is in no way precluded by my thesis and, indeed, the value of doing it is greatly enhanced once that thesis is assented to; once contingency (or ad-hocness or makeshiftness or rhetoricity) is recognized as constitutive of the law's life, its many and various instantiations can be explored without apology and without any larger (that is, grandly philosophical) rationale.

This brings me to my final point. Assuming, for the sake of argument, that I am right about the law and that it is in the business of producing the very authority it retroactively invokes, why should it be so? Why should law take *that* self-occluding and perhaps self-deceiving form? The short answer is that that's the law's job, to stand between us and the contingency out of which its own structures are fashioned. In a world without foundational essences—the world of human existence; there may be another, more essential one, but we know nothing of it—there are always institutions (the family, the university, local and national governments) that are assigned the task of providing the spaces (or are they theaters) in which we negotiate the differences that would, if they were given full sway, prevent us from living together in what we are pleased to call civilization. And what, after all, are the alternatives? Either the impossible alternative of grounding the law in perspicuous and immutable abstractions, or the unworkable alternative of intruding that impossibility into every phase of the law's operations, unworkable because the effect of such intrusions would be so to attenuate those operations that they would finally disappear. That leaves us with the law as it is, something we believe in because it answers to, even as it is the creation of, our desires.

12

Play of Surfaces:
Theory and the Law

Legal theory today circles around a set of distinctions that are more or less
the content of legal theory today: originalism versus nonoriginalism (or
originalism plus), determinacy versus indeterminacy, interpretivism versus
noninterpretivism, historical versus ahistorical interpretation, etc. It is assumed by those who wield these terms (1) that the positions they name are
conceptually discrete and (2) that there is a strong correlation between the
stances one takes (as in "I am an originalist" or "I choose to interpret
historically") and the shape of one's practice. Thus, for example, Michael
Perry cites the recent (mis)fortunes of Robert Bork, who, he says, was
"rejected as unfit for membership on the Supreme Court principally because of his position—originalist—on issues contested in constitutional theory." Perry's point is complex and involves three claims: (1) that Bork's
originalism was regarded by his opponents as a reason for rejecting him;
(2) that in their eyes the relationship between his originalism and his position on specific issues was strong and even inevitable; and (3) that the
relationship was in fact strong and inevitable. ("Given his originalism,"
says Perry, Bork's "deep skepticism about . . . due-process doctrine regarding reproductive rights is easy to understand.") The first and second
claims may or may not be true; what is certainly true is that originalism
was something Bork avowed, *and* (not necessarily "therefore") he was
rejected. The third claim—that given Bork's originalism, his opposition to
abortion rights was a matter of course—is false, and because it is false,

This essay was written as a review of Gregory Leyh, ed., *Legal Hermeneutics: History,
Theory, and Practice* (Berkeley: University of California Press, 1992). References are to
pieces in that volume, reprinted by permission of the publisher, although the issues raised
throughout are those that vex the legal culture in general.

Perry's conclusion that "constitutional theory matters" is false too, at least in the sense he intends it. (I too shall be arguing that constitutional theory matters, but in terms Perry would regard as weak.)

The third claim is false, first of all, for the reason Terence Ball gives: "a judicial strategy of original intent . . . can cut in quite unexpected directions." That is, it is perfectly possible to be in favor of abortion rights and also to label oneself an originalist, as someone who hews to the intentions of the framers. It would just be a matter of characterizing those intentions so that the right to abortion would seem obviously to follow from them. One might, for example, argue (as many have) that even though the Fourteenth Amendment nowhere mentions abortion rights, a correct understanding of its authors' more general intention requires that such rights be protected. (The point is made elegantly by Knapp and Michaels in this volume.) Conversely, it would be perfectly possible to be against abortion rights and think of oneself as a nonoriginalist, as someone who looks elsewhere than to the intentions of the framers for the point of constitutional law. (In a moment I will question the possibility of "looking elsewhere" than to intention, but for now I will stay with the formulations provided by the debate.) One might, for example, look to precedent and contend that whereas the right to privacy was firmly established, the right to an abortion should not be understood as falling under that rubric. In short, there is no necessary relationship between declaring oneself an originalist and coming out on one side or the other of a particular issue; "given [Bork's] originalism," one could not have inferred his "deep skepticism about . . . due process doctrine regarding reproductive rights," or anything else for that matter.

It is wrong, then, to say that Bork was rejected because he was an originalist. He was rejected because he held certain positions (on the right to privacy and other matters) the Senate found repugnant. Of course, it is true that his holding of those positions was associated in his mind and in the mind of the public with originalism, but that association is contingent not causal, and in the end the public (through its representatives in Congress) was quite able to disassociate the rhetoric of originalism—with its attendant claim of responsibility as opposed to the supposed unfettered creativity of nonoriginalists—from policies it disliked. I can easily imagine two votes being taken on the same day in the Senate: in one vote the senators would decide against Bork just as they did, and in a second vote in which the choice was between originalism and nonoriginalism as modes of constitutional interpretation, originalism would win by the same margin that Bork lost. Such a vote would not be contradictory: it would merely reflect the fact that people vote for outcomes not theories, except in those special (and bizarre) circumstances when theories are on the ballot. This is not to deny that theoretical slogans might play a part in one's thinking about outcomes; it is just that those outcomes will not be *determined* by theory.

The assumption that theories determine outcomes is common to originalists and nonoriginalists alike, and indeed the rival parties are mirror visions of one another in that they both think that there is a choice between originalism and something else. Thus, David Hoy, speaking for the nonoriginalist side, grants that one of a text's meanings is the meaning its author(s) intended but insists that the text can yield other meanings to which we might turn for any number of reasons. Since "linguistic processing can be done without attention to intention," there is no need to restrict ourselves to the meanings an author may have intended: "hermeneutics takes the question about the intended meaning as a secondary one that presupposes a prior understanding of 'meaning' in some other sense." Both Hoy and Bork, then, agree in distinguishing between the meaning an author intended and the meaning one can find in a text without reference to intention; they disagree only in the style of interpretation each prefers, Bork saying, "Let's go for intentional meaning," Hoy saying, "Let's not."

What I want to say (along with Knapp and Michaels) is there is only one style of interpretation—the intentional style—and that one is engaging in it even when one is not self-consciously paying "attention to intention." Hoy's case for nonintentional interpretation rests in part on our capacity (acknowledged by Knapp and Michaels) to hear "what is really noise as an articulate utterance." Hoy concludes from this capacity (evidenced in relation to examples such as computer poems or wave poems or monkeys at typewriters producing Shakespeare) that we are able to construe meaning without taking intention into account. The reasoning is that since it is always possible "for arbitrarily produced sounds or marks to be read as a text," we can "hear something as making sense . . . independently of whether it was intended." But his example works against him, for in order to hear sense in arbitrarily produced sounds or marks we have to hear those sounds and marks within the assumption that they have been produced by some purposeful agent; that is, we have to hear them *as not arbitrarily produced,* even if to do so we must attribute purpose and intention to the waves or to the wind or to the great spirit that rolls through all things. Computer poems are read *as* poems (as opposed to random marks) only because we read them as issuing from intentional beings, only because we grant them—*in* the act of construing, not as an addition to it—a mind. Hoy (and others of his party) confuse the activity of self-consciously arguing about intention and the activity (if that is the word) of assuming intention, not a specific one but just the unavoidable fact of one. The first activity may or may not play a part in interpreting; the second *is* interpreting.

I would not be misunderstood as returning to some notion of authorial control as exercised by a coherent self that knows its own projects unproblematically. The thesis that interpretation always and necessarily involves the specification of intention does not grant priority and authority to the author, who is in no more a privileged relation to his own intentions than is anyone else. Each of us has had the experience of walking away from a conversation and asking himself or herself, with respect to something just said, "Now what did I mean by that?" The question is shorthand for "With what kind of motives and in the context of what hopes, fears, anxieties, and desires did those words issue?" Depending on the answer, one will have fashioned an image of oneself that will constitute one's sense of the kind of person one is. Rather than being a stable center of authority and control, the intending self is the continual creation and re-creation of the interpretations it itself performs in relation to its own actions, including the action of intending. We know ourselves no less interpretively (and therefore revisably) than others know us; the fact that it is intentions that we know, or try to know, does not limit the range of interpretation but merely specifies what the object of interpretation will necessarily be. Intentionalism properly understood involves no methodology, no prescriptive direction. It does not, for example, bind us to the dead hand of the past (as some non-originalists fear) because the hand of the past can appear to us only in an interpreted form, in the form of a constructed intention that can always be constructed *again* in the light of whatever evidence from whatever source seems relevant; and therefore the past we will be bound to will acquire its shape within the horizons of the living and lived-in present.

What this means, among other things, is that there cannot be a distinction between interpreters who look to intention and interpreters who don't, only a distinction between the differing accounts of intention put forward by rival interpreters. If Bork and some other legal scholar disagree about what (if anything) the Constitution says about reproductive rights, the opposition is not between intentional interpretation and some other kind (even if that is the opposition both parties would stipulate to) but between two specifications of the intentional conditions of the Constitution's production. Bork's construal of the Constitution and its amendments assumes an agent (the framers) whose purpose is to confine future generations to the possibilities for action conceivable in 1787 or —; that is, he imagines the framers saying something like, "If we can't think of it now, we don't want anyone to think of it in the future." Other construers posit an agent whose purpose is to lay down general rules and standards (of equality, liberty, etc.) that are to be rendered particular in the light of emerging information and changing circumstances; in this vision the framers are saying something like, "Equality and liberty are the standards we want to promote, but whether or not a particular action affirms these values is something future generations, faced as they will be by possibilities we cannot foresee, will have to decide for themselves.'' In either case the act of construing will begin by assuming an agent endowed with purposes, and *then* the words will be read as issuing from that agent; and in the (certain) event of a dispute between the construers, pointing to the words as a way of adjudicating the dispute will be futile because the words will mean differently in the light of the differently assumed intentions.

The relevance of this account of the matter to Perry's argument has already been asserted: "originalism" is not the name of a distinct style of interpretation but the name of interpretation as it is practiced by anyone; since meaning cannot be determined apart from (the prior and simultaneous) assigning of intention, everyone who is an interpreter is in the intention business, and there is no methodological cash value to declaring yourself (or even thinking yourself) to be an intentionalist because you couldn't be anything else.

Perry would reply, as he does in his essay, that there is certainly a distinction between original meaning and what he calls "aspirational meaning," the meaning the Constitution takes on when regarded "as a symbol of a fundamental aspiration of our political tradition." Although I would acknowledge the distinction as a possible one, I would redescribe it (in line with my discussion of the Bork example) as a distinction between two forms or versions of originalism, between two competing accounts of what the framers intended, either to bind future jurists to the conditions obtaining in 1787 or to set those same jurists the task of extending general principles to the particular circumstances thrown up by history. The "aspirational" interpreter who looks to "relevant precedent" is not looking *away* from the framers' intention but *to* the place where he thinks the best evidence of that intention is to be found. He thinks that since the framers intended that we search in present configurations for the course of action most faithful to their aspirations, the writings of those charged with the responsibility of carrying on that search is a good place to look. Others, of course, may think differently, which doesn't mean that they have forsaken intention but that they have a different notion as to where evidence of that intention is to be sought. They may think, for example, that the text provides the best evidence for the author's (or authors') intention (formalism and intentionalism are not necessarily opposed); or they may think to consult the author's biography; or they may research the received opinions of his (her, their, its) culture; or, if they have a certain view (bizarre to some but widely held in the culture) of the influences that produce verbal and physical action, they may consult the author's horoscope. No matter where they look, what they'll be looking for is evidence of intention, and, moreover, since the question of what *is* evidence is a matter of dispute (depending on one's notions of agency, mental processes, linguistic structure, class consciousness —in short, of anything or everything), their common intentionalism will in no way be predictive of the conclusion any interpreter might come to.

To all of this someone committed to the distinction between intentionalist and nonintentionalist interpretation might reply (with Perry) that one can always find meanings in a text other than the ones intended by the author. This statement is undoubtedly true, but the question is, in what sense would that action be an instance of interpreting? Suppose, for example, my method of interpreting a text consists of taking every third word of it and seeing what patterns of significance then emerged. There are at least three understandings within which I might be proceeding: (1) I might believe that the author, in writing the text, employed a code that I have now discovered; (2) I might believe that the author (unbeknownst to him) was controlled (through brainwashing, injection, diabolic possession) by a superior force employing that same code (that force would then *be* the author); or (3) I might believe neither (1) nor (2) nor any version of them but simply want to see what I could make of the text by subjecting it to this procedure. Under the first and second understandings I am interpreting; under the third I am playing with the text, something I could be doing for any number of reasons, serious and unserious. I might be playing with the text in order to amuse myself and my friends; or I might be playing with the text because I was determined to bring it in line with a conclusion I had already reached. In either case the text is the object of my experiment and not of my interpretive attention. I am not trying to figure out what it means but trying to see what meanings it could be made to yield. I have no necessary quarrel with those who want to do that (although when the text is one I have produced or one produced by someone I regard as authoritative, a quarrel is likely to be provoked), but I do not think it should be called interpreting. Interpreting is the act of determining, or trying to determine, what is meant *by* an utterance or a gesture, that is, of what it is intended to be the expression. Of course, as has already been acknowledged, there is no uncontroversial road to that determination and no independent way of confirming that it has been correctly made (intentionalism is not a method); but there is a great difference between saying that the effort to specify what is meant *by* a text can yield multiple and competing results and saying that the competition must admit those who are not making the effort at all.

Once one sees (1) that aspirational and nonaspirational readings of a text are rival specifications of what its author intended and (2) that those who

find patterns in a text but do not claim they were intended are not interpreting, the conclusion I have already reached is inescapable: originalism is
not an *option* for interpreters but the name of what they necessarily do.
And the conclusion to be drawn from that conclusion is that originalism
cannot be the issue it is for Perry and others. Arguments in favor of it are
superfluous since it is what everyone does anyway; arguments against it
(like Terence Ball's) are futile since one cannot avoid it and still be interpreting; the only arguments there can be are about which version of originalism, which specification of the meaning an author intended, is the correct one. Moreover, since looking for evidence of intention is what every
interpreter does, knowing that one is doing it is no methodological advantage, and believing (incorrectly) as Hoy, Perry, and Ball do that one is not
doing it is no methodological handicap. Nor will announcing that you are
an originalist or declaring that you wouldn't be caught dead being one lead
you to look in some particular direction since the direction you look in will
depend on where you think the evidence is, and that will vary independently of whether or not you are a declared originalist. In short, if theoretical pronouncements are consequential insofar as they directly affect practice (Perry's explicit claim and the implicit claim of some others in this
volume), then theory, at least in relation to the originalist question, doesn't
matter.

The same argument will hold for the other issues examined in these essays,
historical versus ahistorical interpretation and determinate versus indeterminate practice. The historical/ahistorical opposition is complicated because
there are two versions of it. In one version, the ahistorical stands for the
immutable laws or principles to which persons should be faithful no matter
what their historical (i.e., local, partial, biased) circumstances; in the other,
history stands for the authority of traditions and public codes as against the
ahistorical preferences of isolated and self-regarding individuals. In the first
version, an ahistorical purity is to be protected from the polluting touch of
merely historical imperatives; in the second, the considered and settled doctrines produced by historical process are invoked as a stay against the willful actions of interpreters who consult only their own wishes.
   The two versions get mixed up and merge into a third when cultural
relativism—the notion that interpreters are limited in what they can see or
discover by the horizons of their own historical moment—enters the picture. To those who are distressed by it, cultural relativism presents the same
liability as does the thesis of interpretation on the basis of personal preference: true, the interpreter in a cultural-relativist world does not consult his
own (single, unique, self-originating) preferences but rather the preferences
made available to him by the norms and standards of his time; still, since

he is confined to those norms and standards, he cannot get a critical distance on them, and therefore practice, as the extension either of an individual agenda or of the agenda of a particular interpretive community, is without a check on its desires and becomes (in the words of Jerry Stone as he paraphrases Habermas) "subjectivized, impressionistic, and incapable of rigorous self-criticism." In the legal culture this debate is folded into the familiar (and durable) opposition between a government of men and a government of laws, which is itself a variant of the even more durable opposition between truth and politics. Shall we follow our own inclinations (either personal or communal), or shall we subject them to the scrutiny of a standard that reflects something larger than our desires or the desires of those who are associated with us? In the eyes of some the question is rendered meaningless by the cultural relativism thesis, which precludes the possibility of any check either on our desires or (it is the same thing) on our politics. As Fred Dallmayr puts it (he is rehearsing not endorsing the argument):

> On radical-perspectivist premises an inescapably human and political factor seems to enter the law. . . . with the intrusion of politics the rule of law is in danger of collapsing into the very "government of men" that it was originally meant to forestall. . . . Does law . . . not become a captive or instrument of arbitrary caprice, of the whim of particular interpreters? Are we not witnessing here the triumph of power over law, of *voluntas* over *ratio* . . . ?

There are in fact two questions here, and they should be disentangled. The first question concerns the (supposed) choice between interpretation from within some local structure (either of the self or of some currently in-place interpretive community) and interpretation that looks for guidance and (possible) correction to a structure beyond the present horizon. In the essays collected here the terms of this choice are most clearly formulated by E. D. Hirsch, Jr., as reported by Stone:

> [T]he interpreter can choose someone else's, perhaps another culture's understanding as [a text's] best meaning—an *allocratic* ("other-governed") choice— then allow this delegated meaning to stand in critical judgment of the interpretation that seems most significant in the present moment. As a contrast to the "realists" who make *allocratic* choices, Hirsch describes the "cultural Kantians" as "idealists" who believe that the only authoritative interpretation is the *autocratic* one, that is, the self-governed one that emerges from within a person's own horizon.

The formulation is neat and apparently clean, but a single question undoes it: where is the interpreter when he makes this choice? The answer can only be that he is situated within some horizon of present understanding and that therefore his ''other-governed'' choice will in fact be governed by his horizon-conditioned sense of where to look for authoritative guidance. It is not that the scenario Hirsch and Stone imagine is an impossible one; we often say to ourselves something like, ''This (text, gesture, event) seems to me to have such and such a meaning, but perhaps I'd better check (by consulting a dictionary, or some historical archive, or some recognized expert, or a psychoanalyst) in order to make sure the meaning I see is the correct one.'' It is just that the check we perform will not be a check *on* our present horizons but will proceed *from* our present horizons (which may then possibly be altered by a mechanism internal to them). Nor does this mean that when we interpret we always do so in a self-rather-than-other manner, for the self that is interpreting will always be constituted in its possibilities by the ongoing traditions through which it must pass in order to have an interpretive direction, in order to *be*. The self, in short, is already formed by that which is other than it—in Derridean language, the self is always already other than itself—and therefore the judgments that issue from it cannot be contrasted to judgments that are ''other-governed.'' It is *all* ''other-governed,'' or if you prefer (and with the proper understanding), it is all ''self-governed.'' What is not possible is a genuine (as opposed to a polemical) choice between them; the only choice will be between the various reservoirs of evidence that present themselves to interpreters for consulting, and that choice will have no direct relationship to an interpreter's declared autocratism or allocratism since both programs of action— being purely self-governed or purely other-governed—are impossible to put into practice.

Does this mean that we are trapped within our present horizon with no hope of a critical perspective on the judgments we feel inclined to make? This is the second question raised by the passages from Dallmayr's and Stone's essays, and the answer is no. The urgency of the question stems from the assumption that criticism must come from the outside in order to *be* criticism, and this assumption itself rests on the assumption that horizons of understanding are monolithic and will go their unconstrained way unless challenged by something external to them. But any horizon of understanding, any agenda, any political arrangement bears within it the dynamics of its own alteration; this is the case even for the ''royal absolutism'' that for Gadamer represents the arrangement least open to change because everything happens ''only through absolute fiat'' (as Dallmayr phrases it). Although royal absolutism may be a claim at the heart of a political theory (of the divine right of kings), it is never a fact because its exercise or apparent exercise depends on a populace that receives its acts within a set of assumptions (Gadamer's ''shared meanings'') that renders them legitimate. Of course, given enough military and economic power royalty can have its way even when those shared meanings are absent; but every time it does so it energizes and authorizes resentments that will in time undo it, as Charles I of England proved to his cost. If we have learned anything from Foucault, it is that power is never concentrated in one place (or person) but is distributed everywhere throughout a system, so that its exercise is always reciprocal, a two- (or more) way traffic in relation to which the action of one person is effective only insofar as some other persons affirm its scope and thereby maintain a balance that can always be altered at almost any point. What this means is that there is never a situation in which there is "no room for hermeneutics" (Dallmayr, quoting Gadamer), that is, no room for the interpretive maneuvering that produces change; since the structure of any situation or of any system of ideas is one of layered dependencies and coordinations, there are innumerable nodal junctures at which a shift in emphasis and pressure can lead to a systemwide readjustment or even to a systemwide breakdown. As I have put it elsewhere *(Doing What Comes Naturally)*, a horizon of understanding is not a monolithic unity of which one asks, how can it change? Rather, it is itself an *engine* of change, a complex mechanism whose every exertion is simultaneously a self-alteration. If we understand criticism as the possibility of correction and reform (a word that should be read with a literal emphasis), then it is a possibility that can never be foreclosed.

The conclusion is the same one we reached vis-à-vis the supposed choice between originalism and nonoriginalism: (1) neither autocratism—the preferring of one's own point of view to the point of view of another—nor allocratism—the deliberate choosing of another's point of view in preference to one's own—is a possible option (although they are options one can *profess,* a point we shall return to); and (2) declaring oneself to be either autocratic or allocratic cannot have the consequence of turning one away from or toward history, or away from or toward individual desire, or away from or toward abiding and eternal truth, or away from or toward politics, and so on. In short, such a declaration cannot of itself turn you toward or away from *anything,* and therefore the entire issue, at least as it has been posed in the literature, is moot.

The choice between determinacy and indeterminacy is similarly moot, although the argument takes a somewhat different form. Indeterminacy has become an issue in legal theory because of the assertion by Critical Legal Studies members that the decisions judges render do not follow from the materials (laws, precedents, evidence, etc.) they invoke, materials that could be made to yield almost any decision one wishes to reach. According to the indeterminacy thesis, judges are not constrained by the rules and texts that supposedly ground the legal process, and this absence of constraint, as

Ken Kress points out, "rais[es] the specter that judicial decision making is often or always undemocratic and illegitimate": undemocratic because the decisions follow from the desires of particular judges rather than from the directions embedded in public texts, and illegitimate because the result is a government not of laws but of men.

Indeterminacy as a property of legal decisions can be urged in different ways. Some claim that indeterminacy is a product of the contradictions embedded in a liberal/bourgeois/capitalist legal system: since the system is full of gaps—its first principles do not cohere with one another—its "application" to a set of facts will yield any number of conclusions. In this version the indeterminacy thesis admits of a remedy: reground the system in coherent first principles and principled conclusions will follow. In another version the thesis is more radical and asserts indeterminacy as a *general* feature of *all* interpretation; no matter what constraints are supposedly in place, they will not check the interpretive will, which can always recharacterize them on the way to pursuing its own agenda. To those who are alarmed by it, this position (often buttressed by references to the polysemous nature of language) amounts to nihilism—the denial of meaning and authority—and various defenses are constructed against it. The more sophisticated of these defenses acknowledge that interpretation is not determinate down to the last detail but insist that at a basic level constraints are always in place, or can be if the drafters of laws and contracts are sufficiently careful. That basic level is sometimes identified with literal (as opposed to metaphorical or analogized) meaning, sometimes with a settled core as opposed to a penumbra of uncertainty (H. L. A. Hart's terms), sometimes with easy as opposed to hard cases. Whatever the label, the argument is that the existence of unproblematical instances of interpretation disproves the indeterminacy thesis in its strongest and most corrosive form.

The counterargument is that unproblematical instances are unproblematic only within interpretive conditions—specifications of what counts as evidence, arguments as to the weight and shape of precedent, etc.—which, while presently settled, can themselves become the object of dispute and so become problematical. I find this counterargument persuasive and therefore am persuaded of the indeterminacy thesis in its strong form; but it is precisely because I am persuaded by the indeterminacy thesis in its strong form that I agree with Kress when he says that the indeterminacy issue is a "red herring." It may be that at a *general* level interpretation and language are radically indeterminate because every interpretation (decision, specification of meaning) rests on a ground that is itself interpretive and therefore challengeable; but since life is lived not at the general level but in local contexts that are stabilized (if only temporarily) by assumptions already and invisibly in place, the inherent indeterminacy of interpretation is without the practical consequences both feared and hoped for it.

That is, although the logic of a decision can always be undone by a deconstructive analysis of it or by the elaboration of a more powerful logic, until that happens (and in some cases it may not happen for a very long time, long enough to feel like forever) the decision is as determinate as one would like and has all the consequences of a decision that was absolutely determinate. People will act on it, be influenced in their calculations by it, cite it, invoke it, believe in it. What this means is that the distinction between determinate and indeterminate does not refer to possibilities anyone could actually experience; no state of interpretive affairs can be determinate in the sense that it is invulnerable to challenge, but no interpretive state of affairs is (within its own challengeable terms) characterized by the instability that subsequent events can retroactively confer. To put it another way, everything is always determinate and indeterminate at the same time: interpretation is always determinate in that within the context of its occurrence the meanings it yields will seem obvious and inescapable; and interpretation is always indeterminate because meanings thus yielded can always be dislodged by successfully recharacterizing the context in which they emerged.

Like originalism and nonoriginalism and historicism and ahistoricism, determinacy and indeterminacy do not name cognitive behaviors that could be actualized; and although they do name positions that one could affirm— as in "I believe in determinacy" or "I believe in indeterminacy"—those affirmations do not translate into interpretive options. One cannot *be* determinate simply by declaring an intention to be so because the determinations one makes can always be upset; and one cannot be indeterminate, in the sense of *doing indeterminacy* or practicing instability, because indeterminacy and instability are features of interpretation as viewed from the long range, and it is in the short range that interpretation is always experienced and practiced. In the short range a situation-specific determinacy provides all the stability one needs. Determinacy and indeterminacy may be polemical identifications in the current epistemological wars, but those who identify with them are not thereby led to interpret in one direction rather than in another. In short, and to repeat myself, in terms of the claims made by Perry and others, theory doesn't matter.

In another sense, however, theory matters very much, for as Kress observes, "[L]egal indeterminacy is of great importance because many scholars think that significant indeterminacy results in illegitimacy." Posing the question "Why do we and should we care about indeterminacy?" Kress answers correctly, "Indeterminacy matters because legitimacy matters," because "many legal scholars hold that the legitimacy of judicial decision making depends on judges applying the law and not creating their own." Of course, the options "applying the law" versus "making one's own" are no more realizable than are any of the other oppositions discussed here (and for the same reasons), but that is beside the point because the point is not theoretical but rhetorical and political. That is, although determinacy and indeterminacy, historicism and ahistoricism, and originalism and nonoriginalism do not name practices one can choose to instantiate, they do name positions (and accusations) one can invoke strategically in the course of engaging in the practice of what I have called "doing what comes naturally" (with "naturally" defined as whatever is felt to be appropriate and necessary given the assumed goals, accepted procedures, notions of evidence, etc., of a particular enterprise). Although one cannot be a nonoriginalist and cannot avoid being an originalist, the announcement that you are or are not one or the other may be crucial to the effectiveness of what you are trying to do. And if, as Kress observes, the assertion of a general indeterminacy would be understood both by jurists and the lay public as undermining the judicial process, then to assert it is to run the risk of undercutting your efforts (unless, of course, the rhetorical structure in which determinacy is thought necessary to legitimacy changes). And if it is important either to ground your position in eternal principles or to claim for it the virtue of being progressive and forward-looking, you will want to distance yourself from historicism and the suggestion that you are being dictated to by "the dead hand of tradition." In all of these cases, and any others that might be imagined, theoretical distinctions name moves that are possible within the practice; one can invoke them (on either side) with a fair (but by no means certain) confidence in their effectiveness not because they ground the practice in the sense of sitting above or below it but because in the current shape of the practice—a shape they do not generate but contribute to—they carry a calculable rhetorical weight.

There are two mistakes to make with respect to the distinctions we know as theory: one can think that they stand to the side of practice in an independent and authorizing relationship, so that the course of practice will follow from their invocation; or one can think that because they are not thus generatively authorizing they are empty and "merely" rhetorical. But if practice is through-and-through rhetorical, the components of its present rhetorical structure cannot be inconsequential, even if the consequences can never be as total or stable as hard-core theorists (the only real kind) desire. If the practice is so constituted that claims to originalism or nonoriginalism stand in an enabling or inhibiting relation to other claims (also components of the practice's rhetorical structure) one might want to make or disallow, then originalism is important to the practice and one cannot afford to ignore it; but its importance is a function of a web of interrelations in which no one point is supervening because all points are mutually constitutive.

What this means is that there is nothing "deep" to know about the legal process, which is, as Gerald Bruns says, a "social rather than propositional discourse," a discourse "without ground": "[The law] is not a system working itself pure but a play of surfaces, a heterogeneous cultural practice that cannot be formally reduced but needs to be studied locally in terms of its position and effects within specific social and political situations." This is so good (and of a piece, I might add, with a very fine essay) that one is surprised to find Bruns then declaring, "The hard part is getting clear about the consequences entailed in such a view of legal study." That is, now that we know that the law is a play of surfaces, what does this knowledge direct us to do? Even to ask this question, however, is to retreat from the insight that the law is a play of surfaces by making that insight into something deep, into something that should generate a new practice. If "the very idea of law itself is in constant revision," then any idea of the very idea of law is part of that revisionary process rather than its impelling force. The notion that the law is social, not propositional, cannot itself be a new proposition from the vantage point of which the law can be reformed and corrected. Like any other proposition, it plays its role in the process by which the law builds and rebuilds its justificatory structure; but that role can never be theoretical in the strong sense; that is, it can never point you in a particular direction or exclude other directions in advance.

Bruns thinks otherwise and cites as the first methodological consequence of his insight that "one ought to unhook the law from such insupportable distinctions as the one between logic and rhetoric, not to mention the one between knowledge and power." But by Bruns's own logic ("the law is without ground") distinctions are neither supportable nor insupportable, but cashable or uncashable; either they serve to carry the discourse along, enabling and authorizing certain movements the argument seems to demand, or they don't and they are discarded. The distinctions that inhabit the space of law do not do so from the outside, where they preside over activities from which they are distinct; rather; they are constitutive of the law not in the sense of producing it but as components of its intelligibility. Take away the distinctions between logic and rhetoric and knowledge and power and you take away the entire basis of the law's current self-presentation. The law cannot be "unhooked" from the distinctions in relation to which its every action is now read and justified and still remain what it is. To be sure, many are dissatisfied with the way the law is and wish to change it; however, change will be achieved not by unhooking the law from distinctions but by replacing the distinctions now in place with others that will be equally insupportable outside of the contexts they themselves help to create. And were the law to be unhooked from *all* distinctions—a suggestion that Bruns does not make, but one that lies in waiting in his thought—it will be nothing at all.

Another way to put this is to say that without these distinctions the law would be without principles; and this is indeed the conclusion reached by some members of the Critical Legal Studies movement, who go from the discovery that the law is without a deep ground and is a play of surfaces to the accusation that the law is a sham. It is this jump in reason that Drucilla Cornell attributes to the "irrationalists" in Critical Legal Studies, those who declare the "the absence of a fully cognizable good leaves us with the irrationality of all legal and ethical choice." The absence of a fully cogniz- able good is the consequence, as Cornell indicates, of the Derridean epis- temology in which the signified (the ideal, the true, the Good) is never apprehended directly but appears to us always under the aegis of the signi- fier; and since signification is a human, social process, the values we meet and invoke are to a great extent (if not wholly) our own constructions. In short, we are not guided by anything we do not ourselves make, and as a result, the argument goes, "there can be no rational limiting principle by which to judge competing interpretations of legal doctrine."

As Cornell points out, by this move the irrationalists "replace the truth of reconciliation with the truth of castration"; that is, having lost faith in the immanence of the true and the good, they declare them to be forever unavailable and empty the world of meaning. To this thesis Cornell opposes the notion of "the appeal to the beyond": although the good "eludes our full knowledge," it "is a star that beckons us to follow." "The Good can never be fully enacted in space. That is why as a prescriptive command it points us toward the future." The principle underlying law may not be fully graspable in the sense that no enacted state of affairs will perfectly capture it, but it can serve as a "light that guides us and prevents us from going in the wrong direction," an "appeal to and enrichment of the 'universal' within a particular *nomos.*" In this vision the legal order is at every moment in- complete but leaning forward in the direction of that which it must finally join in order to be whole; it is the pull of the Good as it exists *outside* the system that prevents its operations from merely reflecting the status quo; it is the pull of the Good that assures forward movement.

Even as she unfolds this position, however, Cornell points (perhaps in- advertently) to another in which the inside/outside distinction need not be invoked because the mechanism of change is interior to the system. As she observes correctly, "no line of precedent can fully determine a particular outcome . . . because the rule itself is always in the process of reinterpre- tation as it is applied." In other words, precisely because legal rules (or any other for that matter) are underdetermined in their content and scope, the very act of applying them—of using them to order a piece of the world—will result in their alteration. It is only if legal rules could achieve the status of "a self-sufficient mechanism"—of directions so complete that they did not require, indeed repelled, interpretation—that their application could preserve a status quo. "Interpretation," Cornell insists, "always takes us beyond a mere appeal to the status quo"—not (I would add) because of a regulative ideal that exerts a magnetic attraction and refuses to allow the system to remain complacent but because every operation of the system at once extends it and reforms it. There is no need for an appeal to something beyond the status quo because any appeal to the status quo—requiring, as it will, a recharacterization of the status quo in relation to the events that provoke the appeal—will always necessarily go beyond "it" (now in quotation marks because the status quo must be understood not as a static entity but as an ensemble of transformative possibilities). If, as Cornell argues, transformation is inevitable, and if, by virtue of the constitutive self-understanding of the enterprise, we can not "escape our responsibility . . . to elaborate principles of justice that can guide us," there is no need for anything outside the system to impel it forward, no need of a "guiding light" brighter than the light the enterprise is already shedding.

What I am suggesting is that one needn't choose between a view of the law as autonomous and self-executing and a view in which the legal process is always unfolding in relation to the pressures and needs of its environment; for autonomy should be understood not as a state of hermetic closure but as a state continually achieved and reachieved as the law takes unto itself and makes its own (and in so doing alters the "own" it is continually making) the materials that history and chance put in its way. The law (or any other enterprise) can display autonomy only in the course of stretching its shape in order to accommodate what seems external to it; autonomy and the status quo are conceivable and achievable only within movement; identity is asserted not in opposition to difference but in a perpetual recognition and overcoming of it.

This process of forging identity and autonomy out of difference and relation is what Peter Goodrich describes as the achievement of English jurisprudence:

> Far from being a technical and internal development the new jurisprudence responded to and was molded by a series of discourses external to law. Jurisprudence was marked by external discourses and desires, and its subsequent reformulations still carry those marks even though the historians of law prefer to recycle the juridical fiction of a true discourse and its authoritative judgments.

My only quarrel with this statement concerns the divorce Goodrich feels obliged to institute between the fact of the law's social construction and its claims of authority and autonomy. It is no doubt true that in forming itself, jurisprudence (English or any other kind) is "marked by external discourses and desires," but in the course of its (self) elaboration those discourses and desires are recharacterized and given a jurisprudential form; that is, they are brought into the discourse of the law and become components of its new (and ever renewing) unity. Insofar as this unified (but always developing) discourse is looked to by those who find themselves engaged in disputes, its judgments (at least insofar as they conform to what is, for the time being, considered legal form) will be authoritative. If the autonomy and authority of jurisprudential discourse is a "fiction," it is so only in respect to a discourse that is (1) self-generated and (2) authoritative irrespective of any political and social arrangement; but the only candidate for that discourse is the word of God, and as the history of biblical interpretation shows, even God's word must appear in some socially constructed and authorized form. Of course, within the social fabric jurisprudential discourse always presents itself, whether explicitly or implicitly, as either substantiating God's word or descending from it. That is why, as Goodrich points out, "law needed first to base itself on a conception of sacral origination . . . [and] needed second to elaborate a rigorously esoteric hermeneutic that would guard the legal missive from profane interpretations." Goodrich intends this as an indictment against an enterprise that must obscure its humble origins lest it be seen not as authoritative but as just one more competing conversation; but if it is the *job* of the enterprise to constitute its own authority in a neverending responsiveness to changes in the body politic, then it makes no sense to indict it unless one is comparing it unfavorably to an enterprise that wears its divinity on its face—that is, to an enterprise that Goodrich in all his writing declares unavailable. Goodrich's (implicit) complaint is that the law's authority is a political achievement, but the complaint is incoherent in the context of the antifoundationalist views he himself professes.

Indeed, it is precisely *because* the law's authority is a political achievement that it cannot be accused of holding itself aloof from the Other and of refusing to take into account what its internal mechanisms do not recognize. A politically earned authority (as opposed to an authority wielded by deity) can emerge only by taking into account the environment over which it then claims sway. A politically earned authority is *always already* in a relation to the Other it is accused of scorning, and the problem (as some see it) of opening the law's self-referential procedures to the pressures of the "real world" is no problem at all because that very self-referentiality (autonomy, unity, integrity, etc.) has been constructed (and reconstructed) in response to those pressures. Everything that is supposedly to be conferred on the law by some guiding light external to it—social awareness, the capacity for chance, the historical sense—is already built into its mechanisms.

Moreover, if this is so, if the law is already socially and historically constituted, then the demand that it be socially and historically responsive is superfluous. Methodologically there is nothing to be gained from the thesis of the law's social construction, no direction in which it points us, no direction it rules out. Gregory Leyh thinks otherwise. In his view the value of the interpretive insight—the insight that without contexts there can be no understanding at all—lies in its fostering of a critical attitude toward our own constitutive contexts. That is, now that we know that the urgencies we feel, the values we resonate to, and the facts we unhesitatingly affirm are contextual constructions, we can go about the demystifying business of critically examining those constructions so that we can be more responsibly aware of what we are doing. The problem is that the thesis of pervasive contextuality bars any such account of its own possible effects. If all understanding is contextual, then any "critical" understanding we achieve will be no less contextual; that is, it will depend for its critical force on a background of values, distinctions, desiderata, etc., that cannot itself be examined critically because its parameters will (for a time) determine what will count as criticism. The "critical attitude" can always be assumed, but its very shape will be a function of a context in relation to which it does not have and could not have any distance. Leyh's mistake (and he shares it with many) is to think that knowledge of contextuality relaxes its constraints, but in fact knowledge of contextuality is a *general* knowledge that is of no force whatsoever in the specific situations in which we find ourselves. You may know (in the sense that you have certain answers to some traditional philosophical questions) that the urgencies you feel, the values you resonate to, the facts you affirm, are contextually produced and therefore revisable, but that knowledge neither loosens the hold of those urgencies, values, and facts nor provides instructions (or even reasons) for their revision. If instructions and reasons do emerge, they will have been produced by the particular circumstances that give rise to the need for particular actions and not by some abstract conviction of circumstantiality in general. Gadamer tells us, "True historical thinking must take account of its own historicality" (quoted in Leyh), but once historical thinking presumes to do that it has ceased to be historical because it is offering itself as a vantage point from which the present historical moment can be surveyed and transcended.

I know that to many it will seem counterintuitive to deny consequentiality to the realization that the legal process is historical and pragmatic rather than foundational, but this opposition is just like the others we have examined between originalism and nonoriginalism, historical and ahistorical interpretation, determinacy and indeterminacy. Indeed, it is the same opposition and is amenable to the same analysis. A foundational practice would be one that derived from absolutely clear and uncontroversial first principles; but whenever such principles are put forward, they turn out to be knowable only within a set of historical circumstances of the kind to which they are supposedly prior. A pragmatic (or radically historical) practice would be one that proceeded on the basis of no principles whatsoever, flying by the seat of its pants; but it is impossible even to conceive of a practice that was not the extension of some notion of what the world is and should be— in short, of some principle. It follows then that foundationalism and pragmatism (or historicism) cannot be the names of alternative modes of being in the world, and it follows further that identifying oneself as one or the other will not, in and of itself, determine one's practice. However, it may very well be that identifying oneself as one or the other is an important strategy given the conditions currently prevailing in one's practical world. That is to say (once again), that although theory does not and could not matter in the way both theorists and antitheorists assume it must, it can matter if the rhetorical structure of an enterprise is sensitive to its pronouncements.

The point is made dramatically by one of Lief Carter's judges, who says of his (her?) own practice, "Sure, you occasionally have results that just don't write, but most of the time you reach the result that's fair and then build your thinking around it." By "building your thinking around it" the jurist means preparing an opinion in which certain theoretical propositions may figure prominently. The mistake is to think that those propositions generate the decision they are brought in to justify. Of course, it would be equally a mistake to think that they had nothing to do with it since in the process of making a decision the judge will pass through and incorporate the forms of thought the culture makes available, and these will certainly include theoretical formulations. It is just that those formulations will be components of the decision-making process rather than its source. They are resources for the judge, whose sense of the enterprise does not derive from them but from the institutional experience in relation to which they have an honored but not a foundational place. Carter reports that he tried to get the judges to talk about the distinction between a government of laws and a government of men as it related to the understanding they had of their own authority, but he found that they understood their authority to flow from the role the society authorized them to play and not from any of the abstract concepts they might employ in the course of playing that role. No wonder that despite his urgings in the direction of theory, Carter found that "the conversation would quickly drift from the theoretical points [he] had introduced.''

Where then does this leave us? Just about in the position indicated by Francis Lieber's *Legal and Political Hermeneutics* of 1837, as discussed by James Farr. We have a generous supply of "rules of thumb"—"There can be no sound interpretation without good faith and common sense"; "The construction ought to harmonize with the substance and general spirit of the text"; "We ought not to build arguments of weighty importance on trifling grounds"—but these are not rules: one must still decide what common sense currently is and what does or does not harmonize with the spirit of the text, and, indeed, what *is* the spirit of the text, as well as what is important and trifling; and in making these decisions, commonplaces like those rehearsed by Lieber can inform us of certain public requirements of judicial decorum, but they cannot tell us how to meet those requirements in particular instances. In short, all the work remains to be done, and that perhaps is the one lesson to be taken away from a collection that would offer us lessons more programmatic and ambitious: the only thing to know about interpretation is that it has to be done every time.

13

Almost Pragmatism: The Jurisprudence of Richard Posner, Richard Rorty, and Ronald Dworkin

I

In *The Problems of Jurisprudence,*[1] Richard Posner announces that he is a pragmatist, by which he means that he rejects many if not most of the goals of legal theory, and especially the chief goal of offering an account of the law that is at once comprehensively abstract, strongly normative, and predictive of outcomes, that is, of decisions and holdings. He begins by declaring that he will "argue against 'artificial reason,' against Dworkin's "right answer' thesis, against formalism, against overarching conceptions of justice such as 'corrective justice,' 'natural law,' and 'wealth maximization' . . . against 'strong legal positivism' " (26), and he ends by proclaiming that the search for "an overarching principle for resolving legal disputes" (302) has failed and that "no keys were found" (455). The process of finding no keys gives the book its structure. In other treatises on jurisprudence the argument is built up step by step into what promises to be a magnificent edifice (or empire), but here "there is no edifice" (69), only the repeated attempt to lay still another foundation that is almost immediately found to be as "rotten" as the last one (392).

Something of the feel of this negative project emerges early on in the discussion of objectivity. Objectivity, Posner tells us, comes in three flavors. First, and most ambitiously, there is "objectivity as correspondence to an external reality" (7); second, the scientific sense of objectivity as a procedure that is replicable independently of the differences between agents who execute it: "A finding is replicable in this sense if different investiga-

Reprinted from *Pragmatism in Law and Society,* Michael Brint, editor, Westview Press, 1991, by permission of Westview Press, Boulder, Colorado.

tors, not sharing the same ideological or other preconceptions . . . would be bound to agree with it'' (7); and, third, there is objectivity in the sense of ''merely reasonable—that is, as not willful, not personal, not (narrowly) political, not utterly indeterminate though not determinate in the ontological or scientific sense, but as amenable to and accompanied by persuasive though not necessarily convincing explanation'' (7).

The first kind of objectivity—the conforming of our procedures to an independent and external truth—''is out of the question in most legal cases'' (31). The second, scientific or replicable and convergent objectivity, ''is sometimes attainable, but given the attitudes of and the constraints on the legal profession, and the character of the problems it deals with, often not''; and the third form of objectivity, named by Posner ''conversational objectivity,'' the objectivity that seems achieved in moments (however temporary) of successful persuasion, ''is attainable—but that isn't saying much'' (31). It isn't saying much because its attainability is not a matter of method or planful design (conversational objectivity cannot be generated by a mechanical procedure; if it could it would be replicable and scientific objectivity) and therefore it is in some sense fortuitous; in any situation it may or may not occur, depending on the degree of homogeneity in the relevant community, the relation of available argumentative resources to skillful advocates, the pressures for generating a conclusion in one direction or another, the routes by which that decision might be reached, and innumerable other *contingencies* that may or may not meet together in a happy conjunction.

In a word, conversational objectivity is a *political* achievement, and therefore an achievement that is the antithesis of objectivity as many understand it, a state of certitude that attends the identification and embrace of bedrock and abiding fact and/or principle. To those for whom objectivity can only come in this (hard) form, the temporary outcomes of an indeterminate and messy institutional ''conversation'' hardly meet the test. Neither is the test met by scientific or replicable objectivity as Posner describes it, because it is distinguishable from the softer, conversational, kind only in degree. ''The only way to make [the law] more objective''—the only way to kick legal objectivity up a notch from the conversational to the replicable—''is to make the courts and the legislatures more homogeneous, culturally and politically'' (32).

In short, the only difference between scientific and conversational objectivity is a difference between a community in which assumptions are widely shared and firmly in place, and a community in which assumptions differ and agreement must be repeatedly negotiated. And since the stability of the first community is itself a contingent matter, a stage in the history of a discipline or a society, it is a stability that can always be upset by an unforeseen circumstance.[2] Scientific or replicable objectivity is therefore no less political than conversational objectivity; it is just a matter of how much homogeneity the powers that be have managed to achieve. "Legal thought cannot be made objective by being placed in correspondence with the 'real' world. It owes whatever objectivity it has to cultural uniformity rather than to metaphysical reality or methodological rigor" (30).

If methodological rigor goes south in a pragmatist wind, can formalism be far behind? Formalism is the hope that legal outcomes can be generated by a procedure that is not hostage to any a priori specification of value: "the only prerequisite to being a formalist is having supreme confidence in one's premises and in one's methods of deriving conclusions from them" (40). However, adds Posner, the formalist's confidence is unfounded since the premises are always contestable and therefore incapable of providing a firm foundation for the reasoning that flows from them. So long as one does not notice the contestability, "decisions will appear to be strongly objective because logically deducible" (48); but once the curtain is lifted the observer will "see that the decisions are no more objective in an ultimate sense than those made under [a] more frankly ad hoc regime" (48). One may intone with "great resonance" the "idea of treating like things alike," but the "idea is empty without specification of the criteria for 'likeness'; and . . . those criteria are political" (42). So much for what H. L. A. Hart calls the idea of justice in its simplest form ". . . the notion that what is to be applied to a multiplicity of different persons is the same general rule, undeflected by prejudice, interest, or caprice."[3]

What is true of large abstractions like the "idea of justice" is no less true of rules; they too are political in their operation, says Posner, because although they may be invoked as formal and universal, they are almost always employed in the service of "ad hoc exceptions and adjustments" (46). Rules of a truly formal kind may perhaps be found in games where the player is not free to decide, for example, that his rook will simply not "be captured by his opponent's queen" (50), but in the law judges can do just that and say that "they are doing so in order to comply with a higher level rule" (50). In games the rules apply to carefully circumscribed and static worlds, but the world in which legal rules function is protean and ever changing: the richness of its phenomena is richly in excess of any attempt to formally contain it. Since "[r]ules make dichotomous cuts in continuous phenomena" (46), a rule "suppresses potentially relevant circumstances of the dispute" (44) and a judge is free to decide what will or will not be suppressed.

Thus, rather than constraining judges, rules offer judges the opportunity to engage in temperamentally preferred activities by allowing them either to confine or expand the judicial gaze. Judges who are tolerant of "untidiness, even disorder'' will be ''highly sensitive to the particulars of each case,'' while judges who are invested in tradition and continuity will defer to already-in-place authorities, ''legislators, the founding fathers, higher or earlier judges'' (49). Although judges of both kinds will employ rules, the rules will function not as checks on personal preferences (the standard account of rules and their value), but as their vehicle: ''judges are not *bound* by the rules to do anything'' (47). Here is the formalist fear writ large, a legal system that is no system at all, but a ramshackle nonstructure made up of bits of everything and held together (when it is held together) by transitory political purposes: ''The common law is a vast collection of judge-made rules, loosely tethered to debatable interpretations of ambiguous enactments'' (47).

Loose tethering, however, turns out to be all the tethering one needs in the Posnerian vision, for while ''exact inquiry'' (71) and ''pure'' reason are unrealizable ideals, practical reason takes up the slack. Practical reason ''is a grab bag that includes anecdote, introspection, imagination, common sense, empathy, imputation of motives, speaker's authority, metaphor, analogy, precedent, custom, memory, 'experience' '' (73). In one sense, as Posner points out, the list is too long, because its components are not all of a kind and are sometimes not discrete; but it is also too short because some of the entries can be divided and subdivided. It is that untidiness that makes practical reason what it is, not a self-enclosed mode of algorithmic or mechanical calculation, but an ever changing collection of rules of thumb, doctrines, proverbs, precedents, folk-tales, prejudices, aspirations, goals, fears, and, above all, beliefs.

In the realm of formal objective reasoning (if there were such a thing) belief (personal preference) is precisely what is kept at bay so that the impersonal logic of the deductive machine can run smoothly, without interference. But in the (real) world of practical reasoning, beliefs—the intuitions ''that lie so deep that we don't know how to question them'' (73)— serve as the premises of all reasoning, and rather than being controlled or trumped by evidence (as they are in the popular picture of ''good'' reasoning) beliefs pass on the usefulness and relevance of different kinds of evidence and put the kinds together in ways that sort with an already-in-place structure. ''Pure'' reasoning generates a basis for the taking up of purposes; but practical reasoning begins with purposes, with inclinations toward the inhabiting and building of this or that world, and it is those inclinations that influence and direct the way evidence is marshalled and even seen.

Posner illustrates the point with the doctrine of precedent. The doctrine is that precedent controls, but, says Posner, what really controls is how one ''chooses to read the precedent''; ''the key to the decision is precisely that choice, a choice not dictated by precedent—a choice as to what the precedent *shall* be'' (95). That choice will not have been logically driven, but driven by the direction in which the judge wanted to go. This does not mean that the judge can decide in any direction he or she pleases; the routes of choice, indeed the alternative forms in which choice can even appear, are constrained by the present shape of practical reasoning, by what arguments will work, what categories are firmly in place, what distinctions can be confidently invoked.

Posner asks if a precedent could be distinguished on the basis that in the earlier case "the plaintiff had been left-handed and in the present one the plaintiff is right-handed," and answers, "it could not—but only because there is no consideration of policy or ethics that would justify so narrow an interpretation" (96). That is, the distinction between left- and right-handedness cannot be grabbed hold of by a judge who wants to arrive at a certain conclusion; the state of the culture, of what it will hear as reasonable (not the force of reason itself) bars him, at least now; but there might come a day (perhaps in the context of a new and persuasive account of criminal behavior) when the left-handed/right-handed distinction carries a legal weight, means something in terms related to the concerns of the legal community. Practical reasoning is not a fixed category and its content will not always be the same, but whatever it contains, its mode of calculation will be rhetorical rather than logical, a matter first of determining or sensing where the lines of authority lie—what previous holdings will strike one as settling a question, what rules can be invoked without challenge or qualification, what maxims ("no one should be permitted to profit from his own wrong") will close down discussion, what analogies have stood the test of time, "what politically accredited source" (82) has issued what citable pronouncements, what goals now go virtually unquestioned in the realm of "rational" deliberation—and then of "working" these "authoritative" materials in the direction of one's purposes, one's inclinations, one's intuitions, one's beliefs.

In this vision authority itself is rhetoricized and politicized; that is, authorities do not come ready made in the form of a pure calculus or a scriptural revelation; rather they are made, fashioned in the course of debate and conflict, established by acts that are finally grounded in nothing firmer than persuasion (another name for practical reasoning) and so finally fashioned and maintained by force: "To be blunt, the *ultima ratio* of law is . . . force—precisely what is excluded even by the most latitudinarian definitions of rationality" (83). Posner here endorses and expands on the view of Holmes which he had earlier quoted: "I believe that force, mitigated so far as may be by good manners, is the *ultima ratio,* and between two groups that want to make inconsistent kinds of world I see no remedy except force" (19 n. 29). The conclusion is of course a shocking one, but it follows inevitably from every other part of Posner's argument and it does so for a reason Holmes's sentence nicely highlights: disputes between "groups who want to make inconsistent kinds of world" could be resolved by rational rather than forceful means only if the content and method of rationality could be stipulated apart from the agenda of any particular group; but it is just that kind of specification Posner rules out as a possibility when he declares unavailable to the law (and to much else) objectivity in the ontological sense and grants to the law only scientific or replicable objectivity ("sometimes attainable") and conversational objectivity (contingently attainable).

As I have already observed the second and third senses of objectivity are actually one and the same, since they are distinguished not by an epistemological, but by a social/political condition. In a discipline that can be said to display scientific objectivity—for example science or at least some corners of it—potentially disputable premises are simply not in dispute for reasons of history, disciplinary politics, societal expectations, etc. In a discipline characterized by conversational objectivity, disputes are everywhere and basic premises are often seen to be "up for grabs" (although as Posner correctly points out, not all of them will be so seen at the same time). In either disciplinary situation—the one of potential but quiescent dispute or the one of pervasive and continuing dispute—the settling of dispute, should it break out, can only be accomplished by political means, by the invoking of some sacrosanct (but itself contestable if anyone dared, or even thought, to contest it) first principle of the enterprise ("if we are to remain a government of laws, not men . . ."), or by the pronouncement of someone in a position to make his or her pronouncements stick or by the taking of a vote as the result of which the dispute has been officially or administratively settled (but is sure to erupt on another day) or by the intervention of an armed force.

In this list (certainly not exhaustive) of possibly "authoritative" actions, only the last is usually given the name "force," but in the absence of any neutral calculus or principle to which disputants might have recourse, the other actions are but softened versions of the last, instances of what Holmes refers to as the mitigation of "good manners." To be sure this is a mitigation not to be lightly dismissed; without good manners—a weak phrase for the willingness to refrain from bashing one's opponent's head in—civilization itself would fail, not because, as some have been telling us recently and others had been telling us even before Juvenal's third satire, we have lost hold of first principles and basic truths, but because, given the unavailability of such principles and truths to limited mortals (the phrase is redundant), we would fall instantly to fratricide (and to matricide, and patricide and genocide and every other cide) did we not invest our energies in procedures and habits designed (as it has become fashionable to say) to keep the conversation going. Force, in short, comes in hard and soft versions, and all things being equal, soft is better than hard (a reversal of the usual masculinist metaphor underlying much of academic discourse). But not always, because all things are not equal. That is, at any moment one is always committed to goals and premises in such a way that certain challenges to them will be perceived as socially, not personally, disastrous; and when those challenges arise, it will seem that a soft response—turning the other cheek, writing another page—is a betrayal of one's values and of one's responsibility to the world.

At that point there will be invoked the distinction between legitimate and illegitimate force, a distinction that, as H. L. A. Hart saw, is basic to the law's claim to be law rather than force in law's clothing, but a distinction that will then be invoked in another form—the line differently but just as sincerely drawn—by those whose depredations you feel compelled to resist. At bottom—and the situation of having to confront "at bottom" is what most of life is devoted to avoiding—what is unreasonable is what the other fellow believes, and illegitimate force is the action he is taking in defense of his beliefs. As Learned Hand put it in a statement Posner also cites: " 'Values are incommensurable. You can get a solution also by a compromise, or *call it what you will*. It must be one that people won't complain of too much; but you cannot expect any more objective measure' " (129 n.10).[4]

You cannot expect any more because of the condition whose strong acknowledgement is the basis of all pragmatist thinking, the condition as Posner names it, of heterogeneity or difference as I would name it. (The fact that pragmatism too has its foundational premise is not a contradiction of its antifoundationalism because this particular premise—the irreducibility of difference—*is* antifoundationalism.) In a heterogeneous world, a world in which persons are situated—occupying particular places with particular purposes pursued in relation to particular goals, visions, and hopes as they follow from holding (or being held by) particular beliefs—no one will be in a situation that is universal or general (that is, no situation at all), and therefore no one's perspective (a word that gives the game away) can lay claim to privilege. In that kind of world, a world of difference, in our world according to Posner and according to me, the stipulation both of what is (of the facts) and of what ought to be will always be a politically angled one, and in the (certain) event of a clash of stipulations, the mechanisms of adjudication, whether in the personal or institutional realms, will be equally political.

How then does the business of law get done? "If two social visions clash, which prevails? . . . How does a judge choose between competing social visions?'' (148). Posner's answer to these questions will be troubling to those who seek a jurisprudence in which policy considerations have been either eliminated or subordinated (à la Dworkin), but it is an inevitable answer given everything that precedes the question: ''Often the choice will be made on the basis of deeply held personal values, and often these values will be impervious to argument'' (148–149). This last is particularly devastating, since argument, in the sense of the marshalling of evidence that will be compelling to any actor no matter what his or her ''personal values,'' is supposedly the very life of the law. This is not to say, Posner hastens to add, that because a judge's personal values are impervious to argument, they are impervious to change. Change can and does occur, not however by a process of ''reasoned exposition'' (149), but through conversions, defined nicely as ''a sudden deeply emotional switch from one nonrational cluster of beliefs to another that is no more (often less) rational'' (150).

And what brings that switch about? Almost anything and nothing in particular. That is, there is no sure route—no sequence of formalizable or even probabilizable steps—to conversion, nor are there means or stimuli that are ''by nature'' too weak to produce it. Conversion can follow upon anything—reading at random a verse from the Bible, falling off one's horse on the road to Damascus, suddenly seeing the first gray hair—for anything, given the right history, psychology, pressuring circumstances, etc., can ''jar people out of their accustomed ways of thinking'' (150).

Posner's example is the women's movement, which he says, has become influential because ''[m]any women and some men'' have been brought to see the role of women ''in a different light,'' not however ''by being shown evidence that this is the way things 'really' are, but by being offered a fresh perspective that, once glimpsed, strikes many with a shock of recognition'' (150). But not all. The metaphors, analogies, revisionist histories, slogans (''the personal is the political'') that have struck some as a revelation (''once I saw through a glass darkly'') have struck others as absurd or irrelevant. If the minds of people, including judges, are changed by conversion rather than by the operation of reason and logic, then change is a *contingent* matter and predictability—both prized and claimed by the law—is a chimera. Of course, contingency can sometimes take hold, not however as the result of a plan or campaign, but as the result of notions or vocabularies that somehow get to be ''in the air'' and effect a ''change of outlook'' which when it is noticed (by a historian or social commentator) will be seen to have been caused by no one in particular and certainly not by any rational process. It is just that something that was once ''virtually unthinkable'' (151) now goes without saying. ''My point,'' concludes Posner, ''is that the great turning points in twentieth-century law (and in law, period) [were] not the product of deep reflection on the meaning of the Constitution . . . but instead reflect changing outlooks'' (152).

II

With statements like this, Posner puts the cap on his anti-essentialist, anti-foundational, antirational (in the strong sense), antimetaphysical and deeply pragmatist view of the law, and it is perhaps superfluous for me to say that I agree with him on almost every point. Indeed, as I look back on the preceding pages, I see little effort to separate my account of Posner's argument from my own elaborations of it. Of course I have some quibbles, but that's what they are, even though I shall now be so ungenerous as to rehearse them.

When Posner says that "[a] judicial holding normally will trump even a better-reasoned academic analysis because of the value that the law places on stability" (95), he seems to accord both a privilege and an independence to "reason" that he elsewhere withholds. Would it not be truer to his larger argument to replace "better-reasoned" with "differently-reasoned" and so recognize that the desire for stability is itself a reason, and one no better or worse than the academic reasons that are put forward in the context of institutional norms? And when Posner criticizes the "plain-meaning approach" of excluding consideration of "the communicative intent and broader purposes" of statutes, he seems to think that such an exclusion is possible, that one could, in fact, read in a way that bracketed purposes and intentions not already "in" the writing; but (as I have argued elsewhere and at length) language is only construable within the assumption of some or other human purpose. No act of reading can stop at the plain-meaning of a document, because that meaning itself will have emerged in the light of some stipulation of intentional circumstances, of purposes held by agents situated in real world situations. The difference between ways of reading will not be between a reading that takes communicative intent into account and a reading that doesn't, but between readings that proceed in the light of differently assumed communicative intents. Formalist or literalist or "four corners" interpretation is not inadvisable (as Posner seems to suggest); it is impossible.

And finally in this short list of occasional but not fatal lapses or slips, it really will not do, in the context of the book's informing spirit, to contrast "persuasion by rhetoric" to "the coolest forms of reasoned exposition" (149). Reasoned exposition (which will have different shapes at different times in different disciplines) is itself just one form of rhetoric, cooler perhaps if the measure of heat is a decibel count, but impelled by a vision as partisan and contestable as that informing any rhetoric that dares to accept that name.

As I have already said (twice), these disagreements with Posner do not amount to much, but that should not be taken to indicate that I have no real quarrels with this book, for there is a strain in it, muted at first but heard more often in its second half, that I believe to be at deep odds with Posner's strongest insights. Let me try to focus my criticism by returning to the moment when Posner declares that of the three kinds of objectivity, the third—conversational objectivity—is attainable in the law, but, he adds, "that isn't saying much" (31). It seems to me, however, that in the following pages he sometimes thinks that too much follows from having said "that." Indeed, in my view *anything* that would be said to follow from the fact of "conversational objectivity" would be too much, for it would be to confuse a pragmatist account of the law with a pragmatist program.

A pragmatist *account* of the law speaks to the question of how the law works and gives what I think to be the right answer: the law works not by identifying and then hewing to some overarching set of principles, or logical calculus, or authoritative revelation, but by deploying a set of ramshackle and heterogeneous resources in an effort to reach political resolutions of disputes that must be framed (this is the law's requirements and the public's desire) in apolitical and abstract terms (fairness, equality, what justice requires). By the standards applied to determinate and principled procedures, the law fails miserably (this is the charge made by Critical Legal Studies); but by the pragmatist standard—unsatisfactory as a standard to formalists and objectivists, as well as to deconstructors—the law gets passing and even high marks because it *works*. A pragmatist *program* asks the question "what follows from the pragmatist account?" and then gives an answer, but by giving an answer pragmatism is unfaithful to its own first principle (which is to have none) and turns unwittingly into the foundationalism and essentialism it rejects.

Posner's answer—his program—takes the form of a pro-scientific, no nonsense empiricism that is obviously related to the tradition of legal realism. Signs of this "realist" stance surface early, when in the course of setting out the book's plan, he questions the utility of interpretation as an explanatory concept and declares, "We might do better to discard the term" (31). Discarding terms and much else is a favorite move in legal realist polemic and no one performed it with more flair than Felix Cohen.[5] Cohen begins by heaping scorn on the notion of a corporation as a legal abstraction, as a fictional entity. "Where is a corporation?" he asks, and replies that it is "not a question that can be answered by empirical observation" (810). "Nor is it a question," he goes on, "that demands for its solution any analysis of political considerations or social ideals. It is, in fact, a question identical in metaphysical status with the question which scholastic theologians are supposed to have argued at great length, 'How many angels can stand on the point of a needle?' "[6] In short, a question directed at a speculative, mythical nonobject which, because it was produced by superstition rather than observation, gets in the way of seeing things as they really are. Unfortunately, the law is not (yet) a science and is therefore susceptible to the appeal of "myths [that] impress the imagination . . . where more exact discourse would leave minds cold."[7] The result, Cohen laments, is a world of circular legal reasoning (815) in which jargon-of-the-trade terms interact with one another "without ever coming to rest on the floor of verifiable fact."[8]

The remedy for this sorry state is implicit in the indictment: sweep away the magical but substanceless words that make up the vocabulary of jurisprudence so that we will have an unobstructed view of the situation and problems to which we could then address ourselves. If notions of " 'property' and 'due process' were defined in non-legal terms" (820)—defined that is in a descriptive vocabulary truly in touch with "empirical social facts" (821)—we might be able "to substitute a realistic, rational scientific account of legal happenings for the classical theological jurisprudence of concepts."[9] If we mean ourselves from "supernatural entities which do not have a verifiable existence except for the eyes of faith" we may at last come into contact with "actual experience."[10] Once this happens, once "statistical methods" have brought us close to the "actual facts of judicial behavior,"[11] the "realistic advocate" or judge will be able to "rise above" all distorting lenses including both the lens of "his own moral bias" and the lens of "the moral bias of the legal author whose treatise he consults."[12] No longer will he be "fooled by his own words"[13] or by anyone else's.

In these quotations (which could have easily been supplemented from the pages of Jerome Frank and other early realists) we see that the basic realist gesture is a double, and perhaps contradictory, one: first dismiss the myth of objectivity as it is embodied in high sounding but empty legal concepts (the rule of law, the neutrality of due process) and then replace it with the myth of the "actual facts" or "exact discourse" or "actual experience" or a "rational scientific account," that is, go from one essentialism, identified with natural law or conceptual logic, to another, identified with the strong empiricism of the social sciences.

The problem with this sequence was long ago pointed out by Roscoe Pound who, while acknowledging the force of many of the realists' observations, declares himself "skeptical as to the faith in ability to find the one unchallengeable basis free from illusion which alone the new realist takes over from the illusion-ridden jurists of the past."[14] Given the realist insistence on the unavoidability of bias and on the value laden nature of all human activities, the recourse to a brute fact level of uninterpretive data seems, to say the least, questionable, as does the assumption that if we could only divest ourselves of the special vocabulary of the legal culture (no longer be fooled by our own words) we could see things as they really (independently of any discursive system whatsoever) are. Cohen and Frank are full of scorn for theological thinking and for the operation of faith, but as Pound sees, they are no less the captives of a faith, and of the illusion—if that is the word—that attends it.

That is, however, *not* the word, for "illusion" implies the availability of a point of view uncontaminated by metaphysical entities or by an a priori assumption of values, and as the realists (and Posner after them) argue in their better moments, there is no such point of view, no realm of unalloyed non-mediated experience and no neutral observation language that describes it. The advocate or jurist who moves from the conceptual apparatus codified in law to the apparatus of statistical methods and behaviorist psychology has not exchanged the perspective-specific facts of an artificial discursive system for the real, unvarnished facts; rather he or she has exchanged the facts emergent in one discursive system—one contestable articulation of the world—for the facts emergent in another. It is not that there is no category of the real; it is just what fills it will always be a function of the in-place force of some disciplinary or community vocabulary; eliminate the special jargon of the law, as the realists urge, and you will find yourself not in the cleared ground of an epistemological reform ("now I see face to face") but in the already occupied ground of some other line of work no less special, no less hostage to commitments it can neither name nor recognize.

Much of what Posner writes in *The Problems of Jurisprudence* suggests that he should be in substantial agreement with the previous paragraph. Steeped as he is in the writings of Peirce, Wittgenstein, Kuhn, Rorty, and Gadamer (not to mention Fish), he should be immune to the lure of empiricist essentialism, but he is not. At the end of the chapter on practical reasoning he complains that the law still carries too much conceptual baggage and avers that "the situation would be improved if law committed itself to a simple functionalism or consequentialism," that is, to a program of adjusting the operations of law to precisely specified social goals in relation to which the law would be self-consciously subordinate and secondary:

> Suppose the sole goal of every legal doctrine and institution was a practical one. The goal of a new bankruptcy statute, for example, might be to reduce the number of bankruptcies . . . and if the statute failed to fulfill [that goal] . . . it would be repealed. Law really would be a method of social engineering, and its

structures and designs would be susceptible of objective evaluation, much like the project of civil engineers. This would be a triumph of pragmatism. (122)

This is a complicated statement that looks forward to several arguments Posner will later elaborate, including the argument (more modest than one might have expected) for law and economics. In his chapter on that approach Posner rehearses the familiar thesis that even though the law may not self-describe its operations in economic terms, its history indicates that those terms or something approximating them are impelling legal actors whether they are aware of it or not. It is as if somewhere deep down, in the realm of tacit rather than explicit knowledge, "judges *wanted* to adopt the rules, procedures and case outcomes that would maximize society's wealth" (356). "We should be no more surprised that judges talk in different terms while doing economics than that businessmen equate marginal cost to marginal revenue without using the terms and often without knowing what they mean" (372–373).

Here is the meeting point of Posner's declared pragmatism and his previous self-identification with the law and economics movement. When Posner says that the goal of every legal doctrine should be a practical one, he means (in good realist fashion) that legal doctrine should be reconceptualized so as to accord with the nitty-gritty facts of social life, and that means reconceptualized in the language of law and economics, since in his view the language of law and economics is the language of real motives and actual goals. If "the object of pragmatic analysis is to lead discussion away from issues semantic and metaphysical and toward issues factual and empirical" (387), then by Posner's lights pragmatic analysis and the pragmatic program will succeed when legal concepts and terms have been replaced by economic ones or when "the positive economic theory of law will be subsumed under a broader theory—perhaps, although not necessarily, an economic theory—of the social behavior we call law" (374).

This could possibly happen, but if it ever does, we will not have escaped semantics (merely verbal entities) and metaphysics (faith-based declarations of what is) but merely attached ourselves to new versions of them. As many commentators have observed, "wealth-maximization," efficiency, Pareto superiority, the Kaldor Hicks test, and the other components of the law and economics position are all hostage to metaphysical assumptions, to controversial visions of the way the world is or should be. A transformation such as Posner seems to desire would not lead to methods "susceptible of objective evaluation" but to methods no more firmly grounded than the wholly contestable premises that "authorize" them. Moreover, and this is the more important point, should that transformation occur, the result would not be a more empirically rooted law, but no law at all. The law, as a separate and distinct area of inquiry and action, would be no more; an enterprise of a certain kind would have disappeared from the world (itself not fixed, but mutable and revisable) of enterprises.

At issue here is the nature of the desire to which law is a response. In Posner's view (although he doesn't put it this way), the law is answerable to a desire that can be pragmatically defined (the desire to prevent bankruptcy or protect the integrity of the family); given this view, it makes sense that its forms and vocabulary should match up with that desire, and that they should be criticized when they do not. But I would describe the desire that gives rise to law differently and more philosophically, in a loose sense of that word. Law emerges because people desire predictability, stability, equal protection, the reign of justice, etc., and because they want to believe that it is possible to secure these things by instituting a set of impartial procedures. This incomplete list of the desires behind the emergence of law is more or less identical with the list of things Posner debunks in the course of his book, beginning with objectivity, and continuing with determinate rules, value free adjudication, impersonal constraints, the right of privacy, freedom of the will, precedent, intention, mind, judicial restraint, etc. Repeatedly he speaks of himself as "demistifying" (184) these concepts in the service of "the struggle against metaphysical entities in law" (185), and he writes deprecatingly of the delusions, pretensions, and false understandings with which actors in the legal culture deceive themselves.

But the result of success in this struggle, should he or anyone else achieve it, would not be a cleaned-up conceptual universe, but a universe deprived of the props that must be in place if the law is to be possessed of a persuasive rationale. In short, the law will only work—not in the realist or economic sense but in the sense answerable to the desires that impel its establishment—if the metaphysical entities Posner would remove are retained; and if the history of our life with law tells us anything, it is that they *will* be retained, no matter what analysis of either an economic or deconstructive kind is able to show.

The curious fact is that Posner knows this with at least part of his mind. Anticipating the objection that the adoption of a behaviorist vocabulary (which would have the advantage, he says, of eliminating "fictitious" entities like minds, intentions, the conscience, and guilt) will "strip the moral as well as the distinctively human content from the . . . law" (178), he replies:

> There are no . . . grounds for fearing that speculations in the philosophy of mind are likely to affect respect for, let alone observance of, law. . . . Philosophers who believe in determinism behave in their personal lives just like other people. If freedom is an illusion, it is one of those illusions . . . that we cannot shake off no matter what our beliefs or opinions are. (178)

This is exactly right, not (as Posner implies) because human beings obstinately cling to their "illusions," but because the set of purposes that will lead one to do philosophy of mind and the set of purposes that lead one to administer or make law are quite different and there is no reason to assume that a conclusion reached in one area will have an effect on the central tenets of the other. (More of this later.) Law is centrally *about* such things as conscience, guilt, personal responsibility, fairness, impartiality, and no analysis imported from some other disciplinary context "proving" that these things do not exist will remove them from the legal culture, unless of course society decides that a legal culture is a luxury it can afford to do without.

What Posner calls the "illusions" with which public actors sustain their roles are in fact the assumptions (no more or less vulnerable than any others) that constitute those roles; take them away (not, as he acknowledges, an easy task) and you take away the role and all of the advantages it brings to the individual and the community. As Posner correctly observes, "most judges believe, without evidence (indeed in the face of the evidence . . .) that the judiciary's effectiveness depends on a belief by the public that judges are finders rather than makers of law" (190). The implication is that the belief would be better founded if independent evidence of it could be cited; but this particular belief is itself founding, and constitutes a kind of contract between the legal institution and the public, each believing in the other's belief about itself and thus creating a world in which expectations and a sense of mutual responsibility confirm one another without any external support. Similarly when judges persuade "themselves and others that their decisions are dictated by law" (193), the act of persuasion is not a conscious strategic self-deception, but something that comes with the territory, with the experience of law school, of practice, of a life in the courts, etc. The result is not, as Posner would have it, a "false sense of constraint" (193), but a sense inseparable from membership in a community from whose (deep) assumptions one takes one's very identity. But as I have said, Posner knows all and he even knows that the fictions he debunks are necessary, that "the belief that judges are constrained by law . . . is a deeply ingrained feature of the legal culture" (194), and that this "situation is unlikely to change without profound *and not necessarily desirable* changes in the political system" (193).

III

Perhaps he knows something even deeper, which is that the call for a pragmatist program, for a demistifying of the legal culture, for a clearing away of the debris of faith-based conceptual systems, is at odds with the insight of heterogeneity, with the recognition that difference is a condition that cannot be overcome by attaching ourselves to a bedrock level of social/empirical fact because that level, along with the facts seen as its components, is itself an interpretive construction, an imaginative hazarding of the world's particulars that is finally grounded in nothing stronger than its own persuasiveness (with persuasiveness a function of the number of desires this particular story about the world manages to satisfy).

I said earlier that once pragmatism becomes a program it turns into the essentialism it challenges; as an account of contingency and of agreements that are conversationally not ontologically based, it cannot without contradiction offer itself as a new and better basis for doing business. Indeed, if the pragmatist account of things is right, then everyone has always been a pragmatist anyway; someone may pronounce in the grand language of foundational theory, but since that theory will always be a rhetoric—an edifice supported by premises that might be contested at any moment—such pronouncing is no less provisional and vulnerable than "those made under [a] frankly ad hoc regime" (48). Nor will their advent of a frankly ad hoc regime—one in which contingency and the heterogeneity of value are publicly announced—make any operational difference; awareness of contingency allows one neither to master it (as if knowledge of an inescapable condition enabled you to escape it) nor to be better at it (a quite incoherent notion). Once a pragmatist account of the law (or anything else) has shown that practice is not after all undergirded by an overarching set of immutable principles, or by an infallible and impersonal method, or by a neutral observation language, there isn't anything more to say ("to say that one is a pragmatist is to say little" [28]), anywhere *necessarily* to go; and you certainly can't go from a pragmatist account, with its emphasis on the ceaseless process of human construction and the endless and unpredictable achieving and reachieving of conversational objectivity, to a brave new world from which the constructions have been happily removed. Indeed, if you take the antifoundationalism of pragmatism seriously (as Posner in his empiricism finally cannot) you will see that there is absolutely nothing you can do with it.

The point is one that puts me in a minority position in the pragmatist camp, for most advocates of pragmatism (I have never been one myself) assume that something must follow from the pragmatist argument, that there are, in the words of Richard Rorty's title, consequences of pragmatism.[15] Rorty himself thinks that, although at times it seems that the consequences he identifies are so loosely related to pragmatism that the claim doesn't amount to much. Nevertheless he does repeatedly attach at least a hope to the possible triumph of pragmatism, not Posner's hope for "a method of social engineering . . . susceptible of objective evaluation" (122), but the hope that if we give up the search for just such a method, write it off as an investment that didn't pan out, we will turn from the (vain) search for "metaphysical comfort"[16] to the comfort we can provide each other as human beings in the same afoundational boat: "In the end, the pragmatists tell us, what matters is our loyalty to other human beings clinging together against the dark, not our hope of getting things right."[17]

The idea is that if people would only stop trying to come up with a standard of absolute right which could then be used to denigrate the beliefs and efforts of *other* people, they might spend more time sympathetically engaging with those beliefs and learning to appreciate those efforts. Those who do this will be improving what Rorty believes to be a specifically pragmatist skill, the "skill at imaginative identification," the "ability to envisage, and desire to prevent, the actual and possible humiliation of others."[18] Moreover, although this ability is in some sense an anti-method in that it involves the proliferation of perspectives rather than the narrowing of them to the single perspective that is right and true, one acquires it, according to Rorty, by a technique that is itself methodical if not methodological: one practices "rediscription," not rediscription in the direction of what is really true, but rediscription as a temperamental willingness to try out vocabularies other than our own in an effort "to expand our sense of 'us' as far as we can."[19] "We should stay on the look out for marginalized people—people we still instinctively think of as 'them' rather than 'us'. We should try to notice our similarities to them."[20] In that way, we may "create a more expansive sense of solidarity than we presently have."[21] In the process, Rorty hopes, philosophy will lose its orientation toward truth and become "one of the techniques for reweaving our vocabulary of moral deliberation in order to accommodate new beliefs (for example that women and blacks are capable of more than we white males had thought, that property is not sacred, that sexual matters are of merely private concern)."[22]

My problem with Rorty's formulations can be surfaced by focusing on this last sentence which suggests, as other sentences do, that there is a general nonspecific skill or ability which, if we hone it, will make us the kind of people likely to see that women and blacks are capable of more than we white men had thought, that property is not sacred, that sexual matters are of merely private concern. But in my view the direction is the other way around: first an issue is raised, by real life pressures as felt by men and women who must make decisions or perform in public and private contexts, and then, in the course of discussion or by virtue of the introduction of a new and arresting vocabulary, or by the pronouncements of a particularly revered figure, or by a thousand other contingent interventions, some of us might come to see the situation and its components in new and different ways. Moreover, this "conversion" experience, if it occurs, will not be attributable to a special skill or ability that has been acquired through the regular practice of redescription—through empathy exercises—but rather to the (contingent) fact that for this or that person a particular argument or piece of testimony or preferred analogy or stream of light coming through a window at the right moment just happened to "take."

My point is that moments like that, which could be described (although inaccurately I think) as expansions of sympathy, cannot be planned, and cannot be planned *for* by developing a special empathetic muscle. This leads me to proclaim Fish's first law of tolerance-dynamics (tolerance is Rorty's pragmatist virtue where Posner's is contextual clarity): *Toleration is exercised in an inverse proportion to there being anything at stake.* If I go to hear a series of papers on John Milton, I listen to them with an attention whose content includes my own previously published work, the place of that work in Milton studies, projects presently in process, etc., and therefore as I listen (not after I listen) I perform *involuntary* acts of approval ("that's right"), disapproval ("Oh, not that tired line again"), anger ("he's got me all wrong"), and others too embarrassing to mention. But if I go to hear a series of papers on George Eliot, whose novels I have read but not written about or even formulated a position on, I listen in quite a different mode, one that allows me to take a relatively cool pleasure at a display of contrasting interpretive skills. Indeed I might even be successively convinced by five speakers and never feel obliged to render a judgment of the kind they are directing at one another. Now obviously I will be exercising more tolerance, engaging more empathetically, in the second scenario as opposed to the first, but that will be because I have no investment in George Eliot whereas my investment in Milton has been growing for more than twenty-five years and is at this point inseparable from my sense of my career and therefore from my sense of myself.

What the example shows, I think, is that tolerance (or, if you prefer, sympathy) is not a separate ability, a virtue with its own context-independent shape, but is rather a way of relating or attending whose shape depends on the commitments one already feels. The Rortian injunction "be ye tolerant" or "learn to live with plurality"[23] or "notice suffering when it occurs" (93) or "expand our sense of 'us' "[24] is like the biblical injunction, "be ye perfect" or the parental injunction "be good"; one wants to respond, yes, but in relation to what? One cannot *just* be tolerant; one is tolerant (or not) in the measure a given situation, complete with various pressures and with the histories of its participants, allows. "Avoid cruelty" is a directive that cries out for contextualization and when put in a more qualified way— avoid cruelty when you can, or avoid cruelty, all other things being equal, or avoid cruelty except when the alternative seems worse—it is even clearer that its force depends on how everything is filled in, on what is already felt to be at stake in the situation. It is this sense of there being something at stake, something not just locally but universally, crucially, urgent, that Rorty would like to see lessened if not eliminated; although, as he reports ruefully, William James himself seemed unable to let go of the feeling that life is "a real fight in which something is eternally gained for the universe by success."[25] It is Rorty's hope—ungrounded, as it must be given his (anti)principles—that if "pragmatism were taken seriously"[26]—if we conceived of ourselves as creatures clinging together in a foundationless world rather than as philosophers in search of a foundation—we might cease experiencing life as a fight and we would be less likely to confront one another across firmly drawn lines of battle.

As Rorty presents it, the vision is certainly an attractive one, but his utopian consequence no more follows from pragmatism than does Posner's empiricist consequence. Both theorists begin by asserting the irreducibility of difference and the concomitant unavailability of overarching principles, but then go unaccountably to the proclamation of an overarching principle, in one case to the principle of undistorted empirical inquiry and in the other to the principle of ever more tolerant inquirers. The two programs differ markedly, but they are similarly illegitimate in having as their source an account from which no particular course of action necessarily or even probably follows. In short, to repeat myself, they confuse a pragmatist account with a pragmatist program and thereby fail to distinguish between pragmatism as a truth we are all living *out* and pragmatism as a truth we might be able to live *by*. We are all living *out* pragmatism because we live in a world bereft of transcendent truths and leak proof logics (although some may exist in a realm veiled from us) and therefore must make do with the ragtag bag of metaphors, analogies, rules of thumb, inspirational phrases, incantations, and jerry-built "reasons" that keep the conversation going and bring it to temporary, and always revisable, conclusions; but we could only live by pragmatism if we could grasp the pragmatist insight—that there are no universals or self-executing methods or self-declaring texts in sight—and make it into something positive, use an awareness of contingency as a way either of mastering it or perfecting it (in which case it would no longer be contingency), turn ourselves (by design rather than as the creatures of history) into something new. But while contingency may be the answer to the question "what finally underwrites the law?", it cannot be the answer to the question, "how does one go about practicing law?" The answer to *that* question is "by deploying all of the resources (doctrines, precedents, rules, magic metaphors, standard concepts) the legal culture offers." As an analyst or observer of the law you may know that those resources cannot finally be justified outside the culture's confines; but as a practitioner justification from the outside is not your business (you are not a philosopher or an anthropologist); as a practitioner, you take your justifications where you can get them.

IV

One place you are unlikely to get them is in the practice of describing the practice for which you are seeking a justification. The mistake both Rorty and Posner make (albeit in different ways) is the mistake of thinking that a description of a practice has cash value in a game other than the game of description. You may find (as Posner, Rorty, and I all do) that it is with pragmatist categories that one can best describe the law; but that doesn't mean that it is with pragmatist categories that one can best practice the law or that with a pragmatist description in place you have a new source of justification; or there is no reason to think that the results of an effort at description can be turned into a recipe either for performance or the act of justifying. Description is *itself* a practice performed, with its own conventions, requirements, and *internal* justifications, all of which are *necessarily* distinct from the terms of the practice that is its object.

The point speaks directly to the other large disagreement I have with Posner's book, his position on the question of legal autonomy. Posner believes that the ragtag eclectic content of legal doctrine means that it cannot be a distinctive thing. The reasoning is simple: since the law manifestly makes use of and invokes as authoritative, materials, doctrines and norms from any number of other disciplines and even nondisciplines (there seems no limit to its indiscriminate borrowings) it must itself be multidisciplinary and therefore not autonomous. "Interdisciplinary legal thinking is inescapable" (439), he says, if only because "there is no such thing as legal reasoning" (459), only a "grab bag of informal methods" (455), which includes rules, but rules that are "vague, open-ended, tenuously grounded, highly contestable, and not openly alterable but frequently altered" (455). But the fact that an area of inquiry and practice incorporates material, concepts, and methods from other areas in a mix that is volatile and variable does not mean that there is nothing—no distinctive purpose or perspective—guiding and controlling the mix. The reasoning that if law is not pure, then law is not law, a discipline with its own integrity, holds only if disciplinary integrity is understood in what Posner calls the "strong" sense, "a field that rather than battening on other fields [is] adequately—and indeed optimally—cultivated by the use of skills, knowledge and experience that owed nothing to other fields" (431).

Posner quite correctly observes that the law cannot meet this strong requirement, but then neither can anything else. No field is self-sufficient in these absolute terms; every field depends both for its legitimacy and operations on assumptions and materials it does not contain but on which it draws, often in the spirit of "it goes without saying" or "if you can't assume this, what can you assume?". What makes a field a field—makes it a thing and not something else (a poor thing but mine own)—is not an impossible purity, but a steadfastness of purpose, a core sense of the enterprise, of what the field or discipline is *for,* of why society is willing (if not always eager) to see its particular job done. The core sense of disciplinary purpose is not destroyed by the presence in the field of bits and pieces and sometimes whole cloths from other fields because when those bits and pieces enter, they do so in a form demanded by the definitions, distinctions, convention, problematics, and urgencies already in place.

This does not mean that a field remains unaltered at its core by the entrance into it of "alien stuff"; a purpose that expands itself by ingesting material previously external to it does not stay the same, but even in its new form, it will still be the instantiation in the world of the enterprise's project, a project whose shape may vary so long as it retains its diacritical identity as the shape now being taken by a particular job of work (ensuring justice, understanding poetry, explaining finance, recovering the past). There is no natural reason that any of these projects should continue forever; the world may one day find itself without economics or literary criticism or history as identifiable disciplinary tasks; but before that happens, the natural conservatism of disciplines—the survival instinct that makes them institutional illustrations of Fish's first law (tolerance is exercised in an inverse proportion to there being anything at stake)—will work to prevent borrowed material from overwhelming the borrower. Despite the recent millenarian calls to interdisciplinarity, disciplines will prove remarkably resilient and difficult to kill.

I will have more to say about the lure of interdisciplinarity in a moment, but for now I want to underline the point I have been making; legal autonomy should not be understood as a state of impossibly hermetic self-sufficiency, but as a state continually achieved and re-achieved as the law takes unto itself and makes its own (and in so doing alters the "own" it is making) the materials that history and chance put in its way. Disciplinary identity is asserted and maintained not in an absolute opposition to difference but in a perpetual recognition and overcoming of it by various acts of assimilation and incorporation.

That is why it is beside the point to complain, as Posner does, that when the law takes up foreign concepts and materials it often does so in crude and sloppy ways, failing to avail itself of the latest up-to-date techniques and formulations. Early on, he observes that although the law is obviously dependent on "ethical insights," it declines "to look for the ethical and political materials of judgment . . . in scholarly materials [and] statistical compendia," relying instead on its own "previous decisions" (94); and later he makes a similar point with respect to the philosophy of language:

> If philosophers mount cogent attacks on simple minded ideas of textual determinacy—ideas that as it happens are the unexamined assumptions of many lawyers and judges engaged in interpreting statutory and constitutional texts—can the legal profession brush aside the attacks with the assertion that what lawyers and judges do when they interpret legal texts is its own sort of thing?" (440).

And by the same reasoning, shouldn't the "pieties of jurisprudence . . . be discarded" so that "at long last" judges would "abandon the rhetoric and the reality of formalist adjudication"? (462).

The answer to these questions was given long ago by Dean Pound in response to the realist program of "beginning with an objective scientific gathering of facts." [27] Facts, declares Pound, "have to be selected and what is significant will be determined by some picture or ideal of the science and of the subject it treats." [28] In other words (my words), the particular form in which materials from other disciplines enter the law will be determined by the law's sense of its own purpose and of the usefulness to that purpose of "foreign" information. Pound notes that there have always been calls for jurisprudence to "stand still" until metaphysicians or ethicists or psychologists concluded their debates and determined an authoritative scientific viewpoint, and yet, as he puts it, "certain general ideas . . . served well enough for the legal science of the last century." [29] Jurisprudence "can't wait for psychologists to agree (if they are likely to) and there is no need of waiting" since we "can reach a sufficient psychological basis for juristic purposes from any of the important current psychologies." [30]

"For juristic purposes." It is those purposes and not the purposes at the core of the quarried discipline that rule. The fact that a notion of the self or of the text or of intention is regarded by psychology or philosophy or literary theory as the best notion going does not mean that it is the notion that best serves juristic purposes. To this Posner might object that it is irresponsible for jurists or anyone else to go about their business in the company of discarded or discredited accounts of important matters; but again one must point out that accounts are specific to enterprises, to projects informed by their own sense of what needs to be done and for what (usually metaphysical) reasons; and it makes no sense to require that projects impelled by a *different* sense of what needs to be done in relation to different (usually metaphysical) reasons be faithful to, or even mindful of, the state of the art in the areas they invade. Law will take what it needs, and "what

it needs" will be determined by *its* informing rationale and not the rationale of philosophy, or literary criticism, or psychology, or economics. (That is why psychoanalysis in its classic form, discarded by mainstream psychology today, is alive and well and productively so in English departments.)

Posner doesn't see this because he has a different view of what enterprises are or should be: he thinks that the point of an enterprise is to get at the empirical truth about something; and that enterprises are different only in the sense that they have been assigned (or assigned themselves) different empirical somethings—the mind, the past, the economy, the stars, the planets—to get at the truth of. The trouble with law is that it has not accepted such an assignment in the true empirical spirit: "Law is not ready to commit itself to concrete, practical goals across the board," but instead keeps prating on about "intangibles such as the promotion of human dignity, the securing of justice and fairness, and the importance of complying with the ideals or intentions of the framers of the Constitution" (123). In short, the law is a rogue discipline that refuses to join the general effort to get things right and that is why the list of legal concepts ("metaphysical entities") that must be discarded is so long as to amount to the discarding of the entire legal culture.

This is not a conclusion that Posner himself reaches (in more than one place he seems to affirm the distinctiveness he elsewhere denies), but it is nevertheless a conclusion that follows inevitably from his strong empiricism. If the "intangibles" he finds "too nebulous for progress" (123)—justice, fairness, the promotion of dignity—are removed in favor of "concrete facts," the disciplinary map will have one less country, and where there was law there will now be social science. Were we to heed Posner's advice to get "rid of" the "carapace of falsity and pretense" that has the effect of "obscuring the enterprise" (469), we would end up getting rid of the enterprise. In doing so we would suffer a loss, not because justice, fairness, and human dignity will have been lost—I believe them to be rhetorical constructions just as Posner does—but because we will have deprived ourselves of the argumentative resources those abstractions now stand for; we would no longer be able to say "what justice requires" or "what fairness dictates" and then fill in those phrases with the courses of action we prefer to take. That, after all, is the law's job—to give us ways of redescribing limited partisan programs so that they can be presented as the natural outcomes of abstract impersonal imperatives. Other disciplines have other jobs—to rationalize aesthetic tastes or make intelligible pasts—and they too have vocabularies that do not so much hook up with the world as declare one.

Disciplines should not be thought of as joint partners cooperating in a single job of work (one world and the ways we describe it); they are what make certain jobs (and worlds) possible and even conceivable (lawyering, literary criticism, economics, etc., are not natural kinds, but the names of historical practices); and if we want this or that job to keep on being done— if we want to use notions of fairness and justice in order to move things in certain directions—we must retain disciplinary vocabularies, not despite the fact that they are incapable of independent justification, but *because* they are incapable of justification, except from the inside.

Posner sees the matter differently because he does think that there is a single job of work to do—the job of getting the empirical facts right—and he thinks of disciplines as either participating in the task or going off into self-indulgent "theological" flights. He is, in other words, at least part of the time, an essentialist (the dismissal of categories as merely "verbal" [468] is a dead giveaway), someone for whom the present state of disciplinary affairs with its turf battles and special claims will in time be "subsumed under a broader theory" (374), that is, under a general unified science. Despite the many acknowledgments of heterogeneity in the book, he is finally committed to a brave empirical future in which heterogeneity will have been, if not eliminated, at least grounded and firmly tethered to something more real.

That is why he is an advocate of interdisciplinary work which has the effect, he says, of "blurring the boundaries between disciplines" (432). Presumably at some future date the "blurring" will amount to a wholesale effacing and the disciplines will number only one. (Thou shalt have no other disciplines before me.) This is the basic premise and declared hope of interdisciplinarity (more a religion than a project) as it has recently been described by Julie Thompson Klein.[31] Klein identifies the interdisciplinary impulse with the desire for "a unified science, general knowledge, synthesis, and the integration of knowledge."[32] Disciplinary boundaries, she tells us, divide fraternal paths of inquiry and also force individuals to develop specialized parts of themselves and ignore the "whole person,"[33] and the result is a society that is itself fragmented and must therefore be "restored to wholeness."[34] Klein deplores the combative vocabulary characteristic of both intra- and interdisciplinary discussions, which, she says, leads to "imperialistic claims"[35] and a sense of disciplines as "warring fortresses,"[36] and she laments the fact that our very "vocabulary—indeed our whole logic of classification—predisposes us to think in terms of disciplinarity."[37]

The suggestion is that if we could only change our vocabulary, no longer speak in a way that created hierarchies and divisions, we would not experience ourselves as locally situated, but as members of a vast interconnected community. The suggestion, in short, is that through interdisciplinarity, we can eliminate difference (ye shall see the interdisciplinary truth and it shall make you free). Difference, however, is not a remediable state; it is the bottom line fact of the human condition, the condition of being a finite creature, and therefore a creature whose perspective is not general (that would be a contradiction in terms), but partial (although that partiality can never be experienced as such and those who think it can be unwittingly reinstate the objective viewpoint they begin by repudiating). The "predisposition," as Klein puts it, to think in terms of selves and others, better and worse, mine and yours, right and wrong, to think, that is, in terms rooted in local experiences, is not one that could be altered unless we ourselves could be altered, could be turned from situated beings—viewing and constituting the world from an angle—into beings who were at once nowhere and everywhere. That is the implicit and sometimes explicit goal of the interdisciplinary program, and it is also the goal whether acknowledged or not, of a program that would tie all disciplinary work to a single universal task. The promise of interdisciplinarity, like the promise of a pragmatism that offers either expanded sympathy (Rorty) or bedrock reality (Posner) is the promise that we shall become as gods. Eve believed it. I don't.

V

In what is presumably an effort to do his part in bringing about a world marked by expanded sympathy and tolerance, Rorty invites everyone he knows to come in under the pragmatist umbrella. The invitation is declined by Ronald Dworkin who, for a moment, sounds almost like me when he declares that "there is nothing in [pragmatism] to accept";[38] but it would seem that the skepticism we both display toward pragmatism as a program is differently inflected. Dworkin means that there isn't enough of a program in pragmatism, whereas I have been saying that pragmatism, at least in the forms put forward by Rorty and Posner, is too much of a program. It is this difference between us, perhaps, that leads Dworkin to pick up the threads of our earlier quarrels.

Dworkin continues to be exercised by my effort to devalue theory in favor of what I call an "enriched notion of practice." It is his view that my account of practice renders it "flat and passive, robbed of the reflective, introspective, argumentative tone . . . essential to its character," and he asserts that I everywhere underestimate "the complexity of the internal structure of practices that people can quite naturally fall into."[39] Needless to say, I disagree, and indeed I would contend that it is Dworkin who underestimates the complexity of the internal structure of practice, and that it is this underestimation that leads him into a futile and unnecessary search for constraints that practice, properly understood, already contains. This is not to say (as Dworkin does) that practice, in at least some forms, is inseparable from theory, but that (1) competent practitioners operate within a strong understanding of what the practice they are engaged in is *for*, an understanding that generates without the addition of further reflection a sense of what is and is not appropriate, useful, or effective in particular situations; and that (2) such a sense is not theoretical in any interestingly meaningful way.

Let me clarify the point by focusing on a single sentence from Dworkin's important and influential essay, "Hard Cases": "If a judge accepts the settled practices of his legal system—if he accepts, that is, the autonomy provided by its distinct constitutive and regulative rules—then he must, according to the doctrine of political responsibility, accept some general political theory that justifies these practices."[40]

The first question to ask is, what would it mean for a judge *not* to accept the settled practices of the legal system? Would it mean that he or she would not recognize the authority of the Supreme Court or would refuse, as an appeals judge, to note or take into account the ruling of a lower court, or would decide a case without reference to any other case that had ever been decided, or would decide a case by opening, at random, to a page in the Bible or in *Hamlet* and taking direction from the first phrase that leapt to the eyes or would simply refuse to decide? The question is rhetorical in the technical sense that it enumerates kinds of action that would be *unthinkable* for anyone self-identified as a judge. I emphasize the word "unthinkable" to indicate that the issue is not one of considering options—to accept or not to accept, that is not the question—but of comporting oneself in ways inseparable from one's position in a field of institutional activity. If one asks, "what does a judge do?", among the answers are, "a judge thinks about cases by inserting them into a history of previous cases that turn on similar problems," or "a judge operates with a sense of his or her place in a hierarchical structure that is itself responsible to other hierarchical structures (a legislature, a Justice Department)," or "a judge is someone who, in the performance of his or her duty, consults certain specified materials which (at least in our tradition) do *not* include (except as occasional embellishments) the Scriptures or the plays of Shakespeare" or even more simply, "a judge is someone who decides."

It must be emphasized that these are not things a judge does in addition to, or in the way of a constraint on, his mere membership in a profession. They are the *content* of that membership; there is no special effort required to recall them, and no danger that a practitioner, short of amnesia, will forget them. Nor was there a moment when they were offered as practices for approval or disapproval. No one ever said to a law student or a young lawyer, "Now these are the ways we usually deal with contracts; do you think you can go along with them?" Rather, one learned about contracts—

about consideration, and breach, and remedies for breach—and as one learned these "facts" one learned too (not also) the appropriate, and indeed obligatory, routes by which outcomes might be produced. The normative aspect of the law is not placed on top of its nuts and bolts (like icing on a cake); it comes along with them and someone who is able to rehearse the points of doctrine in a branch of law knows not only what a practitioner does but what he or she is obliged to do by virtue of his or her professional competence.

In short, there is nothing volitional in the relationship between a professional and the practices in which he is settled and therefore no sense can be given to the notion of accepting or rejecting them. Nevertheless, could it not be said that in the course of internalizing the routines of practice, a prospective judge at the same time internalizes a justification of those routines, internalizes a theory? It depends, I think, on what one means by justification. If you ask a contract lawyer to "justify" his or her recourse to categories like "offer and acceptance," "mistake," "impossibility," "breach," etc., you are likely to be met with a blank look equivalent to the look you would receive from an accountant asked to justify a reliance on statistics or a carpenter asked to justify a reliance on hammers and nails. Simply to *be* a contract lawyer or an accountant or a carpenter is to rely, unreflectingly, on those tools. But if you asked a contract lawyer for a "deeper" justification of his routine deployment of certain vocabularies, you might in some cases, but not all, be told the standard story of "classical" contract doctrine, in which two autonomous agents bargain for an exchange of goods or services whose value they are free to specify in any way they like. Such an answer would be equivalent to a theory, and therefore those who were able to give it could fairly be said to have a theory of their practice.

This, however, does not mean that the theory would be necessary to, or generative of, that practice. A theoretical account produced on demand might be little more than the rote rehearsal of something learned years ago at law school and have no relationship at all to skills acquired in the interim. Self-identified theoreticians would then be practicing in ways that could not be accounted for by their theory and yet their practice would not be impaired by this lack of fit. Moreover, other lawyers and jurists who were either incapable of or uninterested in responding to the demand for theoretical justification might nevertheless be among the most skilled practitioners of a trade. A strong proponent of theory might reply (à la Chomsky) that those who are good at their job but cannot produce a theory of it are nevertheless operating on the basis of theoretical principles; they are just not skilled at articulating them, and if they were to become skilled, if they were able to discourse volubly about the presuppositions of their professional performances, they would be even better performers, approaching perhaps the ideal represented by Dworkin's Hercules. But this would be like saying that a cyclist who would be able to explain the physics of balance would be, by virtue of that ability, a better cyclist, or that a motorist who could calculate the relationship between automobile weight, velocity, and road conditions with respect to braking distances would be a safer or more skillful driver. It is not that the skills of explaining and calculating in these examples are not genuine or admirable; just that they are skills different from the skills necessary to the performances of which they might, on occasion, provide a "theoretical" account.

To be sure, in some practices, as Dworkin points out, the relationship between explanatory and performance skills is closer than in others. Judging is one such practice, for, as I put it in an earlier essay, "judging . . . includes as a part of its repertoire self-conscious reflection on itself."[41] This means that in the course of unfolding an opinion, a judge might well think it pertinent to invoke some legal "principle" such as "a court cannot police the bargains of competent private individuals" or, more abstractly, "the law must be color-blind with respect to the safeguarding of rights"; but at the moment of their invocation, such "principles" would be doing rhetorical, not theoretical, work, contributing to an argument rather than presiding over it. Such principles (and there are loads of them) form part of the arsenal available to a lawyer or judge—they are on some shelf in the storehouse of available arguments (the concept is, of course, Aristotelian)— and the skill of deploying them is the skill of knowing (it is knowledge on the wing) just when to pull them off the shelf and insert them in your discourse. Once inserted, they are just like the other items in the storehouse, pieces of verbal artillery whose effectiveness will be a function of the discursive moment; they do *not* stand in a relation of logical or philosophical priority to more humble weapons, although their force will depend to some extent on the reputation they have for being prior. Aristotle advised his students to lard their orations with maxims because they give a show of wisdom to a speech. Augustine told preachers to quote the Scriptures as often as possible, and his seventeenth-century pupil, the poet George Herbert, recommended expressions of piety because they "show holiness." Similarly, a legal master rhetorician might urge the regular invocation of principle as a way of convincing the appropriate audience of one's high seriousness. I should add that there is no question of insincerity here. The orator or preacher who cites a maxim or a precept will almost certainly believe in it; it is just that the maxim or precept will be only a part of the case he is building and as a part no more or less foundational than the other parts (statistics, precedents, empirical examples) that go to make up a coherent story.

In short, while theory or theory talk will often be a *component* in the performance of a practice, it will not in most cases be the driving force of the performance; rather it will appear in response to the demands of a moment seen strategically, a moment when the practitioner asks himself or herself, "What might I insert here that would give my argument more weight?" Of course many practices are devoid of such moments because the giving of reasons is not a feature of their performance. As Dworkin notes, referring to an earlier essay of mine, "Denny Martinez never filed an opinion,"[42] meaning, I take it, that Martinez's resistance to theory (as evidenced in a conversation with a sportswriter) can be explained by the fact that the activity he is engaged in is not a reflective one, as opposed say, to the activity of lawyering. The distinction is a real one and important, but it is a distinction not between practices informed by theory and practices innocent of theory, but between practices in relation to which the introduction of "theory-talk" will be useful and practices in relation to which the introduction of theory-talk would be regarded as odd and beside the point ("that kid thinks too much to be a good shooter").

In Dworkin's view this distinction marks a hierarchy in which the reflective practitioner is superior, even morally superior, to the practitioner who just goes about his business. Indeed, so committed is Dworkin to this valorization of the reflective temperament (his version of philosophy's age-old claim to be the master art underlying all the other arts) that he finds it even in places where he has just declared it absent: "Even in baseball . . . theory has more to do with practice than Fish acknowledges. The last player who hit .400, fifty years ago, was the greatest hitter of modern times, and he built a theory before every pitch."[43] Now I yield to no one in my admiration for Ted Williams who has been my idol since boyhood (perhaps Dworkin's too, since we are both natives of Providence, R.I., a Red Sox–happy town); for years I carried a picture of the Splendid Splinter in my wallet until it was so worn that it fell into pieces. But the fact that Williams was a student of hitting and wrote a much admired book on the subject does not mean either that his exploits are the product of his analyses or that players (like Babe Ruth or Mickey Mantle) less inclined to technical ruminations on their art were lesser performers. Williams himself testifies (indirectly) to the absence of a relationship between theory and practice when he reports the view of another "thinking man's" hitter, Ty Cobb. "Ty Cobb . . . used to say the direction of the stride depended on where the pitch was—inside pitch, you bail out a little; outside you'd move in toward the plate."[44] Of Cobb's analysis Williams says flatly, "This is wrong because it is impossible," and then goes on to explain that, given the distance between pitcher and batter, and the speed of the ball, "you the batter have already made your stride before you know where the ball will be or what it will be."[45] What this means is that Cobb, possessor of the highest lifetime batting average in baseball history (.367), did not understand—theoretically, that is—what he was nevertheless doing and doing better than anyone else who ever played the game, including Ted Williams (lifetime average, .344). Later in the book Williams calls Cobb the "smartest hitter of all,"[46] and we must assume that his intelligence, as exemplified in his day-to-day performance, had nothing to do with his theoretical skills, which were, as Williams has demonstrated, deficient.

What then was the content of the intelligence Williams so much admires if it was not theory? Perhaps we can infer an answer from Williams's account of his own "theoretical" practice. Early on he identifies the chief constituents of intelligent hitting as "thinking it out, learning the situations, knowing your opponent, and most important, knowing yourself."[47] To academic ears this last may have a Socratic ring and seem to gesture in the direction of a deep introspection; but Williams here means no more (or less) than knowing your own physical abilities, where your power lies, the quickness of your bat, the strength (or lack thereof) of your arms and legs. You must know yourself in the same sense that you must know the situations and your opponent; that is, you must be *alert* to the components of your task and avoid inattention and lack of concentration. The point becomes clear when Williams declares that "guessing or anticipating goes hand in hand with proper thinking."[48] A hitter who observes that "a pitcher is throwing fast balls and curves and only the fast balls are in the strike zone . . . would be silly to look for a curve."[49] The smart hitter, in short, *pays attention* and engages in the kind of thinking Williams images in this hypothetical scenario:

> First time up, out on a fast ball. Looking for the fast ball on the next time up. But you go out on a curve. Third time up, seventh inning, you say to yourself, "Well, he knows I'm a good fast ball hitter, and he got me out on a curve. I'm going to look for the curve, even though he got me out on a fast ball first time up."[50]

It's all there; knowing the situation (it's the seventh inning, and the end of the game is approaching), knowing your opponent (he's smart enough to be thinking this), knowing yourself (I'm a good fast ball hitter), and the skill is to pay attention to all these things at once. "Paying attention," however, is not a skill separable from an experienced player's "feel" for the game, nor is it a skill that is in any meaningful way theoretical (unless it is a matter of theory to attend watchfully to all of the variables in play when you are crossing at a busy intersection; for if it is, then theory is just a name for making your way through life and there's nothing much to say about it or claim for it); it is just the skill that attends being good at what you do.

Now it may be that when Dworkin speaks of the theoretical component of practice he is referring to nothing more exalted than the habits of being alert and paying attention, and if this is so then the difference between us is merely terminological and doesn't amount to very much at all. This is what he claims in the conclusion to the essay written for this volume. But somehow I don't think so, for in every formulation that seems to bring us closer together there is something that reopens a gap, as when he declares that a good judge "will naturally see that he must be, in Fish's terms, a theoretician as well as, and in virtue of, occupying his role as a participant."[51] What troubles me here is the slight equivocation (papered over by a strategically placed "and") between "as well as" and "in virtue of." The first suggests that being a theoretician is something in excess of the role of participant, while the second folds the theory *into* the role. But if the theory is folded into the role, as Williams's alertness to the pitcher and to previous moments in the games is folded into his performance as a batter, there seems to be little reason to call it theory, since it is simply the quality of not falling asleep on the job; and, on the other hand, if theory is used in a more exalted sense and refers to something that must be added to ("as well as") the role of participant, we are back in the realm of meta-commentary and high abstraction despite Dworkin's insistence that he really doesn't live there. A similar equivocation attends the assertion the "theory itself is second nature"[52]; if it is truly second nature, and just comes along with each and every territory, then it simply confuses matters to separate it out and give it an honorific name; and if it is second nature in the sense that it must perfect a first nature that is "flat and passive" without it, then theory is a special project undertaken only by heroic souls like Dworkin's Hercules. That finally is what is at stake in the debates between us, whether or not Dworkin has a project. The question of project also marks my disagreements (much less sharp) with Posner and Rorty, both of whom believe, despite their declared pragmatism, in some benefit to be derived for practice from the pursuit of theory or antitheory. Whatever differences separate Posner, Rorty, and Dworkin, they are alike in thinking that they have something to recommend, something that will make the game better. (Rorty's version of this, as we have seen, is the most tentative and least robust.) All I have to recommend is the game, which, since it doesn't need my recommendations, will proceed on its way undeterred and unimproved by anything I have to say.

14

Being Interdisciplinary Is So Very Hard to Do

I

Interdisciplinary has long been a familiar word in discussions of education and pedagogy, but recently it has acquired a new force and urgency, in part because as an agenda interdisciplinarity seems to flow naturally from the imperatives of left culturalist theory, that is, from deconstruction, Marxism, feminism, the radical version of neopragmatism, and the new historicism. Each of these movements, of course, should be distinguished from the others in many respects, but it is fair to say that they are alike all hostile to the current arrangement of things as represented by (1) the social structures by means of which the lines of political authority are maintained and (2) the institutional structures by means of which the various academic disciplines establish and extend their territorial claims. Often this hostility takes the form of antiprofessionalism, an indictment of the narrowly special interests that stake out a field of inquiry and then colonize it with a view toward nothing more than serving their own selfish ends.

In the antiprofessional diatribe, specialization stands for everything that is wrong with a practice that has lost its way, everything that is disappointing about an educational system that seems out of touch with the values it supposedly promotes. Of course the antiprofessionalist attack on specialization is by no means the exclusive property of the left: it has long been a staple of conservative jeremiads against the decline of culture in a world where all coherence is gone and the center has not held. Indeed, at times it is difficult to distinguish the two ends of the political spectrum on this

Originally printed in: Modern Language Association of America, *Profession* 89, 15–22. Reprinted with permission of the publisher.

question. When Russell Jacoby reports that intellectuals have moved out of the coffeehouses and into the faculty lounge and complains that by doing so they have abandoned their responsibility to the public—"as professional life thrives, public culture grows poorer and older" (8)—we might well be hearing the voice of Lynne Cheney contrasting the vigorous cultural life of the American mainstream to the increasingly narrow and jargon-ridden practices of the academy.

Yet, if both the left and the right can lay claim to an antiprofessionalism that regards with suspicion activities tied narrowly to disciplinary pressures, there is nevertheless a difference in the ways in which the antiprofessional stance is assumed. The difference is one of sophistication and complexity in the presentation: whereas the right tends to issue its call for a general, nonspecialized pedagogy in the same flag-waving mode that characterizes its celebration of the American family, the left urges its pedagogy in the context of a full-fledged epistemological argument, complete with a theory of the self, an analysis of the emergence and ontology of institutions, and a taxonomy of the various forms of pedagogical practice, from the frankly oppressive to the self-consciously liberating.

At the heart of that argument is the assumption that the lines currently demarcating one field of study from another are not natural but constructed by interested parties who have a stake in preserving the boundaries that sustain their claims to authority. The structure of the university and the curriculum is a political achievement that is always in the business of denying its origins in a repressive agenda. Knowledge is frozen in a form supportive of the status quo, and this ideological hardening of the arteries is abetted by a cognitive map in which disciplines are represented as distinct, autonomous, and Platonic. Once knowledge has been compartmentalized, the complaint continues, the energy of intellectuals is spent within the spaces provided by a superstructure that is never critically examined. Disciplinary ghettos contain the force of our actions and render them ineffectual on the world's larger stage. In Michael Ryan's words, the present disciplinary "divisions conceal the relationality" of supposedly independent enterprises and prevent us from seeing "that they are nothing 'in themselves' and that they constitute each other as mutually interdependent determinations or differentiations of a complex system of heterogeneous forces" (53–54). One who uncritically accepts the autonomy of his or her "home enterprise" and remains unaware of the system of forces that supports and is supported by that enterprise will never be able to address those forces and thereby take part in the alteration of that system.

This analysis of our situation implicitly includes an agenda for remedying it. One must first (and here Bruce Robbins is speaking) "affirm . . . that no institution is an island" and that while "exercising our profession, we simultaneously occupy overlapping and conflicting institutions'' (3). We must, that is, become sensitive to what Vincent Leitch calls "the 'made up' quality of knowledge" as our present institutional categories deliver it to us (53); and then, as S. P. Mohanty urges, "we must seek to suspend the process of this continuity, to question the self-evidence of meanings by invoking the radical—but determining—alterities that disrupt our . . . discourses of knowledge" (155). That is to say, and Jim Merod says it, we must learn "to situate texts in the field of institutional forces in which they are historically conceived" rather than rest content with regarding them as the special isolated objects of an autonomous practice (92). Once we do this, the smooth coherences and seamless narratives that form the basis of our present knowledge will be disrupted; artificially constructed unities will fall apart; the totalizing discourse in which discrete and independent entities are put into their supposedly natural places by a supposedly neutral discursive logic will be replaced by discontinuity, disorientation, decentering, transformation, fluidity, relation, process. Moreover, as the bonds of discourse are loosened, the mind will be freed from the constraints those bonds imposed, and the person thus freed will move toward "the full development of all human faculties" (Ryan 49), leaving behind the narrowness of vision that befalls those who remain tied to the confining perspectives of the ideologically frozen divisions of intellectual labor.

Of course the program requires some mechanism of implementation, and it is here that we arrive at interdisciplinary study by a route Jeffrey Sammons charts for us. Sammons points out that American education derives from a German model whose goal is "the cultural formation of the self so that it might reach the fullness of its potentialities" (14). In the context of that model it is the task of particular disciplines to contribute to that fullness and avoid the temptation to become ends in themselves, to become nothing more than training schools for entrance into a trade or profession. It always happens, however, that as soon as disciplines are fully established they come quickly to believe in the priority of their own concerns and turn from their larger mission to the training of professionals for whom those concerns are not only prior but exclusive. In short, the structure of the curriculum, or rather the very fact that it has a structure, works against its supposed end, and therefore something must be built into that structure to counter the tendency to produce nonresponsive spheres of self-contained complacency. By definition interdisciplinary studies do exactly that—refuse to respect the boundaries that disciplines want always to draw—and thus encourage a widening of perspectives that will make possible the fullness education is supposed to confer.

Although Sammons and Ryan share the word *full* as a component in their briefs for interdisciplinary thinking, they mean different things by it. Sammons's fullness is the fullness of the imagination. He writes, however critically, in the tradition of High Humanism, and that is why he can locate the potential for destabilizing activity in the humanities or, as he calls them, "the disciplines of the imagination," of a Coleridgean faculty that sees similarities and differences as constructed. Such a faculty, he claims, is inherently "subversive," and therefore the humanities are inherently subversive because they introduce "people to the inexhaustible alternative options of the imagination" (10). Persons so introduced would be "full" in the sense that their intelligences would not be captured by any one point of view but would, rather, be engaged in exploring points of view other than those authorized by current orthodoxies.

To someone like Ryan (or Robbins or Merod) all this would seem suspiciously familiar, especially when Sammons approvingly quotes Robert Scholes's assertion that "poetic texts are designed to discomfort us" (Scholes 43). Left ears will hear this as just another version of a hoary and suspect *disciplinary* claim, which, instead of decentering the curriculum or exposing its affiliations with political and economic forces from which it thinks itself separate, gives it an even firmer center in the humanities and then has the nerve to call that center "subversive." Insofar as such an agenda envisions a fullness, it is merely a fullness of the reflective intellect, an intellect detached perhaps from any of the particular interests that vie for territory in the academy but an intellect nevertheless confined in its operations—however full—to that same academy. What Ryan, Merod, and Robbins want is a fullness of *engagement,* a mind and person that refuses to segregate its activities, to think, for example, that literary study is one thing, participation in the national political process quite another. They would say that the point is not to determine which of the presently situated fields of study is the truly subversive one but to call into question the entrenched articulations within which the divisions between fields (and knowledge) emerge, and *thereby* (or so the claim goes) to subvert the larger social articulation within which the articulations of the academy are rendered intelligible and seemingly inevitable. In short, for these more radical voices, interdisciplinary study is more than a device for prodding students to cross boundaries they would otherwise timidly respect; it is an assault on those boundaries and on the entire edifice of hierarchy and power they reflect and sustain. If you begin by transgressing the boundaries, say, between literature and economics as academic fields of study, you are halfway to transgressing the boundaries between the academy and its supposed "outside," and you are thus brought to the realization that the outside/inside distinction is itself a constructed one whose effect is to confine academic labor to a neutral zone of intellectual/professional play—a realization that then sends you back to operate in that zone in a way that is subversive not only of its autonomy but of the forces that have established that autonomy for their own unacknowledged purposes. In this vision, interdisciplinary study leads not simply to a revolution in the structure of the curriculum but to *revolution tout court*. In the classical liberal paradigm, interdisciplinary studies seek only to transform the academy while maintaining the wall between it and the larger field of social action; and thus, as Ryan points out, "the radical position of pedagogic activism for the sake of an alternative social construction seem[s] a deviation" (49), an intrusion of the political into precincts it is forbidden to enter. Radical interdisciplinarity begins with the assumption that the political is always and already inside those precincts and that the line separating them from the arena of social agitation is itself politically drawn and must be erased if action within the academy is to be continuous with the larger struggle against exploitation and oppression.

II

It is a stirring vision, but it is finally at odds with the epistemology that often accompanies it. That epistemology is either deconstructive or psycho-analytic or a combination of the two, and in any of its forms its thesis is that "meanings do not exist as such [that is, as freestanding and "natural" entities] but are produced" (Mohanty 15). What they are produced *by* is a system of articulation from which we as either speakers or hearers cannot distance ourselves, because we are situated within it. Since that system (call it *différance* or the unconscious) is the unarticulated ground within which specification occurs, "it" cannot be specified and always exceeds—remains after, escapes—the specifications it enables. What this means, as Shoshana Felman observes, is that knowledge is "a knowledge which does not know what it knows, and is thus *not in possession of itself*" (40). That is, as knowledge it cannot grasp, or name the grounds of, its possibility, and whenever it thinks to have done so, those grounds are elsewhere than they seem to be; they are once again under the would-be knower's feet. It is to this point that Felman quotes Lacan—"the elements do not answer in the place where they are interrogated, or more exactly, as soon as they are interrogated somewhere, it is impossible to grasp them in their totality" (Felman 29)—and she might just as well have invoked Derrida as he explains why *différance*, although it makes presentation possible, can never itself be presented: "Reserving itself, not exposing itself, . . . it exceeds the order of truth . . . , but without dissimulating itself as something, as a mysterious being. . . . In every exposition it would be exposed to disappearing as disappearance. It would risk appearing: disappearing" (122). Or again, "the trace is never as it is in the presentation of itself. It erases itself in presenting itself, muffles itself in resonating . . .'' (Derrida 133). That is to say, the truth one would know has always receded behind the formulations it makes possible, and therefore those formulations are always ignorant of themselves and incomplete. Indeed, ignorance, the forgetting of the enabling conditions of knowledge (conditions that cannot themselves be known), is constitutive of knowledge itself. Thus, Felman declares, ''human knowledge is by definition that which is untotalizable, that which rules out any possibility of totalizing what it knows or of eradicating its own ignorance'' (29). It follows then that if ignorance is the necessary content of knowledge as presented at any particular moment, knowledge is not something that should be preserved or allowed to settle, since in whatever form it appears it will always be excluding more than it reveals; and indeed it is only by virtue of the exclusions it cannot acknowledge that it acquires a (suspect) shape.

Not surprisingly, the pedagogy demanded by this insight is a pedagogy of antiknowledge, of the refusal of knowledge in favor of that which it occludes. There must be a new way of teaching, one that ''does not just reflect itself, but turns back on itself so as to *subvert itself* and truly *teaches* only insofar as it subverts itself'' (Felman 39), a pedagogic style that in Lacan's words is ''the ironic style of calling into question the very foundations of the discipline'' (qtd. in Felman 39). Lacan is referring to the discipline of psychoanalysis but, vigorously pursued, the strategy calls into question the foundations of all disciplines, since those foundations will in every case be made of ignorance and therefore must be first exposed and then removed. The way to do this is to work against the apparent coherences that support and are supported by ignorance and to engage in a kind of guerilla warfare in which the decorums disciplines ask us to observe are systematically violated, so that we proceed, ''not through linear progression, but through breakthroughs, leaps, discontinuities'' (qtd. in Felman 27). Rather than teach meanings, we must undo the meanings offered to us by hidden ideological agendas, poking holes in the discursive fabric those agendas weave, replacing the narcotic satisfactions of easy intelligibility with the disruptive dis-ease of relentless critique. The call to battle is sounded in summary but representative form by Vincent Leitch in the name of Roland Barthes:

> . . . uproot the frozen text; break down stereotypes and opinions; suspend or
> baffle the violence and authority of language; pacify or lighten oppressive pater-
> nal powers; disorient the Law; let classroom discourse float, fragment, digress.
> (51)

And then what? Does the pedagogy of antiknowledge hold out the hope of anything beyond its repeated unsettling of whatever claims us in the name of established knowledge? It is in the answer to this question that the tension between the political and the epistemological arguments for inter-disciplinary studies comes to the surface. In the political argument, which sees us currently inhibited in our actions by lines of demarcation we did not draw, the demonstration that those lines and the distinctions they sub-tend are not natural but historical will remove their power and free us from their constraints. "The classroom," says Jeffrey Peck,

> then becomes a productive rather than a reproductive environment. . . . In the spirit of critical reflection meanings and values of traditional pedagogy can be scrutinized. . . . The intersubjectivity of meaning can be exposed, and educa-tional institutions, the classroom, the discipline, and the university can be seen to construct and condition knowledge. In this way literary study, as the study of textuality, . . . reveals the epistemological structures that organize how we know, how our knowledge gets transmitted and accepted, and why and how students receive it. (51)

To this heady prospect, which will end, Peck predicts, with students be-coming better readers "of their own lives, as well as of texts" (53), the epistemological argument poses a dampening question—from what vantage point will the "structures that organize how we know" be revealed?—and the answer can only be, from the vantage point of a structure that is at the moment *un*revealed because it occupies the position formerly occupied by the structures it now enables us to analyze. The strategy of "making visible what was hidden" can only be pursued within forms of thought that are themselves hidden; the bringing to light of what Edward Said calls "the network of agencies that limit, select, shape, and maintain" meaning re-quires the dark background of a network that cannot be seen because it is within it that seeing occurs (34–35). Partiality and parochialism are not eliminated or even diminished by the exposure of their operation, merely relocated. The blurring of existing authoritative disciplinary lines and boundaries will only create new lines and new authorities; the interdiscipli-nary impulse finally does not liberate us from the narrow confines of aca-demic ghettos to something more capacious; it merely redomiciles us in enclosures that do not advertise themselves as such.

In short, if we take seriously the epistemological argument in the context of which the gospel of interdisciplinary study is so often preached, we will come to the conclusion that being interdisciplinary—breaking out of the prison houses of our various specialties to the open range first of a general human knowledge and then of the employment of that knowledge in the great struggles of social and political life—is not a possible human achieve-ment. Being interdisciplinary is more than hard to do; it is impossible to do. The epistemological argument deprives the political argument of any possible force, because it leaves no room for a revolutionary project. Or, rather, it leaves us with projects that look disconcertingly like the disciplinary projects we are trying to escape. Either (as some contributors to a recent piece in the *Chronicle of Higher Education* complain) the announcement of an interdisciplinary program inaugurates the effort of some discipline to annex the territory of another, or "interdisciplinary thought" is the name (whether acknowledged or not) of a new discipline, that is, of a branch of academic study that takes as its subject the history and constitution of disciplines. Either the vaunted "blurring of genres" (Clifford Geertz's now famous phrase) means no more than that the property lines have been redrawn—so that, for example, Freud and Nietzsche have migrated respectively from psychology and philosophy to English and comparative literature—or the genres have been blurred only in the sense of having been reconfigured by the addition of a new one, of an emerging field populated by still another kind of mandarin, the "specialist in contextual relations" (Alton Becker; qtd. in Geertz 521).

III

Needless to say, this is a conclusion many are loath to reach, but in order to avoid it, the proponents of radical pedagogy must negotiate an impasse produced by one of their own first principles, the unavailability of a perspective that is not culturally determined. Since a perspective from which the determinations of culture can be surveyed is a requirement of the radical project, one must ask how that project can even get started. In general, two answers have been given to this question. The first is to move from Robbins's observation quoted above, that "while exercising our profession, we simultaneously occupy overlapping and conflicting institutions," to the critical practice of allowing the claims made on us by one institution to stand in a relation of challenge to the claims made on us by another. As Samuel Weber puts it, "in interpreting a literary text, an interpreter will not necessarily be limited to confronting those interpretations previously certified as inhabiting the discipline of literary studies"; rather, "he may also invoke interpretations emanating from other regions (philosophy, psychoanalysis, etc.) and these in turn may well challenge the unifying assumptions of the discipline of literary studies in America" (38). That is to say, one's practice within a discipline can be characterized by invocations of and frequent references to the achievements, dicta, emphases, and requirements of other disciplines.

This is certainly true (my own practice, like yours, has often been answerable to such a description), but the question is, does the practice of importing into one's practice the machinery of other practices operate to relax the constraints of one's practice? And the answer I would give is no, because the imported product will always have the form of its appropriation rather than the form it exhibits "at home"; therefore at the very moment of its introduction, it will already be marked by the discourse it supposedly "opens." When something is brought into a practice, it is brought in in terms the practice recognizes; the practice cannot "say" the Other but can only say itself, even when it is in the act of modifying itself by incorporating material hitherto alien to it. As Peter Stearns says of history (and it could be said of any discipline), "What has happened is that social historians have borrowed topics, concepts and vocabulary . . . but they have then cast them in an essentially historical frame," and he adds, "This is something . . . more modest than a 'blurring of genres' " (qtd. in Winkler 14). Just so, and it is hard to see how it could be otherwise: terms and distinctions could arrive intact in the passage from one discipline to another only if they had some form independent of the discipline in whose practices they first became visible; but, in our brave new textualist-historicist world, terms and distinctions are no less socially constructed than anything else, and therefore the shape they appear in will always be relative to the socially constructed activity that has received them and made them its own.

Moreover, if materials, concepts, and vocabularies take on the coloring of the enterprise that houses them, so do practitioners, and that is why the second strategy by which pedagogy will supposedly transcend the disciplinary site of its activities fails. That strategy is a strategy of self-consciousness, and it requires us, while performing within a discipline, to keep at least one eye on the larger conditions that make the performance possible. (This is the implication of Robbins's subtitle: "Toward Productively Divided Loyalties.") While some agents confine themselves to the horizons of a particular profession, others situate themselves in the wider horizons of a general cultural space and therefore manage to be at once committed and not committed to the labors they perform. It is the latter group that keeps faith with a higher vision by not forgetting "the forces and factors" that underlie and give point to local urgencies (Weber 37). They remain aware of "the reader's and writer's immersion in a network of social forces that both grant and limit the possibility of intellectual authority" (Merod 93); and unlike their less enlightened brethren they resist the tendency of any "regime of truth" to deny its "constitutive dependence on what it excludes, dethrones, and replaces" (Weber 38). That is, they contrive to practice a particular craft without buying into the claims of that craft to be self-justifying and autonomous and without allowing the perspective of that craft to eclipse the other perspectives that would come into view were the craft's demands sufficiently relaxed.

The question is, as it was before, is this a possible mode of action? Again the answer is no, and for reasons that will become clear if we rephrase the question: can you simultaneously operate within a practice and be self-consciously in touch with the conditions that enable it? The answer could be yes only if you could achieve a reflective distance from those conditions while still engaging in that practice; but once the conditions enabling a practice become the object of analytic attention (against the background of still other conditions that are themselves unavailable to conscious inspection), you are engaging in another practice (the practice of reflecting on the conditions of a practice you are not now practicing), and the practice you began to examine has been left behind, at least as something you are doing as opposed to something you are studying. Once you turn, for example, from actually performing literary criticism to examining the "network of forces and factors" that underlie the performance, literary criticism is no longer what you are performing.

The point of course is tautological, and it would seem unnecessary to make it, except that in recent years it has been obscured by an illegitimate inference that has been drawn from a legitimate thesis. The thesis is the one we began with: disciplines are not natural kinds; they emerge in the wake of a political construction of the field of knowledge. The illegitimate inference is that since disciplinary boundaries are constructed and revisable, they are not real. But of course they are as real as anything else in a world in which *everything* is constructed (the world posited by those who make this argument); even though the lines demarcating one discipline from another can in time blur and become rearranged, until that happens the arrangements now in force will produce differences felt strongly by all those who live within them. Although it is true that disciplines have no essential core (another way of saying that they are not natural kinds), the identity conferred on them by a relational structure—a structure in which everything is known by what it is not—constitutes (however temporarily) a core that does all the work an essentialist might desire, including the work of telling community members what is and is not an instance of the practice it centers. Someone who says, as I have done in the previous paragraph, "that's not literary criticism" has said something that has a basis in fact, even if that fact itself—the fact of the present shape of a diacritically constituted discipline—is one undergoing continual modification and transformation. Again the lesson is only apparently paradoxical: because the core of the discipline is a historical achievement, it is capable of alteration, but as an achievement, it exerts, if only for a time, a force that cannot be ignored or wished away.

This does not mean that a worker in a discipline knows its core in the sense of being able to hold its differential (nonpositive) identity in mind. Indeed, in order to function in the discipline (as opposed to being a student of its formation), the fragility of that identity is something the worker cannot know or at least must always forget when entering its precincts.[1] The mark of that forgetting is the unintelligibility to practitioners of questions one might put from the outside, questions like (for teaching) "why is it that you want your students to learn?" or (for criminal law) "why should we be interested in the issue of responsibility at all?" or (for history) "why would anyone want to know what happened in the past?" You can't be seriously asking these questions and still be a member of those communities, because to conceive of yourself (a phrase literally intended) as a member is to have forgotten that those are questions you can seriously ask. This is the forgetting that Weber, Robbins, Merod, and others excoriate, but it is also the forgetting that is necessary if action of a particular kind is ever to occur. Denying and forgetting are not reformable errors but the very grounds of cognition and assertion. If one were to remember everything and deny nothing, assertion, directed movement, politics itself would have no possible shape.

Some of those who find magic in the word *interdisciplinary* come very close to making this point but shy away from it at the last moment. Richard Terdiman observes, correctly I think, that while "we attend to the content of our instruction, we are fundamentally, but imperceptibly, molded by its form"—that is, by the disciplinary structures within which the instruction occurs—and that therefore "the ideological representation of the world is involuntarily naturalized even through critique of its specific detail" (221). "Are our ways of teaching students to ask *some* questions," asks Barbara Johnson, "always correlative with our ways of teaching them *not to ask*—indeed to be unconscious of—others?" ("Teaching" 173). The answer is yes, and because the answer is yes our pedagogical imperative, no matter how radical its stance, will always turn out, as Terdiman observes, "to have sources and serve interests other than those we thought" (222). In the act of producing this insight, Terdiman terms it "unhappy." But why? All it means is that we will never achieve the full self-consciousness that would allow us at once to inhabit and survey reflectively our categories of thought, but that incapacity only affects our ability to be gods; and were we indeed to become gods, no longer tethered to the local places within which crises and troubles emerge, we would not feel the urgencies that impel us forward. The fact that we do feel these urgencies and are moved by that feeling to act depends on the very limitations and blindnesses Terdiman and company deplore. It is only because we cannot achieve an "authentic critique" (Terdiman 223)—a critique free of any political or conceptual entanglements—that the critiques we do achieve have force, even if it is the nature of things for the force of those critiques to be as vulnerable and transient as the conditions that give them form.

The impossibility of authentic critique is the impossibility of the interdisciplinary project, at least insofar as that project holds out the hope of releasing cognition from the fetters of thought and enlarging the minds of those who engage in it. The obvious response to this conclusion is to point out that interdisciplinary studies are all around us. What is it that all these people are doing? The answer has already been given; either they are engaging in straightforwardly disciplinary tasks that require for their completion information and techniques on loan from other disciplines, or they are working within a particular discipline at a moment when it is expanding into territories hitherto marked as belonging to someone else—participating, that is, in the annexation by English departments of philosophy, psychoanalysis, anthropology, social history, and now, legal theory; or they are in the process of establishing a new discipline, one that takes as its task the analysis of disciplines, the charting of their history and of their ambitions. Typically the members of this new discipline will represent themselves as antidisciplinary, that is, as interdisciplinary, but in fact, as Daniel Schön points out, they will constitute a "new breed" of "counterprofessionals/ experts" (340). Nor is there anything necessarily reprehensible about these activities. Depending on one's own interests and sense of what the situation requires, the imperial ambitions of a particular discipline may be just what the doctors ordered; and it may equally well be the case that, from a certain point of view, the traditional disciplines have played themselves out and it is time to fashion a new one. For my own part I subscribe to both these views, and therefore I find the imperialistic success of literary studies heartening and the emergence of cultural studies as a field of its own exhilarating. It is just that my pleasure at these developments has nothing to do with the larger claims—claims of liberation, freedom, openness—often made for them. The American mind, like any other, will always be closed, and the only question is whether we find the form of closure it currently assumes answerable to our present urgencies.

15

The Young and the Restless

(This essay appeared at the conclusion of The New Historicism, *a volume in which a group of well-known scholars and critics present the case for and against the movement named by Stephen Greenblatt.)*

As a privileged first reader of these essays I want to comment on the many and varied pleasures they provide. Whatever the New Historicism is or isn't, the energies mounted on its behalf or in opposition to its (supposed) agenda are impressive and galvanizing. In a brief afterword, as this is intended to be, I cannot do justice to the arguments and demonstrations of more than twenty pieces, but some things, while perhaps obvious, should at least be noted. The footnotes to some of the essays (see particularly Newton, Klancher, Pecora, Montrose, Marcus, and Fineman) are alone worth the price of admission; they illustrate even more than the essays themselves the richness and diversity of concerns that cluster around the questions raised by the banner of a New Historicism. One is grateful also for the glimpse into new (and at least for this literary scholar) uncharted territories—the politics of modern Indonesia (Pecora), the political emplotment of Latin American narrative (Franco), the fortunes of feminism in the First World War (Marcus), the history of art's efforts to be historical (Bann). These, however, are pleasures along the way, and even when they are provided, they are more often than not ancillary to whatever pleasure is to be derived from polemical debate, which is to say, from theory. For the most part (and this is a distinction to which I shall return) these essays are not doing New Historicism, but talking about doing New Historicism, about the claims

This essay was originally published in *The New Historicism*, edited by H. Aram Veesner (New York and London: Routledge, 1989). Reprinted with permission of the publisher.

made in its names and the problems those claims give rise to; and ungenerous though it may be, those problems will be the focus of my discussion.

The chief problem is both enacted and commented on in more than a few essays: it is the problem of reconciling the assertion of "wall to wall" textuality—the denial that the writing of history could find its foundation in a substratum of unmediated fact—with the desire to say something specific and normative. How is it (the words are Newton's) that one can "recognize the provisionality and multiplicity of local knowledge" and yet "maintain that it is possible to give truer accounts of a 'real' world"?* On what basis would such a claim be made if one has just been arguing that all claims are radically contingent and therefore vulnerable to a deconstructive analysis of the assumptions on which they rest, assumptions that must be suppressed if the illusion of objectivity and veracity is to be maintained? One can see this "dilemma" with Brook Thomas as a tension between the frankly political agenda of much New Historicist work and the poststructuralist polemic which often introduces and frames that same work. Thomas notes the tendency of many New Historicists to insist that any representation "is structurally dependent on misrepresentation" (184), on exclusions and forgettings that render it suspect, and wonders how their own representations can be proffered in the face of an insight so corrosive: "If all acts of representation are structurally dependent on misrepresentation, these new histories inevitably create their own canons and exclusions" (185).[1] Thomas's chief exhibit is Jane Tompkins who, he says, switches from asking the metacritical question, "Is there a text in this classroom" to asking the political and normative question, "What text should we have in the classroom?" and thereby "abandons her up-to-date poststructuralist pose and returns to old-fashioned assumptions about literature and historical analysis" (185). It would seem, he concludes, that "the very poststructuralist assumptions that help to attack past histories seem necessarily forgotten in efforts to create new ones" (186). We shall return to the idea of "forgetting" which appears more than once in this collection, and I shall suggest that as an action of the mind it is less culpable than Thomas seems to suggest; but for the time being, I shall let his formulation stand since it clearly sets out a problematic to which many in this volume are responding.

One response to this problematic is to refuse it by denying either of its poles. Thus Elizabeth Fox-Genovese, Vincent Pecora, Jane Marcus, and (less vehemently) Jon Klancher simply reject the poststructuralist textualization of history and insist on a material reality in relation to which texts are secondary. For these authors the "dance" of New Historicism substi-

*The quoted passages appear in a longer version: Judith Newton, "History as Usual," *Cultural Critique 9* (Spring 1988), 98.

tutes the ingenuity and cobbled-up learning of the critic for the "history" he or she supposedly serves: "When New Historicism plays with history to enhance the text," objects Marcus, "its enhancement is like the coloring of old movies for present consumption. . . . To learn political lessons from the past we need to have it in black and white" (133). Her effort is to remove from the history of literary women in the First World War the coloring imposed by Sandra Gilbert; she wants, she says, to do "justice to women's history" (144). In a similar vein Pecora contrasts "the kinds of symbolic significance Geertzian anthropology creates" (264) in its so called "thick descriptions" of Indonesian life to the "complex and polymorphous internecine struggles of the Indonesian people for self-determination" and finds that in his analyses Geertz reduces historical experience to "a well-known Western myth" (262), to "anthropological abstraction in spite of [his] claim to greater specificity" (266). "What, for many historians, would be . . . 'basic' categories such as material want and material struggle . . . lose their privileged position" in Geertzian New Historicism and become, like everything else, "merely culturally constructed sign systems" (244). New Historicist accounts, Pecora concludes, "are theoretically and anthropologically given and not historically determined" (269).

Fox-Genovese is even more blunt. History, she declares, is not simply "a body of texts and a strategy of reading or interpreting them"; rather, "history must also be recognized as what did happen in the past—of the social relations and, yes, 'events' of which our records offer only imperfect clues." It may be possible "to classify price series or . . . hog weights . . . as texts—possible, but ultimately useful only as an abstraction that flattens historically and theoretically significant distinctions" (216). "Distinctions" is a key word in this sentence and in many of the essays, and indeed it names a place of emphasis for both New Historicists and their materialist critics, finally reducing the difference between them in a shared devotion to difference. I shall return to this point, but for the moment I want simply to note the materialist position and to observe that its large vulnerability is one of the subjects of Hayden White's essay. While the materialist critics of New Historicism criticize the ideological program of its practitioners and declare that program to be ahistorical, the grounds for this condemnation "are themselves functions of the ideological positions of these critics" (296). The implicit claim of the materialists to be more immediately in touch with the particulars of history cannot be maintained, because all accounts of the past (and, I might add, of the present) come to us through "some kind of natural or technical language" (297) and that language must itself proceed from some ideological vision. "Every approach to the study of history presupposes some model for construing its object of study"; and it is from the perspective of that model, whatever it is, that one distinguishes "between what is 'historical' and what is not" (296). What this means is that "the conflict between the New Historicists and their critics" is not a conflict between textualists and true historians, but "between different theories of textuality" (297). Thus everyone's history is textual—"there is no such thing as a specifically historical approach to the study of history" (302)—and while one can always lodge objections to the histories offered by one's opponents, one cannot (at least legitimately) label them as nonhistorical. In the words of Lynn Hunt, herself a respected historian, "there is no such thing as history in the sense of a referential ground of knowledge."[2]

Of course, such remarks return us to the dilemma with which we began and invite the familiar question, "but if you think *that* about history, how can you, without contradictions, make historical assertions?" How, in other words, can you theorize your own position in a way that escapes the critique you want to make of those who have been historians before you? It would seem that one must either give up the textualist thesis as the materialists urge (which leaves them open to the charge of being positivist at a time when it is a capital offense to be such) or stand ready to be accused of the sin of contradiction. Some New Historicists outflank this accusation by making it first, and then confessing to it with an unseemly eagerness. In this way they transform what would be embarrassing if it were pointed out by another into a sign of honesty and methodological self-consciousness.

In this mode Louis Montrose may be thought a virtuoso. His now familiar formula, "the textuality of history, the historicity of texts," is a succinct expression of the (supposed) problem. By the textuality of history Montrose means "that we can have no access to a full and authentic past, a lived material existence, unmediated by the surviving textual traces of the society in question, . . . traces . . . that . . . are themselves subject to subsequent textual mediations when they are construed as the 'documents' upon which historians ground their own texts, called 'histories' " (20). By the historicity of texts, Montrose means "the cultural specificity, the social embedment, of all modes of writing" (20). The problem is to get from the quotation marks put around "documents" and "histories" to the specificities, or more pointedly, to the strong assertion of the specificities that have now, it would seem, dissolved into a fluid and protean textuality. Once you have negotiated the shift, as Montrose puts it later, from "history to histories," that is, to multiple stories and constructions no one of which can claim privilege, how do you get back?

At one point Montrose seems to say "no problem," as he simply declares the passage easy. "We may simultaneously acknowledge the theoretical indeterminacy of the signifying process and the historical specificity of discursive practices" (23). More often, however, he allows the dilemma to play itself out as he tacks back and forth between the usual claims for the special powers of the New Historicist methodology and the admission that his own epistemology seems to leave no room for those special powers. The climax of this drama late in his essay finds Montrose bringing these two directions of his argument together in a moment of high pathos. In one sentence he asserts that by foregrounding "issues such as politics and gender" in his readings of Spenser and Shakespeare, he is participating both in the "re-invention of Elizabethan culture" and in the "re-formation" of the constraints now operating in our own; and in the next, he acknowledges that his work is "also a vehicle for my partly unconscious and partly calculating negotiation of disciplinary, institutional and societal demands" and that therefore "his pursuit of knowledge and virtue is necessarily impure" (30). The conclusion, which seems inescapable, is that you cannot "escape from ideology," but this insight is itself immediately converted into an escape when the act of having achieved it is said to have endowed the agent (in this case Montrose) with a special consciousness of the conditions within which he lives:

> However, the very process of subjectively *living* the confrontations or contradictions within or among ideologies makes it possible to experience facets of our own subjection at shifting internal distances—to read as in a refracted light, one fragment of our own ideological inscription by means of another. A reflexive knowledge so partial and unstable may, nevertheless, provide subjects with a means of empowerment as agents. (30)

Partiality and instability—the very impediments to achieving any distance from our situation—become the way to that distance when one becomes reflexively aware of them. The questions one might ask of this reflexivity—what is its content, where does it come from?—are never asked. Montrose is content to solve his dilemma by producing (but not recognizing) another version of it in the claim that while he, like all the rest of us, is embedded and impure, he and some of his friends *know* it, and thus gain a perspective on their impurity which mitigates it.

Elsewhere I have named this move antifoundationalist-theory-hope and declared it illegitimate,[3] but here I am interested in it largely as one response to the dilemma New Historicists and materialists negotiate in different (and sometimes opposing) ways. What I want to say in the rest of this essay is that it is a false dilemma ("no problem" is the right answer) that is generated by the conflation and confusion of two different questions:

1. Can you at once assert the textuality of history and make specific and positive historical arguments?

2. Can you make specific and positive historical arguments that follow from—have the form they do as a consequence of—the assertion that history is textual?

The answer to the first question is "yes," and yes without contradiction (whether it is a contradiction one castigates or wallows in) because the two actions—asserting the textuality of history and making specific historical argument—have nothing to do with one another. They are actions in different practices, moves in different games. The first is an action in the practice of producing general (i.e. metacritical) accounts of history, the practice of answering such questions as "where does historical knowledge come from?" or "what is the nature of historical fact?" The second is an action in the practice of writing historical accounts (as opposed to writing an account of how historical accounts get written), the practice of answering questions such as "what happened" or "what is the significance of this event?" If you are asked a question like "what happened" and you answer "the determination of what happened will always be a function of the ideological vision of the observer; there are no unmediated historical perceptions," you will have answered a question from one practice in the terms of another and your interlocutor will be justifiably annoyed. But isn't it the case, one might object, that the two are intimately related, that you will answer the question "what happened" differently if you believe that events are constructed rather than found? The answer to that question is no. The belief that facts are constructed is a *general* one and is not held with reference to any facts in particular; particular facts are firm or in question insofar as the perspective (of some enterprise or discipline or area of inquiry) within which they emerge is firmly in place, settled; and should that perspective be dislodged (always a possibility) the result will not be an indeterminacy of fact, but a new shape of factual firmness underwritten by a newly, if temporarily, settled perspective. No matter how strongly I believe in the constructed nature of fact, the facts that are perspicuous for me within constructions not presently under challenge (and there must always be some for perception even to occur) will remain so. The conviction of the textuality of fact is logically independent of the firmness with which any particular fact is experienced.

I would not be read as flatly denying a relationship between general convictions and the way facts are experienced. If one is convinced of the truth of, say, Marxism or psychoanalysis, that conviction might well have the effect of producing one's sense of what the facts in a particular case are[4]; but a conviction that all facts rest finally on shifting or provisional grounds will not produce shifting and provisional facts because the grounds on which facts rest are themselves particular, having to do with traditions of inquiry,

divisions of labor among the disciplines, acknowledged and unacknowledged assumptions (about what is valuable, pertinent, weighty). Of course,
these grounds are open to challenge and disestablishment, but the challenge, in order to be effective, will have to be as particular as they are; the
work of challenging the grounds will have to be as particular as they are;
the work of challenging the grounds will not be done by the demonstration
(however persuasive) that they are generally challengeable. The conclusion
may seem paradoxical, but it is not: although a conviction strongly held can
affect perception and the experience of fact, the one exception to this generality is the conviction that all convictions are tentative and revisable. The
only context in which holding (or being held by) *that* conviction will alter
one's sense of fact is the context in which the fact in question is the nature
and status of conviction. In any other context the conviction of general
revisability—the conviction that things have been otherwise and could be
otherwise again—will be of no consequence whatsoever.

You may have noticed that by answering my first question—can you at
once profess a textual view of history and make strong historical assertions?—in the affirmative, I have at the same time answered my second
question—can you do history in a way that follows from your conviction
of history's textuality?—in the negative. The fact that the textualist views
of the New Historicists do not prevent them from making specific and polemic points means that those points will be made just as everyone else's
are—with reference to evidence marshalled in support of hypotheses that
will in the end be more or less convincing to a body of professional peers.
In short, in my argument New Historicists buy their freedom to do history
(as opposed to meta-accounts of it) at the expense of their claim to be doing
it—or anything else—differently. But of course that is a price the New
Historicists will not be willing to pay, for, like their materialist critics, they
have a great deal invested in being different, and, again like their materialist critics, the difference they would claim is the difference of being truly
sensitive to difference, that is, to the way in which orthodox historical narratives suppress the realities whose acknowledgement would unsettle and
deauthorize them. Whatever their disagreements on other matters, both New
Historicists and materialists are united in their conviction that current modes
of historiography are (wittingly or unwittingly) extensions of oppressive
social and political agendas, and this conviction brings with it an agenda:
what has been marginalized must be brought to the center; what has been
forgotten or left out must be brought to consciousness; what has been assumed must be exposed to the corrosive operation of critique. The materialists believe that the New Historicists default on this program by aestheti-
cizing it; the New Historicists respond by claiming, as Montrose does, that
by producing readings of Shakespeare that foreground the politics of gender

and the contestation of cultural constraints they participate in the redrawing of the lines of authority and power. Both camps are committed to cultural reformation, and both believe that cultural reformation can be effected by opening up the seams and fissures that a homogenized history attempts to deny.

The idea, although it is never stated in quite this way, is not to allow prevailing schemes of thought and organization to filter out what might be embarrassing to the interests they sustain. The byword or watchword is "complexity," a value that is always being slighted by the stories currently being told. Jane Marcus makes the point with the notion of "forgetting." She quotes Milan Kundera to the effect that "the struggle of man against power is the struggle of memory against forgetting. . . . [M]an has always harbored the desire to . . . change the past, to wipe out tracks, both his own and others" (133). Of course, what tracks are wiped out by are other tracks; for it is the nature of assertion to be selective and the path it lays down will always have the result of obscuring other paths one might have taken. This is what Richard Terdiman means when he observes that "classification always entails symbolic violence" (228). "It is not only that the class you take determines what class you get into. It is that *in classes we learn to class*" (227), that is, learn to draw lines, establish boundaries, set up hierarchies. Classification in all of its forms, says Terdiman (following Bourdieu), "forcefully excludes what it does not embrace" (227).

This is no doubt true, and indeed it is so true that one wonders whether there is anything one can do about it. "Symbolic violence" seems to be just a fancy phrase for what consciousness inevitably does in the act of seeing distinctions, whether they be social, political, moral, or whatever. However, several writers in this volume think that there is something to be done, something that will counter the violence of "received evaluations of fundamental social operators" (Terdiman, 229). Montrose counsels a "refusal to observe strict boundaries between 'literary' and other texts" (26); we should instead "render problematic the connections between literary and other discourses, the dialectic between the text and the world" (24). Newton would have us behave in a manner appropriate to a self that is now thought to be not stable, but "multiple, contradictory and in process," and she praises those feminist theorists who "have embraced multiplicity and provisionality."* These and other contributors urge us to the same course of action: We should reject the exclusionary discourses that presently delimit our perceptions and abrogate our freedom of action in favor of the more flexible and multidirectional mode of being that seems called for by

•

*The quoted passage appears in a longer version of Newton's essay in *Cultural Critique* 8 (Spring 1988), note 33, p. 99.

everything we have recently learned about the historicity of our situated-ness; we should classify less, remember more, refuse less, and be forever open in a manner befitting a creature always in process.

The trouble with this advice is that it is impossible to follow. While openness to revision and transformation may characterize a human history in which firmly drawn boundaries can be shown to have been repeatedly blurred and abandoned, openness to revision and transformation are not methodological programs any individual can determinedly and self-consciously enact. One cannot wake up in the morning and decide, "today I am going to be open,"—as opposed to deciding that today I am going to eat less or pay more attention to my children or get my finances in order. Someone who declares "today I am going to be more open" is in turn open to the question, "open with respect to what?" That is an answerable question and the answer can make sense, as in, "with respect to my habit of dismissing relevant student comments I am going to be open" or "where up to now I have refused to consider sex and race as a criteria for admission to this program, I am henceforth going to be open." But of course *that* kind of openness is nothing more (or less) than a resolution to be differently closed, to rearrange the categories and distinctions within which some actions seem to be desirable and others less so; whereas the openness (apparently) desired by several of the contributors to this volume is something very much more, a *general* faculty, a distinct muscle of the spirit or mind whose exercise leads not to an alternative plan of directed action but to a plan (if that is the word) to be directionless, to refuse direction, to resist the drawing of lines, to perform multiplicity and provisionality.

This is not a new goal. It is, as Catherine Gallagher points out, a familiar component of the left radical agenda, at least since the sixties. She calls it "indeterminate negativity" (interestingly Roberto Unger's name for it is "negative capability"[5]) and characterizes it correctly as the attempt "to live a radical culture" (41), an attempt in which Montrose thinks we partially succeed when we live—that is, consciously employ as part of our equip-ment—"the contradictions . . . of our own subjection" (30). What I am saying is that radical culture—understood as the culture of oppositional ac-tion, not opposition in particular contexts, but just *opposition* as a princi-ple—cannot be lived, and it cannot be lived for the same reason that the textualist view of history cannot yield an historical method: it demands from a wholly situated creature a mode of action or thought (or writing) that is free from the entanglements of situations and the lines of demarca-tion they declare; it demands that a consciousness that has shape only by virtue of the distinctions and boundary lines that are its content float free of those lines and boundaries and remain forever unsettled. The curious thing about this demand, especially curious as a component of something that calls itself the New Historicism, is that what it asks us to be is unhistorical, detached at some crucial level from the very structures of society and politics to which the New Historicism pledges allegiance. My point again is that the demand cannot be met; you cannot not forget; you cannot not exclude; you cannot refuse boundaries and distinctions; you cannot live the radical or indeterminate or provisional or textualist life.

What you can do is write sentences like this:

> English Romantic writings were staged within an unstable ensemble of older institutions in crisis (state and church) and emerging institutional events that pressured any act of cultural production—the marketplace and its industrializing, the new media and their reading audiences, the alternative institutions of radical dissent, shifting modes of social hierarchy. (Klancher, 80)

It is tempting (and it is a temptation many of my generation would feel) to mount a full-scale close reading of this sentence, but I will content myself with pointing out the efforts of the prose to keep itself from settling anywhere: English Romantic writings are barely mentioned before they are said to be "staged," i.e. not there for our empirical observation, but visible only against a set of background circumstances that must be the new object of our attention; but before those circumstances are enumerated they are declared to be "unstable" and also an "ensemble" (not one particular thing); and then this instability itself is said to be "in crisis," but in a crisis that is only "emerging" (not yet palpable); and this entire staged, unstable, emerging and "ensembling" crisis is said to put pressure on "any act of cultural production." At this point it looks, alarmingly, as if there is actually going to be a reference to such an act, but anything so specific quickly disappears under a list of the "institutional events" through which "it" is mediated; and finally, lest we carry away too precise a sense of those events (even from such large formulas as "the new media" and "radical dissent") they are given one more kaleidoscopic turn by the phrase "shifting modes." The question is, how long can one go on in *this* shifting mode? Not for many sentences and certainly not for entire essays. Klancher himself touches down to tell a quite linear (and fascinating) story of the institution of Romantic criticism, and indeed no one of the authors in this volume is able to sustain the indeterminacy of discourse that seems called for by the New Historicist creed.

The problem, if there is one, is illustrated by Jean Franco's criticism of allegorical or homogenizing stories. Repeatedly in her essay Franco opposes allegorical readings to enactments of contradiction and difference (see 208, 210). The last few pages of the piece resound with the praise of that which "def[ies] categorization," of the "uneasy and unfinished," of that which "generic boundaries cannot really contain," of the "unclassifiable," of "pluralism," of clashing styles, of the "kaleidoscope [that] constantly shifts to form different and unreconcilable patterns." "We need," she says in conclusion, "density of specification in order to understand the questions to which literary texts are an imaginary response" (212). But while she opposes "density of specification" to allegorical reading, in her own reading of Latin American novels this same insistence on density and the deferral of assertion becomes an allegory of its own. If density of specification is put forward as an end in itself, it is an allegory as totalizing as any, an allegory of discontinuities and overlappings; and if it is urged as a way to flesh out some positive polemical point, it is an allegory of the more familiar kind. Not that I am faulting Franco for falling into the trap of being discursive and linear; she could not do otherwise and still have as an aim (in her terms an allegorical aim) the *understanding*—the bringing into discursive comprehension—of anything. In the end you can't "defy categorization," you can only categorize in a different way. (Itself no small accomplishment.)

Where then does this leave us? Precisely where we have always been, making cases for the significance and shape of historical events with the help of whatever evidence appears to us to be relevant or weighty. The reasons that a piece of evidence will seem weighty or relevant will have to do with the way in which we are situated as historians and observers, that is, with what we *see* as evidence from whatever angle or perspective we inhabit. Of course, not everyone will see the same thing, and in the (certain) event of disputes, the disputing parties will point to their evidence and attempt to educe more. They will not brandish fancy accounts of how evidence comes to be evidence or invoke theories that declare all evidence suspect and ideological, because, as I have already said, that would be another practice, the practice not of giving historical accounts, but the practice of theorizing their possibility. If you set out to determine what happened in 1649, you will look to the materials that recommend themselves to you as the likely repositories of historical knowledge and go from there. In short, you and those who dispute your findings (a word precisely intended) will be engaged in empirical work, and as Howard Horwitz has recently said, arguments about history "are not finally epistemological but empirical, involving disputes about the contents of knowledge, about evidence and its significance" ("I Can't Remember: Skepticism, Synthetic Histories, Critical Action," *South Atlantic Quarterly* 87: 4 [Fall 1988]: 798).

Another way to put this is to say what others have said before me: the New Historicism is not new. But whereas that observation is usually offered as a criticism—it should be new and it's not—I offer it as something that could not be otherwise. The only way the New or any other kind of histo-

ricism could be new is by asserting a new truth about something in opposition to, or correction of, or modification of, a truth previously asserted by
someone else; but that newness—always a possible achievement—will not
be *methodologically* new, will not be a new (nonallegorical, nonexcluding,
nonforgetting, non–boundary-drawing) way of doing history, but merely
another move in the practice of history as it has always been done.

If New Historicist methodology (as opposed to the answers it might give
to thoroughly traditional questions) is finally not different from any other,
the claim of the New Historicism to be politically engaged in a way that
other historicisms are not cannot be maintained. The methodological difference claimed by New Historicism is the difference of not being constrained
in its gestures by narrow disciplinary and professional boundaries. The reasoning is that since New Historicists are aware of those boundaries and
aware, too, of their source in ultimately revisable societal (and even global)
structures, they can angle their actions (or interventions as they prefer to
call them) in such a way as to put pressure on those structures and so
perform politically both in the little world of their institutional situation and
the larger world of POLITICS. Now in essence this picture of the radiating
or widening out effects of institutional action is an accurate one; for since
all activities are interrelated and none enables itself, what is done in one
(temporarily demarcated) sphere will ultimately have ramifications for what
goes on in others. The question is can one perform institutionally with an
eye on that radiating effect? Can one grasp the political constructedness and
relatedness of all things in order to do one thing in a different and more
capacious way? Given that disciplinary performance depends on the in-place
force of innumerable and enabling connections and affiliations (both of
complicity and opposition), can I *focus* on those connections in such a way
as to make my performance self-consciously larger than its institutional situation would seem to allow?[6]

The answer to all these questions is "no," and for the same reason that
it is not possible to practice openness of a general rather than a context
specific kind. The hope that you can play a particular game in a way that
directly affects the entire matrix in which it is embedded depends on there
being a style of playing that exceeds the game's constitutive rules (I am not
claiming that those rules are fixed or inflexible, just that even in their provisional and revisable form they define the range of activities—including
activities of extension and revision—that will be recognized as appropriate,
i.e. in the game); depends, that is, on there being a form of destabilization
that is not specific to particular practices, but is simply DESTABILIZATION
writ large. If there is no such form—no destabilizing act that does not leave
more in place than it disturbs—the effects of your practice will be internal

to that practice and will only impinge on larger structures in an indirect and etiolated way.

In short, there is no road, royal or otherwise, from the insight that all activities are political to a special or different way of engaging in any particular activity, no politics that derives from the truth that everything is politically embedded. Interrelatedness may be a fact about disciplines and enterprises as seen from a vantage point uninvolved in any one of them (the vantage point of another, philosophical, enterprise), but it cannot be the motor of one's performance. Practices may *be* interrelated but you cannot *do* interrelatedness—simultaneously stand within a practice and reflectively survey the supports you stand on. One often hears it said that once you have become aware of the political and constructed nature of all actions, this awareness can be put to methodological use in the practices (history, literary criticism, law) you find yourself performing; but (and this is the argument about openness, provisionality, and interrelatedness all over again) insofar as awareness is something that can be put into play in a situation it will be awareness relative to the demarcated concerns of that situation, and not some separate capacity that you carry with you from one situation to another.

I do not mean to deny that New Historicist practice may be involved in a politics, only that it could not be involved in the (impossible) politics that has as its goal the refusing of boundaries as in "I refuse to think of literature as a discrete activity" (sometimes you do, sometimes you don't) or "I refuse to think in terms of national or regional identities" (it depends on what you're thinking about), and the exclusion of exclusions, as in "I will remember everything" (which means you will be unable to think of anything) or "I close my ears to no voice" (which means that no voice will be heard by you). Nor am I saying that New Historicist practice has no global implications; only that the global implications of New Historicist practice cannot be operated by its practitioners. Of course, this is not true of all practices. The actions, say, of members of Congress or of officials of the national administration will have far-reaching and immediate consequences and those consequences can be held in mind as those actions are taken; moreover, it is not impossible that literary criticism could in time become a practice with similarly far-reaching effects.

One can imagine general political conditions such that the appearance on Monday of a new reading of *The Scarlet Letter* would be the occasion on Tuesday of discussion, debate, and proposed legislation on the floor of Congress; but before that can happen (if we really want it to happen) there will have to be a general restructuring of the lines of influence and power in our culture; and while such a restructuring is not unthinkable, it will not be brought about by declarations of revolutionary intent by New Historicists or materialists or anyone else. So long as literary studies are situated as they are now, the most one can hope for (at least with respect to aims that are realistic) is that your work will make a difference in the institutional setting that gives it a home.

And that, as Catherine Gallagher points out, is quite a lot. She observes that New Historicist and allied practices have already altered the institutional landscape by influencing "the curricula in the literature department, introducing non-canonical texts into the classroom . . . making students more aware of the history and significance of . . . imperialism, slavery and gender differentiation" (44–45). She also notes that for many on the left these changes are insufficient, and there is evidence in the present volume that in the minds of some they are downright suspicious. Montrose, as we have seen, is uneasy at the thought that the successes of New Historicism may be merely professional, and that his own labors may be "a vehicle for . . . partly unconscious and partly calculating negotiation of disciplinary, institutional and societal demands and expectations." He is nervous, that is, at the thought that his career may be going well. Vincent Pecora is distressed at the alacrity at which the New Historicism has turned into a "new kind of formalism" (272), and he complains that cultural semiotics for all its pretensions remains determinedly literary and that its effect has been to make our activities not "more political . . . but . . . less so." It is hard to know whether such anxieties are a sign of large ambitions that have been frustrated—do these critics want to be the acknowledged legislators of the world?—or a sign of the familiar academic longing for failure—we must be doing something wrong because people are listening to us and offering us high salaries. But whatever the source of the malaise, I urge that it be abandoned and that New Historicists sit back and enjoy the fruits of their professional success, wishing neither for more nor for less.[7] In the words of the old Alka-Seltzer commercial, "try it, you'll like it."

16

Milton's Career and the Career of Theory

More than fifteen years ago I appeared at the MLA to speak on the topic "New Directions in Milton Studies." My fellow panelists, then as now twenty-five years my senior, were Joseph Summers and Louis Martz, and I was expected, presumably, to supply the radical voice. No doubt I disappointed when I predicted that there would be no new directions in Milton studies. My reasons were simple, and they still apply. First of all, the tradition of Milton studies is so strongly articulated, so well equipped with a set of hard questions and intractable problems, so burdened or graced with interpretive traditions that have been declared and elaborated by some of the most celebrated voices in literary history—Addison, Bentley, the Richardsons, Newton, Dr. Johnson, Blake, Shelley, Coleridge, Hazlitt, Macaulay, Arnold, Eliot, Pound, Leavis, Lewis, Ransom, Empson—that the conditions in relation to which a desire for the new would be felt, the conditions of surfeit and boredom, have little chance to develop. I am not making a statement about Milton—about his inexhaustible complexity or ineluctable essence—but about Milton criticism and the extent to which its history (a word we shall return to) constrains those who enter its precincts, even those who enter with the intentions of reform and critique. Not only has that history set us any number of tasks in any number of disciplines—theology, linguistics, military science, astronomy, music, dance, prosody, classics, Italian romance, cosmology, philosophy, rhetoric, zoology, etc.—but each of these tasks comes to us in the context of disputes as to how it is to be framed, in what terms, with what emphases, with what degrees of credulity or incredulity, and as a result any thesis strongly argued opens up many more avenues of inquiry than it claims to close. Add to this the full machinery of the Milton industry with its ceaseless production of authoritative editions, annotated bibliographies, and encyclopedias, with its official and semiofficial organs of publication—*Milton Studies,* the *Milton Quarterly,* the *Friends of Milton's Cottage*—all presided over by the Milton Society of America, which takes as its great project the refurbishing and celebrating of the tradition it continues, and it is no wonder that there seems little room or need for revolutionary gestures.

Of course, one might retort that it is just this tidiness, this apparent saturation of the field, this authoritative and indeed authoritarian agenda that in recent years has provoked the intervention of theory, the oppositional reversals of deconstruction and feminism, the readings against the grain by New Historicists and renewed materialists, the vertigo of Bakhtinian or de Manian decenterings. But here, too, the Milton industry has preempted the field and rendered such eruptions belated: what would be the point, for example, of setting out to destabilize the lines of authority and hierarchy in *Paradise Lost* when Blake and Shelley have already done it and established a tradition continued in our century by Empson, Waldock, Peter, Werblowsky, Watkins, and others? Just how innovative would it be to read Milton's text against its apparent grain, when it has long since been a commonplace to observe that each of the poems undermines the generic categories declared by its title and formal signatures? Who would be unsettled by the suggestion that the Miltonic corpus, rather than being autonomous, is intertextual, the product not of a single voice but of multiple voices and traditions that inscribe themselves within it, given the editorial apparatus of the great eighteenth-century editions? To read a page of Newton's commentary or Todd's *Variorum* is to experience more heteroglossic fecundity than is available in all of Barthes's *S/Z.* Just how forceful would it be to argue that Milton's work is an instrument of power and ideology when Samuel Johnson was already making that argument more than two hundred years ago and when we have long had book-length studies of the uses to which Milton has been put by various political and military agendas on both sides of the Atlantic? What would be the power of asserting that Milton is merely the name for a site traversed by innumerable discursive networks when we are already blessed (if that is the word) by studies of Milton and Virgil, Milton and Plato, Milton and the pastoral, Milton and warfare, Milton and science, Milton and opera, Milton and astronomy, Milton and Puritanism, Milton and kingship, Milton and the Kaballah, Milton and the Italian cities, and on and on and on? And how surprising would it be to hear that the Miltonic text is sexually conflicted, a welter of confusing gender identifications and reversals that are inseparable from issues of patriarchy and political authority, when we have the testimony of Robert Graves in *Wife to Mr. Milton,* not to mention the traditions of scholarship and commentary recently recovered by Joseph Wittreich? And how enlivening would it be to introduce the mysteries of the mirror stage into a conversation that has for so long included debates about the meaning of the moment in which Eve makes her own image in the pool the object of her desire?

I do not mean by posing these rhetorical questions (which could have easily been multiplied) to suggest that the vocabularies and methodological agendas of new theories and approaches have not made their way into Milton studies, only that the effect of their introduction has been very much less dramatic than it would have been had the field been less well furnished than it has proven to be. And indeed in recent years at least two critics, Bill Readings and John Rumrich, have complained that despite more than twenty years of theoretical excitement and supposed revolution in the academic world, Milton criticism has not advanced beyond the stage represented by *Surprised by Sin*. (This is a complaint I hear with divided feelings.) One difficulty of their position is that in order even to articulate it, they must pass through, and therefore become *formed* by, the very institutional discourses they wish to escape.

It is perhaps for the reasons I have been adducing that the New Historicism has made relatively little headway in Milton studies, in contrast, say, to its impact on the study of Shakespeare. Never having been celebrated as a universal genius who appeared unbidden on the world's stage, Milton has not been ripe for a revisionary account in which he is reembedded in the multiple histories of his time. There are, however, at least two other impediments to the success of New Historicism in the Milton world, and they are self-generated. The first is a difficulty, not to say a fatal flaw, in the polemic that often accompanies historicist criticism. In fact the difficulty is the fact that there is a polemic at all, that proponents of a New Historicism (whether or not they are willing to identify themselves as such) or of any historicism strongly conceived imagine a practice that could depart from their master rule, "always historicize." If it is the first thesis of historicism that no human gesture can be either produced or received independently of some context of historical understanding, then the one thing a human being, even a literary critic, cannot be is unhistorical. That is, it is impossible to *not* historicize, impossible to conceive of an event or a text independently of some notion of its origin, genealogical affiliations, generic habitation, etc. The difference between thinking of a pastoral poem in relation to a line of poets stretching back to Theocritus and thinking of the same poem in relation to the agrarian policies of a Jacobean monarch (a Rosemond Tuve reading as opposed to a Raymond Williams reading) is not a difference between a historical interpretation and one that ignores history but between interpretations flowing from differently assumed histories. It is of course possible to debate which of these histories is the more relevant and fruitful, but the debate itself would be between historically situated agents whose notions of what would and would not be properly historical evidence would conflict; and given the fact that *everyone* is historically situated (again the first thesis of historical criticism), one could not adjudicate the conflict by performing some independent calculation that would determine which of the two histories was the *more* historical. As Geoff Bennington and Robert Young have remarked, "History cannot provide an unquestionable ground once the working of difference is appreciated" (*Post-structuralisms and the Question of History*, ed. D. Attridge, G. Bennington, and R. Young [Cambridge, England, 1987], 5), that is, once you realize that any invocation of history will be the invocation of a challengeable interpretive mode and not an appeal to a "last instance" (20), to "a harvest of evidence which contrasts with the endless draught of argument" (130), to "History in the upper case" (135). History, in short, is a discipline, an intellectual praxis, and not a thing.

Two conclusions follow from this point: (1) No one is more historical than anyone else, except by a measure that is itself historically—and therefore vulnerably—constructed; and (2) historicism is not and could not be a methodology; to commit yourself to it—as opposed to committing to some particular *version* of history—is to commit yourself to nothing in particular or (it is the same thing) to anything in general. Since being historical is not an *option* but the condition everyone involuntarily occupies and enacts, nothing follows from the declaration that you will pay attention to history; you have not announced a thesis in relation to which evidence will be first identified and then structured into an argument. If I preface a piece by saying that it is feminist or Marxist or psychoanalytical, you know, within limits, what to expect, what concerns, emphases, vocabulary; but if I advertise my piece as historical, you might expect anything or nothing, because historical is a category without a diacritical edge. Perhaps that is why work that aggressively labels itself historical—especially work of the kind thought of as materialist as opposed to the cultural poetics of a Stephen Greenblatt—is so often random or even haphazard in its procedures. Given that it is committed to a program that has no content or, rather, can have any content whatsoever (because by definition nothing falls outside history's scope), it tends to display brightly colored bits of material independently of any principle of selection or arrangement.

Here, for example, is a passage from David Norbrook's *Poetry and Politics in the English Renaissance* (London, 1984), a book cited almost reverentially by those of the historical persuasion. Norbrook is writing about *Comus*:

Milton probably knew Middleton's . . . "A Game of Chess." In the opening scene a Catholic priest uses love-language similar to Comus's to try to tempt the chaste and Protestant White Queen's pawn; his phrase "the opening eyelids of the morn" reappears, with very different connotations, in "Lycidas."

"Comus" was performed on St. Michael's Day so that apocalyptic associations would have been appropriate. . . . [I]t is interesting that when Milton was later charged by an opponent with loose living he declared that his reading in Dante and Petrarch had made him love chastity. These poets also, of course, provided authority for attacks on the papacy. (125)

The nice touch here is the "of course," which casts a retroactive glow of inevitability and logic over the sequence, but the real work, or rather *non-work*, of the passage is done by other words—"probably," "would have been appropriate," "it is interesting"; for these words act to remove the argumentative pressure from the facts they introduce. Milton *probably* knew Middleton, who wrote a scene in which a Catholic tempts a Protestant; they shared a phrase no doubt shared by innumerable other poets. *Comus* was produced on a day when certain associations would not have been inappropriate. Milton read Dante and Petrarch *and* they had been used in antipapist propaganda.

All true, of course, but what these truths add up to is unclear, since no effort is made to link them in an interpretive structure, and indeed the assumption seems to be that no effort is necessary. The assumption behind that assumption is never spelled out, but one suspects that it is something like this: since *Comus* is a product of its moment in history, any aspect of that history—anything that happened in the 1630s—is interpretively relevant. Moreover, that relevance need not be demonstrated; it is merely assumed, and because it is assumed, one need not establish it by going through the steps that practitioners of other kinds of criticism consider obligatory. Thus when Norbrook observes that a "Puritan in the Forest of Dean was arrested in 1631 for preaching universal equality," he thinks that by merely citing the fact he has connected it with the Lady's plan for the equal distribution of goods. Paradoxical though it may be, the very critics who regularly lambast others for not getting their hands dirty in the archival gardens have contrived to believe something that relieves them of the burden of doing any interpretative work.

A bit more of the work is done by Leah Marcus when she researches the story of Margery Evans, a servingmaid who was assaulted by one Philbert Burghill and his servant and thrown into jail when she accused them. She then appealed for redress to the king, who referred the case to the diligent and apparently sympathetic John Egerton, first earl of Bridgewater. Obviously the facts as Marcus rehearses them are already in the neighborhood of *Comus* in several senses, but when it comes time to bring them home as it were, the effort falls short:

The Lady is confronted with a situation very much like that faced by Margery Evans, whose case had absorbed so much of the earl's and council's attention in the year before the masque. The part of the Lady was performed by an earl's daughter while Margery was only a servingmaid but the two were nearly the same "tender age." . . . As Margery Evans had been, the Lady of *Comus* is traveling westward through the lonely and dangerous border country from England toward Wales. Like Margery Evans she finds herself alone, although her solitude, unlike Margery's, is temporary and accidental. Like Margery Evans, she is accosted by a seducer who is well established in his territory with a network of local connections: Burghill had powerful friends; Comus has his own court and an obedient retinue of monsters. ("A 'Local' Reading of *Comus*," in *Milton and the Idea of Woman*, ed. J. Walker [Urbana and Chicago, 1988], 76–77)

To Marcus's credit her rehearsal of the evidence includes the points of difference that tell against the assertion of similarity—the difference in class, in situation, in outcome, in what is at stake, and even the difference in what is supposedly a similarity, Burghill's friends and Comus's obedient retinue of monsters. In the end, the only persuasively made connection is the one with which she begins: the earl has a 15-year-old daughter who plays the part of an endangered maiden, and in his official capacity he had dealt with a case of a 14-year-old girl who had been raped. I am not saying that the link between these two facts could not have been elaborated into a convincing account of *Comus*, only that the attempt has not been made; the details of the poem have not been brought into line with its purported origin in a local occasion; that occasion has not been tied to the moment of production or conception, to the intention whether of a single author or of a discursive formation, or of forces that are in no one's control but are finally controlling.

Marcus is not unaware of this gap in her argument. At the end of her essay she speculates on the relationship of her reading to Milton's intention given the fact that we do not know whether or not *he* knew anything of Margery Evans. She finds herself inclined to believe that Milton knew, but she then declares that it doesn't matter and that Milton might have been the unwitting instrument of the earl or some of his associates who could have made suggestions designed to manipulate the young poet in the direction of an effect they desired. This extraordinary speculation (not the least extraordinary for the picture of Milton it presupposes) is then supposedly supported by the observation that "there are many other cases of official collaboration in the seventeenth century masque." No doubt, but they do not have the ghostly, spectral quality of this one, of a collaboration so loose that one of the partners doesn't know about it, and the others have been literally invented by a critic in order to shore up a thesis that has never quite come into focus.

Marcus's indifference to the proliferating accounts she offers of the masque's production is of a piece with her failure to press hard on the relationship between the story of Margery Evans and the story apparently told by Milton. I say "apparently" so as to acknowledge the possibility of my being persuaded that Milton is telling a story other than the one I have been accustomed to read. But in the absence of any effort to persuade me, I remain confronted only with a piece of archival research (for which one is always grateful) and the *assertion* of its relevance to an interpretive project. I should emphasize that my concern is not primarily with Marcus's essay but with the methodological assumption that seems to have produced it, the assumption that a commitment to historical, local readings *is* a methodology; whereas in fact, it is no method at all but merely a license for making random observations and then, without any further elaboration, *declaring* them to be meaningful.

This is even more egregiously true of still another piece on *Comus*, Michael Wilding's "Comus, Camus, Commerce: Theatre and Politics on the Border" (in *Dragons Teeth: Literature in the English Revolution*, Oxford, 1987). Here we find the increasingly familiar set of references to the wool trade, border problems between England and Wales, class warfare, and, of course, the *Book of Sports*, which in certain quarters seems to have replaced chastity as the masque's true subject. We also have an interpretive strategy that illustrates better than any other the looseness that attends an untheorized historicism. It is a strategy I recognize from my studies in the law, where it takes the form of making an evidentiary virtue of what has not been said. Wilding employs it when he first observes that what the Elder Brother sees as a bandit, "others might see as a freedom-fighter, a Sampson attempting to deliver his people" (this is presumably the Che Guevara reading of *Comus*), and then declares that this "local socio-political meaning is no less present for being repressed" (35–36). Aside from displaying the dangers of reading a little literary theory, this statement raises all kinds of questions. Repressed by whom? With what consciousness? With what *un*consciousness? These and other questions might very well have answers, and the answers might well give a firm content to the notion of repression and its significance. It is not the practice of reasoning from silences, gaps, and absences that I object to; that is after all a practice that in the work of Derrida and others has yielded spectacular results. But here there is no reasoning, just the blanket assertion that something that has not been mentioned but could have been mentioned by someone (and who that someone or someones might be is never revealed) is, by virtue of that nonmention, present, an assertion that, if taken seriously and in the naked form it here has, would lead to a list—necessarily infinite and inexhaustible—of all the things one *might* see if one were not inclined to see a bandit. Again it is of course possible that we *should* be seeing the "local socio-political meaning" Wilding half urges, but if that possibility is to be realized as something like a conviction, Wilding will have to do a great deal more; he will have to do some work.

Not only do historicists of Wilding's persuasion fail to put sufficient pressure on the historical contexts they identify as crucial, they regularly dismiss as *un*historical the contexts identified by an older criticism. This gesture is incoherent for a reason I have already given; everything is necessarily historical, and the fact that a particular tradition comes freighted with the claim that its materials and concerns are timeless does not disqualify it as historical but rather marks an important feature of its historicity, one that should be taken into account when considering its significance for the interpretive task. In Milton studies the contexts most often dismissed in this way are aesthetics and theology, with results that are very often curious. Thus Wilding several times disparages any attention to the classical and pastoral sources of *Comus* as attention to "mere literary decoration," which he stigmatizes without argument as a "depoliticizing act" (45, 37). And when he correctly observes that in the Lady's first speech, "a social public potentially political theme—charity—is replaced by the private issue of chastity" (57), he is unable to extend the observation, which is of interest to him *only* as an indication of how retrograde the Lady is as opposed (and this is his example) to Christopher Hill. Where others might see an opportunity to press an interpretive and *historical* issue—what meanings do the literary conventions Milton invokes carry at this time? how would the relationship between private virtue and public disorder be seen by a Puritan of Milton's temper?—Wilding sees only an opportunity to indict the poet for not having read *Milton and the English Revolution*. This is *history?*

The spectacle of a historicism that deprives itself of historical materials is also on display in Richard Halpern's "Puritanism and Maenadism in *A Mask*" (in *Rewriting the Renaissance: The Discourses of Sexual Difference in Early Modern Europe*, ed. M. Ferguson, M. Quilligan, and N. Vickers, Chicago, 1986). Halpern begins by simultaneously noting and devaluing the histories that weigh on Milton's performance:

> Milton inherits a field of literary instruments; specifically a mythopoesis that tends to congeal into Neoplatonic and Christian allegory. The syncretic mythology . . . and the ponderous symbolism . . . draw on techniques common to both the Spenserian epic and the Jonsonian masque. (88–89)

It is obvious that for Halpern Neoplatonism, mythology, symbolism, Christianity, allegory, epic, and mask are totalizing instruments, systems of thought and representation that refuse the touch of a materiality that would disrupt their false coherence. And because he sees these discursive formations as *impediments* to historical thinking rather than as the *vehicles* of historical thinking, as the locally available vocabularies within which social and political issues are conceptualized and worried, the only response he can give to the significances they carry is one of impatience and scorn. It is not that he cannot see what Milton may be up to, it is just that he cannot take it seriously because it comes to him in terms he has discounted in advance. Thus when he concludes the Lady "is . . . saved from imprisonment by Comus only in order that she may learn self-imprisonment at the hands of Christ" (96), he regards this statement as prima facie evidence of Milton's inability to imagine the spiritual life as anything but a reproduction of "the worst aspects of life on earth" (97), that is, as anything but another form of patriarchal tyranny. In short, for Halpern there is no difference between submission to Comus and submission to Christ.

One hardly knows how to respond to this except to say that everything in the masque is an exploration of that difference, which is crucial not only to Milton's theology as it unfolds in the many pages of the *Christian Doctrine* but to the various and varied debates that were moving a nation in the direction of civil war. The context Halpern so easily dismisses—the context in which one tries to sort out what belongs to Caesar and what belongs to faith, what belongs to the body and what to the spirit—is the context within which questions of materiality—of war, commerce, social and political organization—were inevitably and *historically* being thought. Of course it would be possible to interrogate that context, to resist the lure of its apparent coherence by exploring the ways it constrains the posing of problems and the formulation of issues, but Halpern doesn't do that, or rather he thinks that he does it merely by labeling the context as theological and declaring it congealed; he can then move immediately to such numbingly abstract (and allegorical) pronouncements as "bourgeois strategies of sexual domination and of class exploitation mutually inform one another" (97), a generalization that is then documented (if that is the word) by four quotations from Christopher Hill and three from Leah Marcus.

Let me say again that my quarrel with these essays is not a quarrel with history or with historical criticism, but with a historicism so enamored of its political correctness that it feels justified in saying *anything,* including things that belong more properly to the so-called timeless discourses it regularly impugns. When David Aers and Bob Hodge complain that if Milton "was sincere in his devotion to justice and individual liberty, he should be equally tender of the rights of women" ("Rational Burning: Milton on Sex and Marriage," *Milton Studies* 13 [1979]: 19), one wonders what to make of their "should." Are justice and liberty universals whose shape is always the same independently of local and historical understandings? Is the logic to which Milton is here being held accountable always the right one to invoke? Where did Aers and Hodge get it? Do they know? Do they care? These and other questions might be less urgent if their own claims and pretensions did not raise them, but once raised they prove, if not fatal, at least embarrassing.

In the context of Milton studies, the embarrassments of the New or Newer Historicism are related somewhat asymmetrically to the embarrassments of theory as they were rehearsed in the first part of this essay. Theory, or more precisely high theory (some might want to call historicism a theory, too), is embarrassed because its work has already been done. Historicism, at least as it is represented by these essays, is embarrassed because it refuses to do the work and indeed doesn't even know what its work *is*. That's the bad news. The good news is that these two failures finally don't matter very much because the layered richness of Milton criticism as noted in my opening paragraphs continues to propel it forward no matter what the deficiencies of various new methods and nonmethods. Milton's career is thus on a much surer footing than the career of theory (which according to many observers is more or less over and is certainly played out), not because Milton himself (a phrase finally without a sense) exceeds the enterprise of Milton studies, but because that enterprise is so well established, so much a feature of any presently imaginable scene, that one can say of it that it is something the world will not willingly let die.

17

Milton, Thou Shouldst Be Living at This Hour

In a career marked by what some might describe as a succession of acts of hubris, let me remain in character by suggesting a connection between this moment and the moment of our profession as a whole. This is, to be sure, a glorification and a romanticization of the present occasion; but as my wife will tell you, it is an occasion I have been hoping for and preparing for ever since I saw the first televised presentation of the American Film Institute's Lifetime Achievement Award. You know the one I mean: Jimmy Stewart or Jimmy Cagney or Bette Davis or Kirk Douglas is honored by a series of embarrassing tributes by Jimmy Stewart or Jimmy Cagney or Bette Davis or Kirk Douglas. It turns out, every time, that this actor is the actor's actor, the true professional, the one that every other actor wants to work with, the one every director wants to direct, the one without jealousy or temperament or unreasonable demands. These obligatory and unlikely characterizations of a virtue no one possesses are punctuated by film clips from the honoree's greatest performances—loving close-ups of a face known to every American, moments of high drama cut and edited so as to produce an intensity even greater than that achieved in the "original" ten or twenty or forty years ago. And finally, the centerpiece of it all: the newly canonized or newly recanonized luminary is ushered to the podium amidst tumultuous applause; there he or she stands, head slightly bowed (you have to be able to see enough to extract from the occasion all the pleasure it affords), burdened by praise, overcome by emotion (one thinks of Julius Caesar three times breaking down in tears as he—because he?—refuses the crown), and faced with the sternest challenge of a long career, the challenge of responding to hyperbolic adulation with the appropriate measure of humility, an impossible task that is usually accomplished, when it is accomplished, by the substitution for humility of wit.

267

As I watched the edited-for-television versions of these presentations—
"watched" is too passive a word; I was practically jumping into the screen—
I mentally transposed their content and their ceremonies into the content
and ceremonies of this organization, since it was the only one likely ever
to afford me the opportunity to be theatrically humble. Some of the trans-
position was easy: instead of Jimmy Stewart, Kirk Douglas, and Gregory
Peck I imagined Northrop Frye, Christopher Hill, Arthur Barker, Marjorie
Nicholson, C. S. Lewis, E. M. W. Tillyard, Sir Herbert Grierson, James
Holly Hanford, a list (only partial of course) of names at least the equal of
any that have ever been emblazoned on a movie marquee; and, along with
the names, accomplishments of more durability and resonance than we usually associate with a cinematic career. The year 1910 may be, for all I
know, a date of some significance for the film industry, but the productions
of that year are valued for their historical significance; they do not work
today in the minds either of filmmakers or audiences. In contrast, James
Holly Hanford's essay of the same year, *The Pastoral Elegy and Milton's
Lycidas,* is as vital to the consideration of its topic today as it was when it
first appeared more than eighty years ago; and to mention only one other
example, we have yet to take the measure or exhaust the insights of Arthur
Barker's *Milton and the Puritan Dilemma,* a book published in 1942 but
written largely in the middle and late thirties.

It was easy and pleasant to imagine the members of a large and glittering
audience—not unlike this one—rising to pay tribute to the example and
influence of such men and women, each perhaps recalling a moment of
startled but not astounded illumination when, upon first looking into Lewalski's *Brief Epic* or Lewis's *Preface,* vistas of arguments to be pursued
and books to be written opened up onto the future of a suddenly energized
career. It was not so easy to think of something that would match the film
clips: responsive readings from a seminal article? dramatic re-creations of a
celebrated lecture? recollections of a stern rebuke delivered thirty years ago
to a then-callow youth now a gray-haired or no-haired eminence who solemnly declares that the event, perhaps the most humiliating he has ever
experienced, was the best thing that ever happened to him? Somehow none
of these seemed adequate, but then I recalled a newly comforting fact; for
more than twenty-five years now I have tape-recorded every class I have
taught. These tapes would provide the historical accuracy no recollection
could ever convey. What bliss it would be to reexperience the give and take
of a particular seminar, perhaps the very seminar when I first became aware
of just how surprised one could be by sin or how consumed one could be
by an artifact. The tape would be presented and prefaced by a student who
was present at the creation, and those in the audience would nod knowingly

to one another, as if to say with their looks and glances, "Ah, so that's how it all came about!"

For as you will have guessed long since, the imaginative exercise of translating the American Film Institute celebrations into the language and context of the Milton Society of America was always impelled by a single hope, the hope that I would someday be standing before you as I stand before you now, an uncouth swain entertained, if not by "all the Saints above," then certainly by the "solemn troops and sweet societies" of all of those who have wandered with him in the perilous flood of academic life. In that vision, less sweeping perhaps than that enjoyed (or suffered) by Adam when he ascended to a "Hill/of Paradise . . . from whose top/ The hemisphere of Earth in clearest Ken/Stretch out to the amplest reach of prospect lay," I was nevertheless able to call the roll of gods or demigods, those men and women who would have preceded me and with whom I have had any number of encounters, some epic, more comic. I imagined myself hailing my predecessor at Duke, Alan Gilbert, if only because he was still jogging daily up to the time of his death at the age of 100; if, as has been said, Miltonists rule the academic world, it may be because they live longer— at least I hope so. I hoped to be able to remind John Diekhoff of our first conversation at a meeting of this very society in the sixties when he introduced to me his prize graduate student and asked me to look out for him; his name was Joseph Wittreich, and he turned out to be quite capable of looking out for himself and for many others as well. I would have loved to have saluted C. S. Lewis, not only for his enormous contribution to medieval and Renaissance studies, but for the extraordinary fact of his prose style, on which I have modeled my own with results that fall far short of his example. I looked forward to reminiscing with Balachandra Rajan about a glorious weekend in London, Ontario, when a stellar cast of Renaissance and Romantic scholars celebrated his stellar self in ceremonies as moving as any I ever attended. And I am sad to be deprived of the opportunity once again to identify Dean Patrides, the convener of that celebration, as the person—or is it culprit?—responsible for my becoming a Miltonist; for in 1963 when Dean received a much deserved grant late in the year, the chair of the English department at Berkeley asked a young medievalist to pick up his Milton course. That young scholar, one year out of graduate school, was reluctant to admit that he had never—either as an undergraduate or in graduate school—taken a Milton course or *any* course in seventeenth-century literature for that matter; and so he cheerily accepted the assignment and began the semester by announcing a four-week "research" period during which he frantically read Milton and Milton scholarship so that he might be capable of keeping at least twenty minutes ahead of his students. It took some students *less* than twenty minutes to see through him. And finally I anticipated with a pleasure I cannot now experience reliving with Irene Samuel that moment when we were introduced, I think by Barbara Lewalski, and she fixed me with that piercing glance and said, in that even more piercing voice, "Young man"—the year was probably 1966—"young man, you are perverse," an opinion that I know her never to have relinquished.

The perversity of which Irene spoke was embodied for her in the articles of 1964 and 1966 that a short time later became the first two chapters of *Surprised by Sin*. Irene never elaborated her judgment into a full-scale critical account, but perhaps her view would not have been far from that expressed recently by Dinesh D'Souza, who, after rehearsing the thesis of *Surprised by Sin*—"Fish suggested . . . that in stumbling from line to line, forming premature opinions and then having to revise them, [readers] would be constantly reminded of [their] fallen state"—declares it to be "implausible" (175). But were he to consult Miltonists of the 1990s, D'Souza would find another judgment being made, no less severe in its way, but different. It is a judgment nicely articulated, for example, by John Rumrich, who last year complained that the trouble with *Surprised by Sin* is that it has too strong a hold on the Milton industry; indeed, even those who might wish to strike out in directions it seems to exclude seem obligated to defer to it. Promising interpretive freedom with one hand, the argument of the book sets limits, with the other, to what it purports to enable and is thus profoundly constraining: "*Surprised by Sin* . . . accomplished the theoretical liberation of Milton studies by placing a destabilizing hermeneutics in the service of conservative ideology." Once perverse and implausible, now a conservative ideologue: how the former enfant terrible has fallen or, rather, *risen* to a bad eminence from whose height he spies upstart children and, like a pagan god, strangles them at birth.

Now, I will not be so foolish as to adjudicate between these two negative characterizations of my work. I merely want to point out that when D'Souza weighs in with his opinion, it is an opinion twenty-five years out of date. There are any number of nasty things one could reasonably say about *Surprised by Sin* in 1991, but one of them is not that the book is subversive of orthodoxy and common sense; for, as Rumrich unhappily reports, within the world of Milton studies *Surprised by Sin is* orthodoxy and common sense, and it is understood to have been produced not by a clever and audacious newcomer (D'Souza's words) but by someone whose status as an insider is so well established that he can now be added to a list of nihilistic radicals like Helen Darbisher and Raymond Dexter Havens. The point would hardly be worth making were it relevant only to D'Souza's characterization of me; but in fact that characterization is typical of what passes for trenchant criticism of the humanities on the part of various self-appointed spokespersons of the neoconservative right. Whether issued by D'Souza or Kimball or Sykes, the call to alarm—along with the prediction that unless we act now, Western civilization will unravel—is issued in the face of a threat so domesticated that anthologies memorializing its history are assigned to high school seniors. Can 1991 be the date when the world awakes to the dangers of reader response criticism? Is it in 1991 that we must gird our loins and fight off deconstruction, a technique of analysis and interpretation that was old hat by 1975 and whose death has been announced regularly (and prematurely) since 1980? And is it in 1991 that we must be alerted to the perils of feminism, which has been around at least since Aristophanes and probably since Genesis? The spectacle of grown (and often old) men hurling jeremiads at enemies whose acts of subversion are limited largely to publishing articles in *Representations* and *Signs* would be comical were it not that many even less well informed than they have been persuaded that the sky is really falling.

The truth is that reader response criticism has come and gone and the republic still stands; deconstruction has done its worst and people still manage to mean things that other people still manage to understand; and feminism's assault on the American way of life has somehow left more than 80 percent of the senior positions in the academy securely in the hands of men. I don't rehearse these matters with satisfaction or approval, but only to make once again a point I have been making for some years: institutional life is more durable than the vocabulary of either dissolution or revolution suggests. The neoconservative right rails against change in the name of tradition and continuity and doesn't realize or doesn't want to know that change is the means by which continuity is achieved and reachieved. Tradition does not preserve itself by pushing away novelty and difference but by accommodating them, by conscripting them for its project; and since accommodation cannot occur unless that project stretches its shape, the result will be a tradition that is always being maintained and is always being altered *because* it is always being maintained. As time goes by, almost everything may change, but from moment to moment, almost everything will be the same, and it will only be the Rip Van Winkles of the world— those who like our modern Jeremiahs fall asleep for twenty years—who look around them and cry that all coherence is gone and the center will not hold.

As it is with institutions and traditions, so it is with careers. Like most of you, I now teach and write on and think about things that were never dreamed of in my days as a graduate student or young instructor; nevertheless, a set of apparently disparate concerns has been held together by a thread so strong that it amounts to an obsession. In my early years of teaching, I gave the same exam no matter what the course or its subject matter.

I asked my students to relate two quotations. The first was by J. Robert Oppenheimer: "Style . . . is the deference that action pays to uncertainty." What that means, I take it, is that in a world where certain grounds for action are unavailable, one avoids the Scylla of prideful self-assertion and the Charybdis of paralysis by stepping out provisionally, with a sense of limitation, with a sense of style. That much said, the other quotation needs no gloss, and neither does it need, at least for this audience, an introduction: "Now faith is the substance of things hoped for, the evidence of things not seen." I wanted my students to see that while the moral life cannot be anchored in a perspicuous and uncontroversial rule, golden or otherwise, we must nevertheless respond to its pressures; and indeed it is only *because* the moral life rests on a base of nothing more than its own interpretations that it can have a content; for were there a clearly marked path that assured the safety of pilgrims and wanderers, we would have no decisions to make, nothing to hazard, nothing to wager. The uncertainty of which Oppenheimer and Saint Paul speak is not a defect in our situation but the very ground and possibility of meaningful action. As I said, I wanted my students to see that in 1962, and I still want it today. With them I have pursued this lesson through many traditions and texts, in literature, linguistics, philosophy, and, most recently, the law—but the text to which I always return is named Milton, for Milton teaches the lesson again and again, exploring its glorious difficulties in figures like Abraham who goes out "not knowing to what land, yet firm believes." It is this continuity in my own life that provided me with one of my favorite moments of the last year and one half. I was in a restaurant called Chili's in Durham, North Carolina, in the company of my nephew, David Fish, and I had just given my Visa card to our young waitress, whom I had been admiring in a way natural to a boy who had grown up in the fifties and who therefore had never grown up. In a minute she came back with the card and asked, "Are you Stanley Fish the Miltonist?" Not, "Are you Stanley Fish, the guy who believes that words mean anything you want them to mean?"; not, "Are you Stanley Fish, the person who is destroying Western civilization?"; but "Are you Stanley Fish the Miltonist?" As my nephew looked on in astonishment at the entire sequence, I gave with pleasure the answer I give now, "Yes, I am Stanley Fish, the Miltonist."

18

The Unbearable Ugliness of Volvos

On a day in the mid-seventies—it may have varied in different parts of the country and at different universities—American academics stopped buying ugly Volkswagens and started buying ugly Volvos, with a few nonconformists opting for ugly Saabs. Now on the surface there would seem to be an obvious explanation for this shift in preference: on the one hand, graduate student stipends gave way to the more generous salaries of assistant and associate professorships; on the other, growing families required more than a rudimentary back seat. But the question remains, why Volvos? why not Oldsmobiles, or Chryslers, or Mercury station wagons? The answer, I think, is that Volvos provided a solution to a new dilemma facing many academics—how to enjoy the benefits of increasing affluence while at the same time maintaining the proper attitude of disdain toward the goods affluence brings. In the context of this dilemma, the ugliness of the Volvo becomes its most attractive feature, for it allows those who own one to plead innocent to the charge of really wanting it. There must be another reason for the purchase, in this case a reason provided conveniently by the manufacturer in an advertising strategy that emphasizes safety. We don't buy these big expensive luxurious cars because we want to be comfortable or (God forbid) ostentatious; we buy them because we want to be safe. (I can only guess how many academics are now gobbling up overpriced Michelin tires reassured by a recent advertising campaign that they are purchasing family security rather than performance or glamour.) The ugliness of the automobile makes the cashing in of its negative value a straightforward and *immediate* transaction. Were the car not ugly, a Volvo owner might be in danger of hearing someone say, "My, what a stunning Volvo," to which he or she would have to respond, "Well, perhaps, but I really bought it because it is safe." But no Volvo owner will ever face the challenge of an unwanted compliment, and therefore the disclaimer of nonutilitarian motives need not even be made. Of course this entire economy is now threatened by the revelation that tests demonstrating Volvo's superior safety were faked; but since they are mostly academics, Volvo owners will no doubt be resourceful in the search for rationalizations and will probably take comfort from *Consumer Reports* or some other publication that breathes the proper anticommercial virtue.

Now it might be said that the relationship between academics and their Volvos is not exactly central to the life of the academy; but in fact it seems to me emblematic of a basic academic practice, the practice of translating into the language of higher motives desires and satisfactions one is unable or unwilling to acknowledge. If I can put the matter in the form of a rule or rule of thumb: whenever you either want something or get something, manage it in such a way as to deny or disguise its material pleasures. Nor is this a rule simply of *personal* behavior; it can generate the behavior of the entire profession. Consider, for example, the very material pleasures of the lecture and conference circuit, something that was not in place when I was a graduate student in the late fifties and early sixties. The flourishing of the circuit has brought with it new sources of extra income, increased opportunities for domestic and foreign travel, easy access to national and international centers of research, an ever-growing list of stages on which to showcase one's talents, and a geometric increase in the availability of the commodities for which academics yearn: attention, applause, fame, and, ultimately, adulation of a kind usually reserved for the icons of popular culture. In the face of such a cornucopia of benefits the academic world is in danger of at once providing and experiencing unadulterated gratification, but (never fear) the danger is warded off by a set of practices that exacts a payment, usually in advance, for every potential pleasure.

Nine times out of ten, when you arrive in a strange city, there will be no one to meet you despite elaborate and repeated promises; or, if you are met, you will be taken to a hotel only to find that the reservations were never made and every room is taken; or, if the reservations have been made and there is a room, the hotel is in the course of being rebuilt (this seems to be a requirement for hotels at the MLA convention) and there is no possibility at all of either sleep or study. And if these difficulties are somehow overcome or avoided and you actually get to the campus at the appointed time, more or less prepared, the room you are supposed to speak in will be locked and no one will know who has the key; and if the room isn't locked, it will be occupied by a scheduled class; or, if it is unoccupied, it will be either impossibly small or embarrassingly large; or, if by some accident it is the right size, it will be the only room in the university, if not the entire state, that is not air-conditioned. You will then be introduced by someone who takes pride in either forgetting or mispronouncing your name, who associates you with work you haven't done in ten years, who attributes to you opinions the reverse of those you actually hold, and who believes that it is only by a perverse turn of fate that he or she is not being introduced by *you*. The audience you then address will be a fraction of what it might have been had the lecture or panel been properly advertised; there will be insufficient time for questions and discussion; and when it is all over, and you have endured a dinner only slightly less adversarial than your Ph.D. orals, you are then left alone, at 7:45 P.M., to pass the long night in a town that offers you only a parade of unfamiliar faces.

Transposed to another institutional key, the same scenario plays itself out in the context of the job market. Here the visit is to a potential employer, but it often seems that you are interviewing for a place in a penal colony. Again, the arrangements are systematically botched; planes not met, reservations mixed up; rather than being introduced by someone who barely knows your name, you are interrogated by deans who mistake you for the candidate from neurosurgery; your schedule is a brutal one, in the course of which you meet potential future colleagues who warn you away from their territory even as they try to enlist you in wars now going into their twentieth year; you have coffee with students who inform you that you were the department's fourth choice; later you will have an intimate dinner with department members whose inability to talk to you is exceeded only by their inability to talk to each other. At your presentation or, rather, audition, you will be asked many questions, but the content of all of them will be the same: what makes you think that you're good enough to join us in the first place? Finally you leave town without ever having had a chance to find out very much about the things that would really be of concern to anyone contemplating relocating from one place to another; and in general it seems that no one thought to ask the most obvious question: if I were coming to see whether I wished to exchange my present situation for a new one, what would I want to know and how would I want to be treated?

The result is perfectly in line with the general rule I announced earlier: the pleasure and satisfaction of landing a new and better job are blunted and even overwhelmed by the programmatic unpleasantness of the process. Nor is this an accident that could be avoided if a few obvious truths about social and human obligations were pointed out to the responsible parties; for the inattentiveness, like the ugliness of Volvos, is *purposeful* and is valued both by those who perform it and by those who receive it. Were the process to be efficient and sensitive both to the personal feelings of the participants and to the realities of the marketplace, it would be too much like the world of business, and the investment in being distinguishable from business is so great that academics will pay any price to protect it. In the collective eye of the academy, sloppiness, discourtesy, indifference, and inefficiency are *virtues,* signs of an admirable disdain for the mere surfaces of things, a disdain that is itself a sign of a dedication to higher, if invisible, values.

Once one understands this, an otherwise puzzling feature of academic life makes perfect sense. I am referring to the remarkable uniform incompetence of academic administrators. It is tempting to say, as many have, that academics get the administrators they deserve, but it would be more accurate to say that academics get the administrators they *want*. What they want is administrators who are either so weak that they provide no protection against the pressures exerted by higher-level administrators or so tyrannical that there is no protection against the pressures *they* exert. In either case, by getting the administrators they want, academics get what they *really* want—they get to be downtrodden; and by getting to be downtrodden, academics get what they really *really* want—they get to complain. If one listens to academics, one might make the mistake of thinking they would like their complaints to be remedied; but in fact the complaints of academics are their treasures, and were you to remove them, you would find either that they had been instantly replenished or that you were now their object.

The reason that academics want and need their complaints is that it is important to them to feel oppressed, for in the psychic economy of the academy, oppression is the sign of virtue. The more victimized you are, the more subject to various forms of humiliation, the more you can tell yourself that you are in the proper relation to the corrupted judgment of merely worldly eyes. Were you to be rewarded in accordance with what you took to be your true worth, that worth would immediately be suspect. The sense of superiority so characteristic of the academic mentality requires for its maintenance continued evidence of the world's disdain, evidence that takes the form of poor working conditions, the elimination of so-called privileges like offices and telephones, increases in course loads, decreases in salary, and public ridicule. As each of these misfortunes is visited upon the academic, he or she acquires a greater measure of that pained sensitivity that makes so many academic faces indistinguishable from the faces of medieval martyrs.

Now if martyrdom and self-denial, even self-hatred, are the true treasures of the academic life, it follows that the generous academic will be eager to share those treasures with others. That is the purpose of tenure decisions and other rites of academic passage. Here the skill is once again to manage the bestowing of rewards in such a way as to render them bitter to the taste. The strategies include delay, ritual humiliations, unannounced shifts in standards, procedures that are either frustratingly secret or painfully public. The meetings themselves are an exercise in virtuoso captiousness where the art is to find the formulation that will best express the proper combination of regretful compassion and condemnation: "a great teacher, an excellent colleague, but . . ." "The work is good and there is an impressive amount of it, but . . ." "We have no better undergraduate teacher, but . . ." And if there seems to be nothing to fault, there is always the invocation of the standard that no one, least of all the invoker, could meet: "Is she taking the field in an entirely new direction?" And if the answer should be "yes," it is no trick at all to turn it into a negative. "Is her work so ahead of the field that no one will be inclined to follow her?"

You get the idea. You've been getting and giving it all your life, when you've been blocked from teaching a course for so long that when you are finally assigned it, you are no longer interested in the material; when you've waited for years for an office you're given three months before the building is torn down; when you've been an associate professor for so long that promotion can only be experienced as an insult. Whatever else they are, academics are resourceful, and when they set their minds to it, there are no limits to the varieties of pain they can inflict on one another.

Originally I had intended to write this paper differently, not as an unfolding argument, but as a simple enumeration of the aphorisms I have been fashioning since I entered the profession in 1962. Here is a sample:

—*In the academy, the lower the act, the higher the principle invoked to justify it.* This aphorism underlies the analyses I offered earlier; it speaks to the academic inability to acknowledge desire unless it is packaged as altruism. It also speaks to the bizarre but strangely logical outcome of this transformative practice: pettiness, which might be held in check were it acknowledged for what it is, instead becomes raised to a principle and is renamed eccentricity or even individualism so that it can then be defended in the name of academic freedom. In this way acts of incredible cruelty can be licensed and even admired. The sequence leads us to another aphorism:

—*Academics like to keep their eyes on the far horizon with the result that everything and everyone in the near horizon gets sacrificed.* The curious thing is that academics like to sacrifice themselves, that is,

—*Academics like to feel morally culpable, especially in relation to those who would give anything to be in their place;* and also,

—*Academics like to feel morally superior, which they manage by feeling morally culpable.* Together these aphorisms illuminate a curious history in which already enfranchised academics, largely male, gazed with envy and strangely mediated desire at the disenfranchised, first at Jews, then at women, then at blacks, and then at Native Americans, and now at gays and Arabs.

Let me emphasize again that these aphorisms describe a two-way commerce, victim and victimizer, trashers and trashees, each not only needing but desiring the other. The essence of it all is contained in the very first aphorism I ever formulated, in 1964 as I watched my colleagues at Berkeley turn from abasing themselves before deans and boards of trustees to abasing themselves before students: —*Academics like to eat shit, and in a pinch, they don't care whose shit they eat.* Of course had I known enough at the time, I could have saved myself the trouble and simply quoted from Freud's essays on masochism. For the masochist, Freud explains, "it is the suffering itself that matters; whether the sentence is cast by a loved one or by an indifferent person is of no importance; it may even be caused by impersonal forces or circumstances, but the true masochist always holds out his cheek whenever he sees a chance of receiving a blow" ("The Economic Problem in Masochism"). (A friend suggested to me that a better title for this essay might be "An Academic Is Being Beaten.")

In the past year many blows have been struck and many cheeks have been proffered. The always dormant strain of anti-intellectualism in American life has been reawakened by a virulent union of disgruntled neoconservatives and ignorant journalists. In nearly every newspaper and magazine in the country, teachers of literature have been credited with the destruction of Western civilization and with other nameless crimes. Academic-bashing has become the national spectator sport, and, predictably, some academics are among the best players. They may soon be outdistanced, however, by a growing chorus of politicians led by George Bush who have realized anew what Ronald Reagan discovered in the 1966 California gubernatorial election, that academics present an irresistible target, not simply because they are highly visible, but because, by and large, they will not fight back. The election year of 1992 promises to provide a perfect match of an attacking army and a target population that finds its pleasure in punishment. Before it is all over, both sides may be moved to echo Wordsworth: "bliss was it in those times to be alive."

Of course, I may be wrong; it may not go that way. Indeed, everything that I have said may be wrong or worse, which is why, despite the fifty-year tradition of the English Institute, I would prefer not to entertain questions at the conclusion of this paper. Instead, and in accordance with the spirit of academic practice as I have described it, I will cheerfully plead guilty to all charges in advance. I acknowledge that the statements I have made are too sweeping and admit of innumerable exceptions: that some Volvos are beautiful, that no one here now owns or has ever owned a Volvo; that the life you experience in your various departments is characterized by amity and generosity; and that your relationship to the rewards and privileges of the profession is straightforward and healthy. I further acknowledge that I am necessarily (and multiply) implicated in the critique I have presented; that I have been a member of the academy for thirty years, in which I have been an eager participant in its economy, often providing, as I have here, the desired beating for those who have assembled to receive it; that every sin of which I have accused others is *writ large* in my own performance. And finally I acknowledge that there is no justification whatsoever for that performance, that it is irresponsible, self-indulgent, self-aggrandizing, and entirely without redeeming social or intellectual value. It is just something I have always wanted to do.

APPENDIX

Fish Tales: A Conversation with "The Contemporary Sophist"

GARY A. OLSON

Perhaps one reason why Stanley Fish influences so many of us in rhetoric and composition is that he has always insisted that rhetoric is *central,* that it's the "necessary center," that "substantial realities are products of rhetorical, persuasive, political efforts." As Fish says in the interview that follows, once you "begin with a sense of the constructed nature of human reality," then rhetoric is "reconceived as the medium in which certainties become established." It's no wonder, then, that Fish feels comfortable being called a social constructionist. Nor is it surprising that he finds "perfectly appropriate" Roger Kimball's label for him: "the contemporary sophist." In fact, Fish sees an affinity between sophism and the antifoundationalist project he has so long championed. He credits his work with Milton, his first love and still a driving passion in his intellectual life, as the genesis of his struggle against essentialist, foundationalist philosophies: as an antinomian Christian and an "absolutely severe anti-formalist," Milton was "rather far down the anti-foundationalist road."

Another reason for Fish's influence in rhetoric and composition is his continued interest in and support of composition. He remains conversant with the discipline's intellectual developments, and he even goes so far as to say that much of his thinking about theory and antifoundationalism was formed in the early 1960s when he taught composition classes using Walker Gibson's *The Limits of Language:* "The essays in that book were perhaps the most powerful influence on me." As always, Fish is outspoken about intellectual trends he disapproves of, and certain developments in composition are no exception. He is skeptical of attempts to "teach people that situational experience is in fact always primary" because he believes this "theoretical" lesson will not produce any generalizable result. On the other

Source: Journal of Advanced Composition 12, no. 2 (Fall 1992). Reprinted with permission.

hand, he favors training in which composition students are placed in real-istic scenarios and are asked to write to the scenario. The difference, in Fish's view, is that the first is an attempt to teach students a "theoretical" perspective in the hopes that they can then apply that perspective to partic-ular situations—something that just cannot happen, according to Fish. The second, however, is experience or practice in specific contexts—for Fish the only "real" kind of knowledge. He repeats, "The practice of training students to be able to adjust their verbal performances to different registers of social life requires no theoretical assumptions whatsoever."

Clearly, this position is consistent with his larger campaign over the years "against theory." Says Fish, "I'm a localist. . . . I believe in rules of thumb." That is, he believes intensely in here-and-now situationality; to believe otherwise would be to subscribe to "the fetishization of the unified self and a whole lot of other things that as 'postmodernists' we are sup-posedly abandoning." Thus, he discounts attempts to cultivate critical self-consciousness, another type of *theoretical* capacity: "Insofar as critical self-consciousness is a possible human achievement, it requires no special ability and cannot be cultivated as an independent value apart from partic-ular situations."

Fish also comments on other issues in composition scholarship. Retreat-ing somewhat from his earlier criticism of Kenneth Bruffee, Fish acknowl-edges that collaborative learning *can* be productive. But we must not as-sume, he cautions, that somehow it is inherently superior to other modes of instruction; it is simply "different," each pedagogical strategy having its own "gains and losses." And while he refuses to embrace radical peda-gogy, he sees it as "*a* wave of the future." He himself prefers a more traditional arrangement: perceiving the classroom as "a performance occa-sion," he enjoys "orchestrating the class," noting that no one would ever mistake one of his classes for "a participatory democracy." He quips that he would never adopt liberatory techniques for two very good reasons: "too much egocentrism, too much of a long career as a professional theatrical academic."

In addition, Fish expresses genuine respect for feminism and the influ-ence it has exerted on the intellectual life of society because, for Fish, it has passed the key test that indicates the "true power of a form of inquiry": when "the assumptions encoded in the vocabulary of a form of thought become inescapable in the larger society." He believes that the questions raised by feminism "have energized more thought and social action than any other 'ism' in the past twenty or thirty years." Nevertheless, he does not support feminists "who rely in their arguments on a distinction between male and female epistemologies." Such feminists, he feels, fall prey to the same epistemological difficulties as those who champion critical self-consciousness: a belief that "you can in some way step back from, rise above, get to the side of your beliefs and convictions so that they will have less of a hold on you."

Fish addresses numerous other issues, such as the nature of "intentional structures" and "forceful interpretive acts," the bankruptcy of the liberal intellectual agenda, and the obligation of academics to engage in what Noam Chomsky has called "more socially useful activities." He is particularly concerned about how the larger societal turn toward conservatism is affecting higher education, and he predicts a period of curtailment and purges so long as the well-financed neoconservative political agenda continues to be "backed by huge amounts of right-wing foundation money." The solution is for academics to speak out to audiences beyond the academy, to help explain intellectual developments to the general public in order to counter narrow conservative perspectives: "I think we *must* talk back."

It may seem something of a paradox that Stanley Fish, who argues so vociferously *against* theory, is becoming more influential in composition studies precisely at a time when the field, or at least part of it, is busily engaged in *theory building*. Yet while the role of theory in rhetoric and composition may still be uncertain, it *is* clear that Fish is in many ways an ally. Especially as more compositionists explore social construction and the role of rhetoric in epistemology, Fish's work becomes increasingly relevant. Responding to criticism from both the intellectual right and left, Fish insists that harkening to him will not "lead to the decay of civilization," nor will it "lead to the canonization of the status quo." Harkening to Fish, however, *may* well lead to productive avenues of inquiry for many of us in composition.

Q. In *Doing What Comes Naturally,* you speak of this as the age of rhetoric and the "world of *homo rhetoricus*." You yourself are frequently called a rhetorician *par excellence*. But do you consider yourself a writer?

A. I do in some ways. Last night at the Milton Society of America banquet I spoke of the influence on me of C. S. Lewis. I think of C. S. Lewis and J. L. Austin as the two stylists I've tried to imitate in a variety of ways, and so I'm very self-conscious about the way I craft sentences. I always feel that once I get a particular sentence right I can go on to the next, and I don't go on to the next until I think it's right. In the sense that this is not just superficially but centrally a concern, I consider myself a writer. In other senses—for example, whether I expect people to be studying my works long after my demise—the answer is that I do *not* consider myself a writer. But the craft I think of myself as practicing is the craft of writing, and my obsession there is a very old-fashioned one, a canonical one, a traditional one—and that is clarity.

Q. What you describe is exactly how Clifford Geertz described *his* writing process recently in *JAC*. Would you tell us more about your writing process? Do you revise frequently? Use a computer?

A. I do not use a computer and I do not revise. I now use one of those small electronic typewriters that you can move around and take on so-called vacations. That's about as far as I've advanced in the age of mechanical reproduction. Since my writing practices are as I just described, I don't tend to revise. I go back occasionally and reposition an adverb, and I often go through my manuscripts and cross out what I know to be some of my tics. For example, I use the phrase *of course* too much, I often double nouns and verbs for no particular reason, and I have other little favorite mannerisms that I've learned to recognize and eliminate. But very rarely do I ever restructure an essay or even a paragraph.

Q. That's surprising considering your polished style. I should have thought you spent countless hours revising.

A. Well, I write slowly. My pace is two pages a day when I'm writing well, when I have a sense of where the particular essay is or should be going. That's often when I sit for six to eight hours and am continually engaged in the process of thinking through the essay. Also, I do this often (not always, but often) while watching television. This is a very old habit. Actually, this is a talent (if it is a talent) that more people of the younger generation have today than people of my generation. But I've always been able to do it. To this day when I reread something I've written I can remember what television program I was watching when I wrote it. I remember once when I was in Madrid and went to the bullfights, I wrote a passage about Book Six of *Paradise Lost;* every time I look at it I remember that I was watching the bullfights when I wrote it.

Q. As an English department chair, what are your thoughts about the future of rhetoric and composition as a discipline? What role will it (or should it) play in the modern English department?

A. I don't know because I don't know whether there will be something called "the modern English department" in the next twenty years. I had thought, in fact, that there would be a more accelerated transformation of the traditional English department than there has yet been. My prediction ten years ago had been that by the year 2000 the English department in which we were all educated would be a thing of the past, a museum piece, represented certainly in some places but supplanted in most others by departments of literature, departments of cultural studies, departments of humanistic interrogation, or departments of literacy. That hasn't happened in the rapid way I thought it would; there are *some* places, like Syracuse University, and earlier modes of experimentation, like the University of California at San Diego, Rensselaer Polytech, and others, but not as many as I thought there would be. If the change, when it comes, goes in the same direction that Syracuse has pioneered, then it might be just as accurate to call the department "the department of rhetoric," with a new understanding of the old scope of the subject and province of rhetoric. That's a possibility, but I'm less confident now than I was ten years ago about such predictions. For one thing, the economic difficulties we've been experiencing lately have had a great effect on the academy. Two years ago the job market looked extraordinarily promising, and certain kinds of pressures that departments had always felt seemed to be lessening. There would therefore have been an atmosphere in which experimentation and transformation might have been more possible. But now that we have had a return of a sense of constricted economy and constricted possibilities and everybody is talking retrenchment (an awful word, but one that you hear more and more), it may be that the current departmental sense of the university structure may continue because the protection of interests that are now in place becomes a strong motive once a threat to the entire structure is perceived. And certainly many people are now perceiving a threat to the entire structure.

Q. So for progress, we need prosperity.

A. Absolutely, and especially in the humanities. Two or five or ten years ago, none of us would have predicted the current *political* assault on humanities education and the attempt to—and this is perhaps the least plausible scapegoating effort in the history of scapegoating, which is a very long history—blame all the country's ills on what is being done in a few classrooms by teachers of English and French. It is truly incredible that this story of why the moral fiber of the United States has been weakened has found such acceptance, but it has and now these consequences are ones we have to deal with in some way.

Q. During a talk at the University of South Florida, you repeated on several occasions that you are a "very traditional" teacher who uses "very traditional methods." In *JAC*, Derrida characterized his own teaching in much the same way. Yet in English studies now, especially in rhetoric and composition, there's a movement toward "liberatory learning," radicalizing the classroom and breaking down its traditional power structures by attempting to disperse authority among all participants. What are your thoughts about radical pedagogy in general and its altering of the teacher-student hierarchy specifically?

A. Well, my thoughts about radical pedagogy are complicated by the fact that my wife, Jane Tompkins, is a radical pedagogue and moves more and more in that direction; she writes essays and gives talks that command what I would almost call a cult following. I've seen some of these performances. *That* at least tells me that there's something out there to which she and others are appealing. I have *some* sense—which one might call an anthropological sense—of what that something is. But I'm simply too deeply embedded in and too much a product of my own education and practices to make or even to want to make that turn. I would first have to feel some dissatisfaction with my current mode of teaching or with the experiences of my classroom, and I don't feel that. For me the classroom is still what she has formally renounced: a performance occasion. And I enjoy the performances; I enjoy orchestrating the class in ways that involve students in the performances, but no one is under any illusion that this is a participatory (or any other kind) of democracy in a class of mine. However, having said this I should hasten to add that my own disinclination to turn in that direction does not lead me to label that direction as evil, wayward, irresponsible, unsound, or any of the usual adjectives that follow. It seems to me quite clear that this is, if not *the* wave of the future, *a* wave of the future. In fact, I listened to some of the interviews for our assistant director of composition position yesterday, and every one of the interviewees I talked to identified himself or herself as a person interested in just this new kind of liberatory, new age, holistic, collaborative teaching. So I think it is the wave of the future, and I would certainly welcome those who are dedicated. But I'm sure that I would never do it myself—too much egocentrism, too much of a long career as a professional theatrical academic.

Q. You disagree with Patricia Bizzell and those who encourage us to teach students the "discourse conventions" of their disciplines, arguing that "being told that you are in a situation will help you neither to dwell in it more perfectly nor to *write* within it more successfully." Surely, though, you don't really advise compositionists *not* to teach students that there are numerous discourse communities, each with characteristic discourse conventions? Wouldn't we be remiss to ignore such considerations in our pedagogies?

A. Of course, I quite agree. My objection in that essay—an objection I make in other essays in slightly different terms—is to the assumption that if we teach people that situational experience is in fact always primary and that one never reasons from a set of portable and invariant theses or propositions to specific situations (that is, one is always within a situation in relation to which some propositions seem relevant and others seem out in left field), if you just teach that as a theoretical lesson and walk away from the class and expect something to happen, the only thing that will happen is that the next time you ask that particular question, you'll get that particular answer. However, I do believe in training of a kind familiar to students of classical and medieval rhetoric—training, let's say, of the Senecan kind, in which one is placed by one's instructor in a situation: you are attempting to cross a river; there is only one ferry; you have to persuade the ferryman to do this or that, and he is disinclined to do so for a number of given reasons—what do you then do? That kind of training, transposed into a modern mode, is essential. I don't think it need be accompanied by any epistemological rap. What I was objecting to in Pat's essay—and I was to some extent being captious because in general I am an admirer of her work—was the suggestion that the theoretical perspective on situationality itself could do work if transmitted to a group of students. I think that one could teach that way, and many have—that is, they've taught situational performance and the pressures and obligations that go with being in situations—without ever having been within a thousand miles of a theoretical thought.

Q. But isn't this inconsistent with everything you say about rhetoric in the larger sense, that to be a good rhetorician is to know situatedness?

A. It depends on what you mean by "to know situatedness." There's one sense in which to know situatedness is to be on one side of a debate about the origins of knowledge, to be on the side that locates knowledge or finds the location of knowledge in the temporal structure of particular situations. That's to know situatedness in the sense that one might call either theoretical or philosophical. Now to know situatedness in the sense of being able to code switch, to operate successfully in different registers, is something else, and you don't even have to use the vocabulary that accompanies most theoretical discussions of these points. I was a teacher of composition long ago (relatively—thirty years ago) before any of us in the world of literary studies knew the word *theory* (Ah, for those days! Bliss was it in those times to be alive!), and many people taught what we would now call situational performance; and there were many routes to that teaching. So again, my point always is that the practice of training students to be able to adjust their verbal performances to different registers of social life requires no theoretical assumptions whatsoever. They are required neither of the instructor, and certainly not of the student.

Q. Kenneth Bruffee draws heavily on you and Richard Rorty to formulate his version of collaborative learning theory. Rorty has already distanced himself from Bruffee's project, and you criticize it because it "becomes a new and fashionable version of democratic liberalism, a political vision that has at its center the goal of disinterestedly viewing contending partisan perspectives which are then either reconciled or subsumed in some higher or more general synthesis, in a larger and larger *consensus*." Given one of the major (and typical) alternatives—a teacher-dominated classroom and an information-transfer model of education—and given the fact that much of your own life's work has been devoted to illustrating how interpretive communities work, wouldn't you agree that Bruffee's collaborative learning is productive despite his own naive liberalism?

A. Yes, it could be. I think that's an excellent point. I don't know whether it is, but there's nothing to prevent it from being productive. That is, collaborative learning is a mode of knowledge production *different* from other modes of knowledge production. In my view, differences are always real; however, differences should never be ranked on a scale of more or less real. So to refine what might be a point of contest between Bruffee and me, I would agree with him that if we move into a mode of collaborative learning, *different* things will happen, things which probably would not have been available under other modes. Also, *some* possibilities will be lost. I tend to think of pedagogical strategies as strategies each of which has its gains and losses. I also believe that there are times in the history of a culture or a discipline when it's time to switch strategies, not because a teleology pulls us in the direction of this or that one, but because the one in which we've been operating has at the moment taken us about as far as we can go and so perhaps we ought to try something else—which, as you will have recognized, has a kind of Rortian ring to it.

Q. Yes, it does. In fact, let me ask you a related question. Your essay "Change" is a detailed discussion of how interpretive communities change their beliefs and assumptions, and in making the argument that "no theory can compel change" you say that we should think of the community not as an "object" of change but as "an engine of change." Although you don't use Rorty's vocabulary of "normal" and "abnormal discourse," like Rorty you seem to be arguing against the notion that substantive disciplinary or intellectual change transpires as a result of persuasive abnormal discourse. What are your thoughts on the role of abnormal discourse as a catalyst of change?

A. I think that abnormal discourse *can* be a catalyst of change, and that's because I think that *anything* can be a catalyst of change. This goes back to a series of points I've been making "against theory" for a number of years now. One of my arguments is that strong-theory proponents attribute to theory a unique capacity for producing change and often believe (and this is perhaps a parody) that if we can only get our epistemology straight, or get straight our account of the subject, then important political and material things will follow. It's that sense of the kind of change that will follow from a new theoretical argument that I reject. However, theory—or as I sometimes say tendentiously in these essays, "theory talk"—can like anything else be the catalyst of change, but it's a contingent and historical matter; it depends on the history of the particular community, the kinds of talk or vocabularies that have prestige or cachet or are likely to trump other kinds of talk. And if in a certain community the sense of what is at stake is highly intermixed with a history of theoretical discourse, then in *that* community at *that* time a change in practices may be produced by a change in theory.

Of course, "abnormal" discourse comes in a variety of forms. For example, if one takes the term "discourse" in its larger senses, one can think that a recession is an abnormal form of discourse, that suddenly one's ordinary ways of conceiving of one's situation are complicated by facts that a year or two ago would have seemed to belong in another realm. Abnormal discourse can always erupt into the routine structures of an interpretive community, but there's no way to predict in advance which ones will in fact erupt and with what effects.

Q. Well, Rorty claimed in *JAC* that abnormal discourse is "a gift of God," while Geertz prefers a less grandiose notion he calls "nonstandard discourse." For Rorty this happens rarely; for Geertz it occurs all the time. Obviously, these are two different conceptions of abnormal discourse. Does either one seem more useful than the other?

A. Not really. I'm not sure whether Rorty and Geertz are making this assumption, but it could be that one or both of them is assuming that abnormal discourse is itself a stable category, and it seems to me that what is or is not abnormal in relation to a discourse history will itself be contingent. For example, in some literary communities that I know about and that I'm a participant in, it now becomes "abnormal" to begin a class by saying, "Today we will explicate Donne's 'The Good-Morrow.' " *That* would be, at least in some classrooms at Duke, a *dazzling* move—not, I hasten to add, in *mine,* because that's a practice I've never ceased to engage in.

Q. Both you and Rorty have been cited as two of the principal intellectual sources of social construction; however, when I asked Rorty in an earlier interview if he considered himself a social constructionist he seemed baffled by the appellation. Despite your understandable resistance to limiting categories, and given your continual insistence that everything is rhetorical and situated, would you consider *yourself* a social constructionist?

A. In a certain sense I would say, "Sure." If I were to be asked a series of questions relevant to a tradition of inquiry in which several accounts of the origin of knowledge or facticity were given, I would come out on the side that could reasonably be labeled "social constructionism." I myself have not made elaborate arguments for a social constructionist view—though I've used such arguments at points in my writing—but I have no problem being identified as someone who would support that view.

Q. In the essay "Rhetoric" you examine the history of anti-rhetorical thought and the unchanging "status of rhetoric in relation to a foundational vision of truth and meaning." You state, "Whether the center of that vision is a personalized deity or an abstract geometric reason, rhetoric is the force that pulls us away from the center and into its own world of ever-shifting shapes and shimmering surfaces." You contrast this mainstream tradition with a counter tradition, represented in classical times by the sophists and today by the anti-foundationalists, whom you credit with helping to move "rhetoric from the disreputable periphery to the necessary center." First, in establishing rhetoric as a kind of master category, don't you run the risk of what Derrida has warned of in *JAC* and elsewhere: rhetoricism, "thinking that everything depends on rhetoric"?

A. What is his point? What's the risk?

Q. He claims that rhetoricism leads us down an essentialist path. His definition is literally that rhetoricism is "thinking that everything depends on rhetoric." It seems to me very different from what *you* say.

A. I'm surprised to hear that answer from Derrida because it seems to buy back into a view of the rhetorical that would oppose it to something more substantial, whereas in my view substantial realities are products of rhetorical, persuasive, political efforts. When discussing these matters with committed foundationalists, of whom there are still huge numbers, one always is aware that for them the notion of rhetoric only makes sense as a category of inferiority in relation to something more substantial. For someone who listens with a certain set of ears, the assertion of the primacy of rhetoric can only be heard either as an evil gesture in which "the real" is being overwhelmed, or as a gesture of despair in which either a hedonistic amorality or paralysis must follow. All of these responses to the notion of the persuasiveness of rhetoric are, of course, holding on for dear life to a paradigm in which the rhetorical only enters as the evil shadow of the real. If, on the other hand, you begin with a sense of the constructed nature of human reality (one leaves the ontological question aside if one has half a brain), then the notion of the rhetorical is no longer identified with the ephemeral, the outside, but is reconceived as the medium in which certainties become established, in which formidable traditions emerge, are solidified, and become obstacles (not insurmountable ones, but nevertheless obstacles) to the force of counter-rhetorical movements. So I would give an answer like that to what might seem to be one reading of Derrida's warning, though I am loathe to put *him* anywhere near the camp of those whose thoughts I was describing, since he's a man, as everyone knows, of extraordinary power of intelligence.

Q. Then do you conceive of the project of the anti-foundationalists as an extension or resurgence of sophism?

A. I think that's one helpful way of conceiving of it, and it's helpful in a rhetorical sense. Roger Kimball, in *The New Criterion,* wrote an essay that I think later became part of *Tenured Radicals: How Politics Has Corrupted Our Higher Education.* I don't know what the title of the essay is in *Tenured Radicals,* but in *The New Criterion* the essay on me was entitled "The Contemporary Sophist." He meant that as a derogatory label, but I thought it was perfectly appropriate. To call oneself a sophist is rhetorically effective at the moment because you seem to be confessing to a crime. If you begin by saying, "I am a sophist," and then begin unashamedly to explain why for you this is not a declaration of moral guilt, it's a nice effective move; it catches your audience's attention. So I think that right now there's some mileage (although it's mileage that's attended by danger, too) in identifying the new emphasis on rhetoric with the older tradition of the sophists.

Q. Recently in *JAC* Clifford Geertz said that Kenneth Burke was one of two thinkers who had the most influence on him intellectually, the other being Wittgenstein. You yourself have referred to Burke's work from time to time. What is your assessment of Burke's contribution to our ways of thinking about language and rhetoric?

A. I don't have a strong assessment. I've read Burke only sporadically and only occasionally and have never made a sustained study of his work and therefore could not say that I have been influenced by it directly. I'm sure that I've been influenced by it in all kinds of ways of which I am unaware because of the persons that I've read or talked to who have themselves been strong Burkeans. I can think of two such people that I've talked to and read a great deal: my old friend Richard Lanham at UCLA and Frank Lentricchia, my colleague at Duke, both of whom are committed Burkeans. No doubt lots of things that they have said to me over the years have passed on a heritage of Burke to me, but I've never myself studied him in an intensive way.

Q. Who *has* had a major influence on you?

A. Well, that's difficult to say. Of course, Milton has been a major influence on me. That would be inescapable having spent thirty years studying his work.

Q. Milton, the anti-foundationalist?

A. Yes, in a way. Milton is an antinomian Christian. That is, he's an absolutely severe anti-formalist. Everyone has always known that about Milton. He is continually rejecting the authority of external forms and even the shape of external forms independently of the spirit or intentional orientation of the believer. In his prose tract called *The Christian Doctrine,* which was only discovered many years after his death, Milton begins the second book, which is devoted to daily life, to works in the world, by asking the obvious question, "What is a good work?" He comes up with the answer that a good work is one that is informed by the working of the Holy Spirit in you. That definition, which I've given you imperfectly, does several things. It takes away the possibility of answering the question "What is a good work?" by producing a list of good works, such as founding hospitals or helping old ladies cross the street. It also takes away the possibility of identifying from the outside whether or not the work a person is doing is good or bad, since goodness or badness would be a function of the Holy Spirit's operation, which is internal and invisible. Milton then seals the point by saying a paragraph or so later that in answer to the question "What is a good work?" some people would say the ten commandments, and therefore give a list. Milton then says, "However, I read in the Bible that *faith* is the obligation of the true Christian, not the ten commandments; therefore, if any one of the commandments is contradictory to my inner sense of what is required, then my obedience to the ten commandments becomes an act of sin." Now, if within two or three paragraphs of your discussion of ethics, which is what the second book of *The Christian Doctrine* is, you have dislodged the ten commandments as the repository of ethical obligation, you are rather far down the anti-foundationalist road. And Milton is a strong antinomian, by which I mean he refuses to flinch in the face of the extraordinary existential anxiety produced by antinomianism. So, much of my thinking about a great many things stems from my study of Milton.

Also, I've been strongly influenced as a prose stylist, as I've already mentioned, by C. S. Lewis and J. L. Austin. In fact, I've been very much influenced by J. L. Austin in my thinking about a great many things in addition to my thinking about how to write certain kinds of English sentences. I've also been influenced by Augustine. It's a curious question to answer because many of the people whom I now regularly cite in essays are people that I read *after* most of the views that found my work were already formed. That is, I hadn't read Kuhn before 1979. I'm fond of citing Kuhn, as a great many other people are. I have found support again and again in the pages of Wittgenstein, but I cannot say that it was a study of Wittgenstein that led me to certain questions or answers.

Let me say one more thing. When I was first starting out as a teacher, I gave the same exam in every course, no matter what the subject matter. The exam was very simple: I asked the students to relate two sentences to each other and to the materials of the course. The first sentence was from J. Robert Oppenheimer: "Style is the deference that action pays to uncertainty." I took that to mean that in a world without certain foundations for action you avoid the Scylla of prideful self-assertion, on the one hand, and the Charybdis of paralysis, on the other hand, by stepping out provisionally, with a sense of limitation, with a sense of style. The other quotation, which I matched and asked the students to consider, is from the first verse of Hebrews Eleven: "Now faith is the substance of things hoped for, the evidence of things not seen." I take that to be the classically theological version of Oppenheimer's statement, and so the question of the relationship between style and faith, or between interpretation and action and certainty, has been the obsessive concern of my thinking since the first time I gave this test back in 1962 or 1963. I think there is *nothing* in my work that couldn't be generated from those two assertions and their interactions. They came from a book I used in my composition teaching from the very beginning, and I don't even know how I came to use it. The essays in that book were perhaps the most powerful influence on me. It's a book edited by Walker Gibson, and it's called *The Limits of Language*. It had this essay by Oppenheimer; essays by Whitehead, Conant, and Percy Bridgman, the Nobel Prize winning physicist; Gertrude Stein's essay on punctuation (which is fantastic); and several others that I used in my classes and that informed my early questionings and giving of answers. That book was an extraordinarily powerful influence. Of course, the quotation from Hebrews Eleven came in from my Milton work.

Q. You just mentioned finding support in the pages of Wittgenstein. In "Accounting for the Changing Certainties of Interpretive Communities," Reed Way Dasenbrock suggests that your debt to Wittgenstein is far greater than you have yet acknowledged. Do you think Wittgenstein was a major influence on your work?

A. No I don't, because I don't know him well enough. Reed was a student of mine. I have a bunch of students out in the world who make what I hope is a very good living writing essays that point out my limitations and flaws, and he's one of them. He no doubt knows Wittgenstein much better than I do and has learned a great deal from him; he therefore probably assumes that I *must* have been influenced by him. Now it *is* true that back in about 1977 or 1978 I was for a semester in a reading group with two or three philosophers from Johns Hopkins: David Sachs, George Wilson, and my friend the art historian, Michael Fried. We read Wittgenstein and talked about him for a period of months. Somewhat earlier—and here's another influence that I'd forgotten to acknowledge but should have acknowledged—there's probably a larger influence from Heidegger as transmitted to me in a series of courses I attended given by Hubert Dreyfus, a philosopher at Berkeley whose notes on *Being and Time* have just been published and have been long awaited, and whose early book *What Computers Can't Do* was another strong and powerful influence on me. That's a great book, both in its first and second editions. Through my friendship with Dreyfus, who is a magnificent teacher, and because of the pleasure and illumination I gained from his courses, there is probably some kind of Heidegger-Wittgensteinian circuit (there *is* a relationship, though a tortured one, between Heidegger and Wittgenstein) that has had more power in my work than I consciously acknowledge. Therefore, I guess I end up saying that in a way Reed may be right.

Q. Last October, over dinner, you and I discussed various issues with Dinesh D'Souza, and I remember your eloquent and impassioned plea for him to believe that feminism is "the real thing," that significant and substantial developments are occurring within and because of feminism. Exactly what of importance is happening in feminism?

A. I couldn't answer that question because feminism has become as a discipline and a series of disciplines so complicated, such a map with so many different city-states or nation-states, that it would be foolish of me to start pronouncing. What I was trying to convey to Dinesh (the question of whether or not one conveys *anything* to Dinesh is an interesting one) is that the questions raised by feminism, because they were questions raised not in the academy but in the larger world and that then made their way into the academy, have energized more thought and social action than any other "ism" in the past twenty or thirty years, including Marxism, which may have been in that position in an earlier period but is in our present culture no longer in that position. Now what is that position? It is the position that in my view marks the true power of a form of inquiry or thought: when the assumptions encoded in the vocabulary of a form of thought become inescapable in the larger society. For example, people who have never read a feminist tract and would be alarmed at the thought of reading one are nevertheless being influenced by feminist thinking in ways of which they are unaware or are to some extent uncomfortably aware. Such influence often exhibits itself in the form of resistance: "*I'm* not going to fall in with any of that feminist crap," thereby falling in headfirst as it were. My benchmark comparison here is with Freudianism. Freudianism's influence on our society is absolutely enormous and in the same way. People who have never read Freud, and who would not think of reading Freud, nevertheless have a ready store of Freudian concepts about the unconscious, repression, slips of the tongue, a vague sense that there's something called the "Oedipus Complex," and so on. *That's* when a form of thought has genuine power; it becomes unavoidable in our society. Feminism, I think, has that status and will continue to have that status (especially if there are more things like the Clarence Thomas/Anita Hill hearings).

Q. You have argued that "feminists who rely in their arguments on a distinction between male and female epistemologies are wrong, but, nevertheless, it may not be wrong (in the sense of unproductive) for them to rely on it." Currently, the most influential version of feminism in composition is concerned primarily with such a distinction. Would you explain, first, why such a distinction is problematic, and also how it nonetheless might be productive?

A. Well, it's problematic in relation to my own notion of the way belief and conviction work. My stricture on that particular piece of feminist theory follows from my general position on critical self-consciousness. Critical self-consciousness, which was my main object of attack for a number of years (now I see that the true object of attack all along was liberalism in general) is the idea that you can in some way step back from, rise above, get to the side of your beliefs and convictions so that they will have less of a hold on you than they would had you not performed this distancing action, thereby enabling you to survey the field of possibilities relatively unencumbered by the beliefs and convictions whose hold has been relaxed. This seems to me to be *zany* because it simply assumes but never explains an ability to perform that distancing act, never pausing to identify that ability and to link the possession of that ability with the thesis that usually begins discussions that lead to this point—the thesis of the general historicity of all human efforts. That is, most people who come to the point of talking about critical self-consciousness or reflective equilibrium or being aware of the status of one's own discourse are also persons who believe strongly in the historical and socially constructed nature of reality; but somehow, at a certain moment in the argument, they are able to marry this belief in social constructedness with a belief in the possibility of stepping back from what has been socially constructed or stepping back from one's own self. I don't know how they manage this. I think, in fact, that they manage it by not recognizing the contradiction.

The feminist version of this, at least in the strain of feminism to which you were referring, is to identify the ability to step back and not be gripped in a strong and almost military way by one's convictions, to identify that softer relationship to one's beliefs as "feminine" while perceiving the aggressive assertion of one's beliefs as "masculine." Well, if I'm right about the impossibility in a strong sense of that stepping back, then there could not be such a distinction between ways of knowing. There could be, however, as I do go on to say in that essay, different *styles* in relation to which one's beliefs are held and urged and introduced to others. And those different styles will have different effects, although, again, contingent on particular situations. It's not always the case that proceeding in a soft and relatively mild way to forward a point of view will produce effect X while a brusk and peremptory declaration of one's point of view will produce effect Y. It depends. I like to think of these not so much as a difference in female and male ways of knowing but a difference in modes of aggression. So, finally, it's whether or not you favor or at the moment find useful garden-variety aggressiveness, or whether you take refuge in passive aggressiveness— which can often be the most aggressive form of aggressiveness. This is the difference often at the root of these discussions, and it also gets into discussions of collaborative learning and of attempts to decenter classroom authority.

Q. You just mentioned your distaste for liberalism . . .

A. Yes, I never tire of it.

Q. I remember your saying not long ago that you see conservatives today as behaving "like a bunch of thugs" and liberals as "foolish and silly," but given a choice between the two, you'd side with silliness over thuggishness. You've been openly contemptuous of liberals, both within the field and in society at large. What is it about the liberal intellectual agenda that you find so repugnant?

A. What distresses me about liberalism is that it is basically a brief against belief and conviction. I understand its historical origins in a weariness with theological battles that were in the sixteenth and seventeenth centuries and earlier (and still today in parts of the world) real battles: people bled, died, mutilated one another, and so on. As every historian has told us for many years, the passions of seventeenth-century sectarian wars, especially in England, led to a sense of weariness, to a lack of faith in the ability of persons ever to be reconciled on these points, and therefore to a desire to diminish their centrality to one's life. That's one of the sources, not the only source, of liberalism's appeal. Liberalism takes the inescapable reality of contending agendas or of points of view or, as we would now say in a shorthand way, of "difference" and tries to find an overarching procedural structure which will accommodate difference and will at least defer the pressure to decide in a final way between strongly differing points of view. Liberalism is a way not so much to avoid conflict (because liberalism is born out of the unhappy insight that conflict cannot be avoided) but to contain it, to manage it, and therefore to find some form of human association in which difference can be accommodated and persons can be allowed the practice and even cultivation of their points of view, but in which the machinery of the state will not prefer one point of view to another but will in fact produce structures that will ensure that contending points of view can coexist in the same space without coming to a final conflict.

The difficulty with this view is that it assumes that structures of a kind that are neutral between contending agendas can in fact be fashioned. What I wish to say, and I'm certainly not the only one or by any means the first one to say it, is that *any* structure put in place is *necessarily* one that favors some agendas, usually by acts of recognition or nonrecognition, at the expense of others. That is, any organization that one sets up already is based on some implicit ordering of possible courses of action that have been identified or recognized as being within the pale. Then there are other kinds of actions that are simply not recognized and are therefore, as it were, written out of the program before the beginning. Now, this has not been a *conscious* act because for it to have been a conscious act it would have to have been produced in the very realm of reflective self-consciousness that I am always denying. Nevertheless, it is an inescapable fact about organization, from my point of view. So what liberalism does in the *guise* of devising structures that are neutral between contending agendas is to produce a structure that is far from neutral but then, by virtue of a political success, has claimed the right to think of itself as neutral. What this then means is that in the vocabulary of liberalism certain kinds of words mark the zone of suspicion—words like *conviction, belief, passion,* all of which are for the liberal mentality very close to fanaticism.

You could have noted a nice instance of this in the Gulf War frenzy of 1990 and 1991 when the charge that was made again and again about Saddam Hussein and his followers was that they *believed* something so strongly that they wouldn't "listen to reason." The February 1991 issue of *The New Republic* was devoted largely to the situation on campus but had three or four essays on the still-evolving Gulf War situation, and it became quite clear that for the editors and writers of *The New Republic* the danger represented by Saddam Hussein and the danger represented by multiculturalism or ethnic studies were exactly the same danger. This was the danger of persons passionately committed to an agenda, a set of assumptions—on the one hand a bunch of nutty Iraqis and on the other hand a bunch of nutty English teachers. In both cases, the obvious and compelling power of reason and rationality somehow had been overwhelmed by passion and conviction. In a way, liberalism, under this description, could be seen as a post-eighteenth-century variation of an old Judeo-Christian account of the nature of man in which man is composed of two parts: willful, irrational passion on the one hand and on the other hand something still residing in the breast, that spark of true intuition left us after the fall. So in many Christian homiletic traditions, human life is imagined as a battleground between the carnal self controlled by its appetites and something else, often called "conscience" or the "word of God," within. Now what happens in the Enlightenment is that the theological moorings of this view are detached, and in place of things like the conscience or the memory of God or the image of God's love, one has Reason. But in the older tradition (and here's the big difference), that which was contending with the carnal, because it was identified with the divine, had an obvious theological valance to it. You take away that and substitute for it Reason and then you have something as your supposed lodestar which, by the Enlightenment's definition of Reason, is independent of value. It seems to me that out of this many of the problems of liberalism, as described by a great many people, arise. So I think that liberalism is an incoherent notion born out of a correct insight that we'll never see an end to these squabbles and that therefore we must do something, and the doing something is somehow to find a way to rise above the world of conviction, belief, passion. I simply don't think that's possible.

Q. What would be an intellectual agenda that is *not* silly or thuggish?

A. I'm a localist, which is already almost a dangerous thing to say. By that I mean I don't have an intellectual agenda in any strong sense, or to put it in deliberately provocative terms: I don't have any principles. If I believe in anything, I believe in rules of thumb, in the sense that in any tradition there are certain kinds of aphorisms or axioms which encode that tradition's values, purposes, and goals; and people who are deeply embedded in that tradition are in some sense, often below the threshold of self-consciousness, committed to those values, purposes, and goals which, however, *can* in the course of the history of a tradition or profession, change. Therefore, as I say quite often (and it's true) my forward time span is generally two hours. By that I mean I tend not to think about or worry about anything more in the future than two hours hence. From a negative point of view, one might characterize my vision, therefore, as severely constrained and limited. I walk into a situation and there's something wrong sometimes, but my sense of what is wrong is very much attached to the local moment, the resources within that moment that might be available to remedy the wrong, and the possibility that my own actions might in some way contribute to that remedy. Then if someone starts commenting, "You act this way in situation *A* and three weeks ago in situation *B* I saw you act in ways that would under a general philosophical description be thought of as a contradiction," I answer, "Don't bother me. Give me a break. I am not in the business of organizing my successive actions so that they all conform to or are available to a coherent philosophical account." A lot of people assume that this is what action in the world should be: you strive from some mode of action that, if viewed from outside over a period of time, would be seen as consistent in philosophical terms. Again, I don't see that. That seems to me to go along with the fetishization of the unified self and a whole lot of other things that as "postmodernists" we are supposedly abandoning but that keep returning with a vengeance.

Q. In *Doing What Comes Naturally,* you discuss at length the role of "intention" in the production and reception of discourse. As a check on both those who "ignore" authorial intention as well as those who defer to it, you explain that "there is only one way to read or interpret, and that is the way of intention. But to read intentionally is not to be constrained relative to some other (nonexistent) way of reading." You say this is so because any meaning is "thinkable only in the light of an intentional structure already assumed." Would you elaborate on the nature and role of "intentional structure"?

A. Sure. I would back off for a moment and consider what the alternative picture would be. The alternative picture would be intention as something added on to a meaningful structure. In other words, those people who wish either to avoid or ignore intention believe that it is possible to speak of the meaning of something independently of a purposeful human action. I do not so believe. Another way to put this is that linguists (some linguists, not all) often talk about what words mean "in the language" as opposed to what they might mean in particular situations. I don't believe that the category "in the language" has any content whatsoever. I do believe, of course, in dictionaries and in grammars or accounts of grammar, but I always assume that dictionaries and accounts of grammar are being written from within the assumption of a range of possible human intentions as realized in particular situations, and that the fact that this range of possible human intentions as realized in particular situations is not on the surface, is not a part of the surface accounts of words given in a dictionary or in a grammar, is simply to be explained by the deep assumption of intentionality which is so deeply assumed that some people think they can in some particular situations get along without it. I always say to my students, "Just try to imagine uttering a sentence that is meaningful and, not as an afterthought but already in the act of thinking up such a sentence, imagine some intentional situation—that is, a situation with an agent with a purpose in relation to the configurations of the world that he or she wishes in some way to alter or announce—imagine doing without that and I say that you won't be able to." It's always the case that when you're attempting to determine what something means, what you are attempting to do is to penetrate to, to identify the intention of, some purposeful agent.

Now having said that, what methodological consequences follow? The answer is "none whatsoever," because (this is usually my favorite answer to almost any question) having now been persuaded that to construe meaning is also to identify intentional behavior, you are in no better position to go forward than you were before because all the problems remain. You mùst yourself decide what you mean by an "agent." Are you talking about the "liberal individual" formulating thoughts in his or her mind? Are you talking about the agency of a "community," of a group in *my* sense, or of a paradigm member, in Kuhn's sense? Are you talking, in an older intellectual tradition of the history of ideas, of the *Zeitgeist*—the spirit of an age within whose intentional structure everyone writes? Or in theological terms are you talking of a tradition in which my hand held the pen but it was the spirit of the Lord that moved me—a tradition I myself in no way denigrate? These are not decisions to which you will be helped by having decided that the construal of meaning is inseparable from the stipulation of intention. You then will also have to decide what is evidence for the intention that you finally stipulate, and that too is a question that was as wide open and as difficult before you came upon the gospel of intentionalism as it is now that you *have* come upon the gospel of intentionalism. So for shorthand purposes and in terms that most of your readers and mine would recognize, E. D. Hirsch was right when he asserted the primacy of intention back in 1960 and 1967, and he was simply wrong to think that having done so he had provided a methodological key or any kind of method whatsoever. This is also the argument, made brilliantly in my view, of Knapp and Michaels's essay "Against Theory."

Q. In "Going Down the Anti-Formalist Road" you write, "There is no such thing as literal meaning, if by literal meaning one means a meaning that is perspicuous no matter what the context and no matter what is in the speaker's or hearer's mind, a meaning that because it is prior to interpretation can serve as a constraint on interpretation." You conclude, "Meanings that seem perspicuous and literal are rendered so by forceful interpretive acts and not by the properties of language.'' Exactly what is a ''forceful interpretive act''? What lends it its ''force''?

A. A forceful interpretive act needn't be committed or performed by any one person; in fact, usually it is not, except in extraordinary cases. The forceful interpretive act takes place over time, and the agencies involved in it are multiple. Its effects are more easily identified than the process that leads to them. The effects are the production of a situation in which for all competent members of a community the utterance of certain words will be understood in an absolutely uniform way. That *does* happen. It is a possible historical contingent experience. When that happens you have, as far as I'm concerned, a linguistic condition that it might be perfectly appropriate to characterize as the condition of literalism. That is, at that moment you can with some justice say that these words, when uttered in this community, will mean only this one thing. The mistake is to think that it is the property of the words that produces this rather than a set of uniform interpretive assumptions that so fill the minds and consciousness of members that they will, upon receiving a certain set of words, immediately hear them in a certain way. Of course, that can always be upset by a variety of mechanisms, but it need not be upset; this condition can last a long, long, long time.

I'll tell you a story I've told many times. When my daughter was six years old, we were sitting at the dinner table one evening. We then had two small black dachshunds. My daughter Susan was doing something with the dachshunds under the table, and it was experienced at least by me as disruptive. So I said to her, ''Susan, stop playing with the dachshunds.'' She held up her hands in a kind of ''Look, Dad, no hands'' gesture and said, ''I'm not *playing* with the dachshunds.'' So I said, ''Susan, stop *kicking* the dachshunds.'' So I said, forgetting every lesson I had ever learned as a so-called philosopher of language, ''Susan don't do *anything* with the dachshunds!'' She replied, ''You mean I don't have to feed them anymore?'' At that moment I knew several things. First, I knew I was in a drama called ''the philosopher and the dupe'' and that she was the philosopher and I was the dupe. I also knew that this was a game that she could continue to play indefinitely because she could always recontextualize what she understood to be the context of my question in such a way as to destabilize the literalness on which I had been depending, which she too—within the situation of the dinner table, our relationship, our house—recognized in as literal a way as I did. That story, which can be unfolded endlessly, encapsulates for me this set of issues that you were asking about.

Q. In your essay on critical self-consciousness, you take issue first with

Stephen Toulmin because he "advocates self-conscious reflection on one's own beliefs as a way to neutralize bias immediately after having asserted the unavailability of the 'objective standpoint' that would make such reflection a possible achievement." Then you criticize the tradition of critical self-consciousness on the left as being "frankly political," as "rigorously and relentlessly negative, intent always on exposing or unmasking those arrangements of power that present themselves in reason's garb." Finally, you pronounce the critical project "a failure." Granting your argument that we are never free of constraints and therefore there never are truly free actions, would you not agree that the project of critical self-consciousness—whether conservative or radical— is nonetheless productive and beneficial, that we'd be poorer without it?

A. No. I do not agree because I sense you venturing into the regulative ideal territory—that is, we can never do this but it's a good thing to try. The bad poetic version of this is given in a line (that's even bad for him) by Browning (in my view the worst major poet): "A man's reach should exceed his grasp or what's a heaven for?" That is really the philosophy or point of view behind regulative ideal arguments, whether they're Kantian or Habermasian or any other "ian." I have no truck with them; I just don't see their point. It's just a form of idealism.

Q. Sure, it's idealistic to think that we can be truly self-conscious in a critical way, but doesn't the process of trying to get there turn out to be productive?

A. It depends on what you mean by "the process of trying to get there." You may be surprised or even distressed to hear this, but there is about to be published another Fish/Dworkin debate. I participated last year in a conference at Virginia, Pragmatism in Law and Society, which was in some ways appropriately centered on the work of Richard Rorty. The organizer, a professor of political science, assembled a really interesting cast of characters to speak about these questions. A few weeks before the conference, I received what I thought was a strange call from the convener of the conference who said that Ronald Dworkin wished to know which of the participants in the conference were going to write about him and what they were going to say. I said *I* wasn't going to write about him, that I was writing about Posner and Rorty, which I did. What I didn't know at the time is that for some reason Dworkin had been asked to be a commentator on the proceedings. His idea of being a commentator was to find out what essays would be directed either wholly or partly at him so that he could in the G. E. Moore tradition write a reply to his critics yet again. I would have seemed to have, at least with respect to me, foiled this intention because I didn't say *anything* about Dworkin. But when Dworkin came to write his commentary on the conference papers, he ignored this small difficulty and simply picked up the threads of earlier quarrels as if I *had* written my paper on him. When I saw that, I became distressed, and so I wrote a reply to Dworkin.

Now, Dworkin was arguing against the "theory has no consequences" position and for critical self-consciousness and for critically reflective stances on one's own assumptions—for a strong relationship, in short, between critical theory and practice. He chose as his example (this was a huge mistake) Ted Williams. Ted Williams was my hero as a boy. I had carried a picture of him around in my wallet for many years until it just fell apart. What Dworkin said was, as a kind of knock-down argument in his view, "The greatest hitter of modern baseball built a theory before every pitch." His source for this was Ted's book, *The Science of Hitting.* I got the latest edition of *The Science of Hitting*, read it carefully, annotated it, and pointed out several things. First of all, in *The Science of Hitting* Ted has an account of Ty Cobb's theory of hitting which he examines in detail—Cobb thought this and thought that in relation to velocity, to the way the foot moved, what you did with the bat, and so forth. Then after doing this, Ted absolutely demolishes it. He says, in effect, that what Cobb was advising is not possible for the human body to perform. Five pages later, Ted describes Cobb quite reasonably as the greatest hitter in baseball history. The conclusion is inescapable: the greatest hitter in baseball history had a theory of how he did it which had no relationship whatsoever, and could not have had any relationship, to what he did. Ted then goes on in another section of the book to describe what he thinks of as the mode of action of a great hitter. He goes on to hypothetical (but not really hypothetical; you have a sense that he's reconstructing moments in his own career) accounts of what a good hitter is doing as he stands up at the plate. What a good hitter is doing, according to Ted, is thinking things like this: "Well, last time he threw me a fast ball and there were two men on base and it was the fourth inning; now it's the eighth inning and there's no one on base but the score is four to three; I know that he doesn't like to rely on his fast ball so much in the later innings, and so forth and so on." Now what can one say about thoughts like that? First, one wants to say that they're highly self-conscious. They're self-conscious in the sense that there is a definite reflection not only on the present moment of activity but on the relationship between the present moment of activity and past moments which are now being "self-consciously" recalled. However, my point is that this self-consciousness really is not another level of practice but in fact is, how shall I describe it, itself a component in practice and that what Ted was saying to the would-be hitter was something like, "Be attentive to all dimensions of the situation." Now, is there a separate capacity called the "being attentive capacity" or what we might call the "critically reflective capacity"? Answer: No. Is it the case that you can develop a muscle or a pineal gland or something such that you could in any variety of different situations involving different forms of action activate that muscle? The answer is no. What you in fact do, when you do it well, is become *attentive* to the situation. The shape of your attentiveness is situation specific and dependent, so that—returning to your question—insofar as one is ever critically reflective, one is critically reflective *within* the routines of a practice. One's critical reflectiveness is in fact a function of, its shape is a function of, the routines of the practice. What most people want from critical reflectiveness is precisely a distance on the practice rather than what we might call a heightened degree of attention while performing in the practice. I haven't given you the argument as elegantly as I gave it in my reply to Dworkin. I guess in the end what I would want to say is that insofar as critical self-consciousness is a possible human achievement, it requires no special ability and cannot be cultivated as an independent value apart from particular situations: it's simply being normally reflective. It's not an abnormal, special—that is, *theoretical*—capacity. Insofar as the demand is for it to be such—that is, special, abnormal—it is a demand that can never be fulfilled.

Q. In "Profession Despise Thyself" you say that we in literary studies have made ourselves "fair game" to criticism "by subscribing to views of our enterprise in relation to which our activities can only be either superfluous or immoral ("How can you study Milton while the Third World starves?")." Noam Chomsky has said in *JAC* that he does not find most academic questions "humanly significant," suggesting that to be humanly useful academics should devote some of their time to social activism. Do you believe we in English studies should turn our attention to more socially useful activities?

A. I think it depends. English studies cannot itself be made into a branch of inquiry that has direct and immediate social and political payoffs, at least not in the way the United States is now structured. In other countries and other traditions, it would have been more possible for there to be a direct connection between literary activity and social and political activity, and perhaps in some transformation of our society that has not yet occurred it could be the case that the kinds of analyses we're performing in class could have an immediate impact on the larger social and economic questions being debated in society. As for the question (which I now will understand in a way that Chomsky would probably find trivial), "Should English teachers devote their energies to social causes?'' my answer is, "Why not?'' It's like *pro bono* work in the legal world: you decide what it is you're interested in doing, working in political ways, in social ways, and you volunteer. In a way, what Chomsky is saying is very congenial to the academic mentality—a mentality that has a deep interest in diminishing its own value. Just why this is so is worthy of many pages of analysis. The academic generally participates in the devaluation of his or her own activities to a much greater degree than the practitioners in other fields do. It seems to me that academic activity is a human activity. As a human activity, like any other activity, it has its constraints and therefore its areas of possible effectivity as well as many areas in which it will not be effective because it will not touch them. This makes it no better or no worse on some absolute scale (that doesn't exist) than any other human activity. However, at a particular moment in history a legitimate question is, "Do we want to put our energies in this human activity that has this structure of plus and minus in terms of gains and losses and opportunities, or this one?'' That's a perfectly reasonable question to ask as long as one doesn't think that one is asking a question that has a Platonic structure, in the old sense, or a surface/deep structure opposition, in the Chomskian sense. I'm temperamentally opposed to those who wish to regard the academic life as an inferior, unauthentic form of human activity. It's *another* form of human activity. It should neither be privileged—as some romantic humanists privilege it so that only those who "live the life of the mind" are really living—nor should it be denigrated as the area of the trivial in relation to which getting one's hands authentically dirty is the true counterweight. I think both of those characterizations are bankrupt.

Q. In *Doing What Comes Naturally,* you speculate that the immense popularity of books like E. D. Hirsch's and Allan Bloom's signal that "the *public* fortunes of rationalist-foundationalist thought have taken a favorable turn'': "One can expect administrators and legislators to propose reforms (and perhaps even purges) based on Bloom's arguments (the rhetorical force of anti-rhetoricalism is always being revived).'' Do you predict massive (and counterproductive) state intervention in the educational system?

A. The current political situation (by "current'' I mean at this moment) suggests that that would be an unhappily canny prediction. Secretary of Education Lamar Alexander is poised to implement some of the ideas he inherited from William Bennett and which are being given continuing vitality in the administration's thinking by Lynne Cheney, as advised by people like Chester Finn and Diane Ravitch, among others. In all of these instances, the tendency is to label as disruptive and subversive—almost in a sense that returns us to the 1950s—all forms of thought that question the availability of transcendental standards and objective lines of measurement so that these forms of thought are regarded by the persons that I have named not as possible contenders in an arena of philosophical discussion but as Trojan horses of evil, decay, destruction of community, and so on. So long as these persons hold important positions in the government, positions connected to the administration of the educational world and the dispensation of funds, I think we do face a period in which there will be (at least on the national level, and in some cases on local levels) moves to curtail and purge. We're already seeing this in the activities of organizations like the National Association of Scholars and in the extensive network of student journalism that began with the *Dartmouth Review* but that has now extended far beyond the confines of Hanover, New Hampshire, allied with a number of prominently placed journalists in the national news media: people like Dorothy Rabinowitz and David Brooks of the *Wall Street Journal;* Jonathan Yardley at the *Post;* Charles Krauthammer and John Leo; political/popular writers like Dinesh D'Souza, Roger Kimball, and Charles Sykes; Nat Hentoff at the *Village Voice*—a whole series of people who can be relied upon to be mouthpieces for this very neo-conservative political agenda which is backed by huge amounts of right-wing foundation money provided by William Simon and others. I think that's a real force at the moment and a force to which many in the academy are only just now waking up.

Q. You said we should be socially active. What measures can we take to prevent such reactionary trends?

A. The MLA panel I'm about to attend is entitled "Answering Back." Though I'm not a member of the panel, I'll be in the audience and I think we *must* talk back. I think that academics too often disdain communication with people outside the academic world and believe that attempting to speak to the public must necessarily be a diminution of our normal mode of discourse and that in order to speak to the public we must gear down and simplify our usual nuanced perspectives. In fact, I know from experience that speaking to the general public is indeed a task equally complex and difficult, but *differently* complex and difficult, as speaking to one's peers in learned journals or at conferences. There is a set of problems of translation and rhetorical accommodation that one comes upon when attempting to talk to audiences outside the academy which is absolutely fascinating and difficult. So unless we set our mind to this task, the capturing of the media pages and airwaves will continue as it has continued in the past year and a half so that up until four or five months ago it was difficult to find a view widely published *other* than the view being put forward by what we might call "Cheney and Company."

Q. Certainly you have had your share of critics and detractors from both the left and the right. Are there any criticisms or misunderstandings of your work that you'd like to take issue with at this time? Anything to set straight?

A. No, not in any sense that hasn't been attempted before. As I say in *Doing What Comes Naturally* and elsewhere, there are basically two criticisms of my work; they come from the right and the left. The criticism from the right is that in arguing for notions like interpretive communities, the inescapability of interpretation, the infinite revisability of interpretive structures, I am undoing the fabric of civilization and opening the way to nihilist anarchy. The objection from the left is that I'm *not* doing that sufficiently. My argument to both is that on the one hand the fear that animates right-wing attacks on me is an unrealizable fear because one can never be divested of certainties and programs for action unless one believed that the mind itself could function as a calculating agent independently of the beliefs and convictions which supposedly we're going to lose; and on the other hand (or on the same hand), therefore, a program in which our first task is to divest ourselves of all our old and hegemonically imposed convictions in order to move forward to some new and braver world is an impossible task. On the one hand, hearkening to me will not lead to the decay of civilization, and on the other hand hearkening to me will not lead to the canonization of the status quo. In fact, on *these* kinds of points—and this is what most of my critics find most difficult to understand—hearkening to me will lead to *nothing*. Hearkening to me, from my point of view, is *supposed to* lead to nothing. As I say in *Doing What Comes Naturally* in answer to the question "What is the point?" the point is that there *is* no point, no yield of a positive programmatic kind to be carried away from these analyses. Nevertheless, *that* point (that there is no point) *is* the point because it's the promise of such a yield—either in the form of some finally successful identification of a foundational set of standards or some program by which we can move away from standards to ever-expanding liberation—it's the unavailability of such a yield that *is* my point, and therefore it would be contradictory for me to have a point beyond *that* point. People absolutely go bonkers when they hear that, but that's the way it is.

Notes

Chapter 2

1. Brad Leithauser, review, *The New Yorker* (21 August 1989): 90.

2. Ibid., 93.

3. Robert Alter, *The Pleasures of Reading in an Ideological Age* (New York, 21989), 76, 54, 270.

4. Review of R. Alter, *The Pleasures of Reading in an Ideological Age, The New Yorker* (21 August 1989): 94.

5. George Steiner, *Real Presences* (Chicago, 1989), 48.

6. *The New Yorker*, 21 August 1989, 94.

7. Stephen White, "Educational Anomie," 2; William J. Bennett, "Why Western Civilization?", 3; Lynne V. Cheney, "Canons, Cultural Literacy, and the Core Curriculum," 3; Lynne V. Cheney, "Canons, Cultural Literacy, and the Core Curriculum," 11; Chester E. Finn, Jr., "A Truce in the Curricular Wars?", 16, 17; Elizabeth Fox-Genovese, "The Feminist Challenge to the Canon," 34, the *National Forum* (Summer 1989).

8. Finn, the *National Forum*, 17.

9. Bennett, *National Forum*, 4.

10. R. T. Smith, "Canon Fodder, the Cultural Hustle, and the Minotaur," the *National Forum* (Summer 1989): 26.

11. Gerald Graff and William Cain, "Peace Plan for the Canon Wars," the *National Forum* (Summer 1989): 9.

12. Betty Jean Craige, "Curriculum Battles and Global Politics," the *National Forum* (Summer 1989): 31.

13. R. Alter, *The Pleasures of Reading*, 28, 27.

14. Ibid., 54.

15. Ibid., 38.

16. Wayne C. Booth, *The Company We Keep: An Ethics of Fiction* (Berkeley, 1988), 223.

17. When I say that everyone is an ethicist I mean more than that everyone is already engaging with books—canonical and noncanonical—in ways that involve ethical commitments; for regarding books as the training ground for moral activity—which is what Alter and Booth mean by ethical reading—is only one of the many relationships one can have with books, all of them ethical in the only sense of the word that makes sense. Polemical ethicists tend to contrast their life with books with the impoverished life led by those who look to books for leisure (we shall see a stunning example of this in a moment) or information, for advice or political inspiration or sociological data; but these and many other forms of engagement are no less ethical than the engagement with Great Questions demanded by our severe moralists, that is, no less a matter of purposeful behavior in relation to some goal considered (by the actor) to be good. Not only is the ethics of reading a problematical concept because there are any number of possible ethical yields from innumerable literary and nonliterary texts, but the very idea of requiring an ethical yield from our reading is itself a debatable notion, and the alternatives (only partially listed above) are to be found not outside but *within* ethics, which, like difference—indeed the ethical *is* difference—is a realm no one can grasp for one is always and already within it.

18. R. Alter, *The Pleasures of Reading,* 16.

19. Anne Barbeau Gardiner, *ADE Bulletin* (Fall 1989): 24.

20. Martha C. Nussbaum, "Reading for Life," *Yale Journal of Law and Humanities* 1, no. 1 (December 1988): 171.

21. Ibid., 176.

22. Lynne V. Cheney, *Humanities in America: A Report to the President, the Congress, and the American People* (Washington, D.C., 1988), 14.

23. Ibid., 32.

24. William J. Bennett, *National Forum,* 4.

25. Chester E. Finn, Jr., *National Forum,* 18.

26. Sidney Hook, "Civilization and Its Malcontents," *The National Review* (13 October 1989): 31.

27. George Steiner, *Real Presences,* 8, 60.

28. Ibid., 67, 32.

29. Ibid., 33.

30. "Derisory Tower" (Editorial), *The New Republic* (18 February 1991): 5.

31. *Academic Questions* (March 1986): 8.

Chapter 10

1. Carter, "Evolution, Creationism, and Treating Religion as a Hobby," *Duke L. J.* (1987): 977, 995.

2. *Id.*

3. *Id.* at 985.

4. *Id.* at 992.

5. *Id.* at 995.

6. *Id.* at 978.

7. *Id.* at 995.

Chapter 11

1. *Manwill v. Oyler,* 11 Utah 2d 433, 361 P.2d 177 (1961).

2. P. 192. Trans. Max Knight from the 2d (rev. and enl.) German ed. (Berkeley: University of California Press, 1967).

3. *Trident Center v. Connecticut General Life Insurance Company,* 847 F.2d 564 (9th Cir. 1988).

4. Ibid., 566.

5. Ibid.

6. Ibid.

7. Ibid., 567.

8. Ibid., 566.

9. Ibid.

10. Ibid., 568.

11. Ibid.

12. Ibid., 567 n.1.

13. Ibid., 568.

14. Ibid.

15. 68 Cal.2d 33, 442 P.2d 641 (1968).

16. *Trident Center v. Connecticut General,* 568 (citing 69 Cal.2d 38).

17. Ibid., 569.

18. Ibid., 569–570.

19. Ibid., 569.

20. Ibid. (citing 69 Cal. 2d 37).

21. Ibid., 569.

22. *Uniform Commercial Code,* 10th ed. (St. Paul, Minn.: West Publishing, 1987), 71.

23. Gordon D. Schaber and Claude D. Roher, *Contracts in an Nutshell,* 2d ed. (St. Paul, Minn.: West Publishing, 1984), 243.

24. *Warren's Kiddie Shoppe, Inc. v. Casual Slacks, Inc.,* 120 Ga. App. 578, 171 S.E.2d 643 (1969).

25. *Dekker Steel Co. v. Exchange National Bank of Chicago,* 330 F.2d 82 (1964).

26. 451 F.2d 3 (1971).

27. Ibid., 9.

28. Ibid., 7.

29. Steven Emanuel and Steven Knowles, *Contracts* (Larchmont, N.Y.: Emanuel Law Outlines, 1987), 160.

30. *Columbia Nitrogen v. Royster Company,* 451 F.2d 3 (1971), 9.

31. Ibid., 9–10.

32. *Uniform Commercial Code,* 71.

33. 65 Cal. Rptr. 545, 436 P.2d 561 (Cal. Sup. Ct. 1968).

34. Ibid., 562.

35. Ibid., 565.

36. *Trident Center,* 569.

37. *Uniform Commercial Code,* 72.

38. *Masterson,* 731–32.

39. *Mitchill v. Lath,* 247 N.Y. 377, 160 N.E. 646 (1928).

40. *Masterson,* 731n.

41. Corbin, *Corbin on Contracts,* one-volume edition (St. Paul, Minn.: West Publishing, 1952), 496.

42. Ibid., 497.

43. Ibid., 515.

44. 196 Minn. 60, 264 N.W. 247 (1935).

45. Ibid., 433.

46. Ibid., 431.

47. Ibid.

48. Ibid., 433.

49. Ibid., 431.

50. Ibid., 433 (citing *City of Marshall v. Gregoire,* 193 Minn. 188, 198–199, 259 N.W. 377, 381–382).

51. Ibid., 433.

52. Emanuel and Knowles, *Contracts,* 72.

53. *Restatement of the Law Second, Contracts Second,* vol. 1 (St. Paul, Minn.: American Law Institute Publishers, 1981), sec. 79.

54. *Wolford v. Powers,* 85 Ind. 294 (1882), 303.

55. E. Allan Farnsworth, *Contracts* (Boston: Little, Brown, 1982), 6.

56. *Continental Forest Products, Inc. v. Chandler Supply Company,* 95 Idaho 739, 743, 518 P.2d 1201 (1974), 1205.

57. Stanley Henderson, "Promises Grounded in the Past: The Idea of Unjust Enrichment and the Law of Contracts," *Virginia Law Review* 57 (1971): 1141.

58. *Restatement Second, Contracts Second,* 233.

59. Grant Gilmore, *The Death of Contract* (Columbus: Ohio State University Press, 1974), 74–75.

60. Ibid.

61. Henderson, "Promises Grounded in the Past," 1127.

62. *Restatement Second, Contracts Second,* 233–234.

63. Henderson, "Promises Grounded in the Past," 1122–1123.

64. Ibid., 1122.

65. *Pull v. Barnes,* 142 Colo. 272, 350 P.2d 829 (1960).

66. 27 Ala. App. 82, 168 So. 196 (1935).

67. Ibid., 196–197.

68. Ibid., 198.

69. Ibid.

70. Ibid., 197.

71. Ibid.

72. Ibid., 198.

73. Ibid.

74. Ibid., 199.

75. Clare Dalton, "An Essay in the Deconstruction of Contract Doctrine," *Yale Law Review* 94 (1985): 1007.

76. Ibid., 1035.

77. Ibid., 1066.

78. Ibid., 1084.

79. Ibid., 1091.

80. Ibid., 1087.

81. Harry Scheiber, "Public Rights and the Rule of Law in American Legal History," *California Law Review* 72 (1984): 236–237.

82. Steven J. Burton, "Rhetorical Jurisprudence: Law as Practical Reason" (unpublished manuscript), 69.

83. Ibid., 63.

84. Ibid., 9.

85. Ibid., 2.

86. Ibid., 9.

87. Ibid., 69.

88. Ibid.

89. James Boyd White, *Heracles' Bow: Essays on the Rhetoric and Poetics of the Law* (Madison: University of Wisconsin Press, 1985), 33.

90. Ibid., 34.

91. Ibid., 42.

92. Ibid., 39.

93. Ibid., 47.

94. Ibid., 45.

95. Ibid.

96. Ibid.

97. Ibid., 124.

98. Ibid.

99. Ibid., 125.

100. Peter Goodrich, *Legal Discourse: Studies in Linguistics, Rhetoric, and Legal Analysis* (New York: St. Martin's Press, 1987), 5–6.

101. Ibid., 27.

102. Ibid., 61.

103. Ibid., 77.

104. Ibid., 58.

105. Ibid., 57.

106. Ibid., 147.

107. Ibid., 88.

108. Ibid., 90.

109. Ibid.

110. Ibid., 175.

111. Ibid.

112. Ibid., 176.

113. Ibid., 173.

114. Ibid., 132.

115. Ibid., 183.

116. Ibid., 204.

117. Stanley Fish, *Doing What Comes Naturally* (Durham, N.C.: Duke University Press, 1989), 397.

118. Goodrich, *Legal Discourse*, 212.

119. Ibid.

Chapter 13

1. Richard A. Posner, *The Problems of Jurisprudence* (Cambridge, Mass.: Harvard University Press, 1990). All page numbers in the text refer to this book.

2. This is the great lesson of Thomas Kuhn's *The Structure of Scientific Revolutions* (Chicago: Chicago University Press, 1962).

3. H. L. A. Hart, *The Concept of Law* (Oxford: Oxford University Press, 1961), 202.

4. Quoting Learned Hand, "A Personal Confession," in *The Spirit of Liberty: Papers and Addresses of Learned Hand*, 3d ed. (New York: Knopf, 1960), 307 (emphasis added).

5. Felix S. Cohen, "Transcendental Nonsense and the Functional Approach," *Columbia Law Review* 35 (June 1935): 809–848.

6. Ibid., 809–810.

7. Ibid., 812.

8. Ibid., 814.

9. Ibid., 820–821.

10. Ibid., 822.

11. Ibid., 833.

12. Ibid., 841.

13. Ibid.

14. Roscoe Pound, "A Call for a Realist Jurisprudence," *Harvard Law Review* 44 (March 1931): 699.

15. Richard Rorty, *Consequences of Pragmatism* (Northfield: Minnesota University Press, 1982).

16. Ibid., 166.

17. Ibid.

18. Richard Rorty, *Contingency, Irony, and Solidarity*, 93 (Northfield: Minnesota University Press, 1982).

19. Ibid., 196.

20. Ibid.

21. Ibid.

22. Ibid.

23. Ibid., 67.

24. Ibid., 196.

25. Rorty, *Consequences of Pragmatism*, 174.

26. Ibid.

27. Pound, "A Call for a Realist Jurisprudence," 700.

28. Ibid.

29. Ibid., 706.

30. Ibid.

31. Julie Thompson Klein, *Interdisciplinarity: History, Theory, and Practice* (Detroit, Mich.: Wayne State University Press, 1989).

32. Ibid., 19.

33. Ibid., 23.

34. Ibid., 41.

35. Ibid., 79.

36. Ibid., 78.

37. Ibid., 77.

38. Ronald Dworkin, "Pragmatism, Right Answers, and True Banality," 369 (Chapter 19 in this volume). In *Pragmatism in Law and Society,* Michael Brint and William Weaver, eds. (Boulder, San Francisco: Oxford, 1991), 359–388.

39. Ibid., 382.

40. Ronald Dworkin, *Taking Rights Seriously,* 105.

41. Stanley Fish, *Doing What Comes Naturally* (Durham, N.C.: Duke University Press, 1989), 378–379. Simultaneously published by Oxford University Press.

42. Dworkin, "Pragmatism, Right Answers, and True Banality," 382.

43. Ibid.

44. Ted Williams and John Underwood, *The Science of Hitting* (New York: Simon & Schuster, 1986), 47.

45. Ibid.

46. Ibid., 84.

47. Ibid., 14.

48. Ibid., 29.

49. Ibid.

50. Ibid., 32.

51. Dworkin, "Pragmatism, Right Answers, and True Banality," 381.

52. Ibid., 382.

Chapter 14

1. Stephen Booth tells me that this formulation may be too strong, and he reminds me of an experience many of us will be able to recall, knowing while watching a horror movie that certain devices are being used to frighten us and yet being frightened nevertheless despite our knowledge. In experiences like this an analytical understanding of what is happening exists side by side with what is happening but does not affect or neutralize it. It would be too much to say, then, that when engaging in a practice (and watching horror movies is a practice) one must forget the analytical perspective one might have on the practice at another time. It would be more accurate to say that an analytical perspective on a practice does not insulate one from experiencing the practice in all its fullness, that is, in the same way one would experience it were the analytical perspective unavailable.

Works Cited

Adams, Hazard, and Leroy Searle, eds. *Critical Theory since 1965*. Tallahassee: Florida State UP, 1986.

Cheney, Lynne V. *Humanities in America: A Report to the President, the Congress, and the American People*. Washington: NEH, 1988.

Derrida, Jacques. "Differance." Adams and Searle 120–136.

Felman, Shoshana. "Psychoanalysis and Education: Teaching Terminable and Interminable." Johnson, *Pedagogical Imperative* 21–44.

Geertz, Clifford. "Blurred Genres: The Refiguration of Social Thought." Adams and Searle 514–523.

Jacoby, Russell. *The Last Intellectuals*. New York: Basic, 1987.

Johnson, Barbara, ed. *The Pedagogical Imperative. Yale French Studies* 63 (1982): 1–252.

———. "Teaching Ignorance: *L'Ecole des Femmes*." Johnson, *Pedagogical Imperative* 165–182.

Leitch, Vincent. "Deconstruction and Pedagogy." Nelson 45–56.

Merod, Jim. *The Political Responsibility of the Critics*. Ithaca: Cornell UP, 1987.

Mohanty, S. P. "Radical Teaching, Radical Theory: The Ambiguous Politics of Meaning." Nelson 149–176.

Nelson, Cary, ed. *Theory in the Classroom*. Urbana: U of Illinois P, 1986.

Peck, Jeffrey M. "Advanced Literary Study as Cultural Study: A Redefinition of the Discipline." *Profession 85*. New York: MLA, 1985. 49–54.

Robbins, Bruce. "Professionalism and Politics: Toward Productively Divided Loyalties." *Profession 85*. New York: MLA, 1985. 1–9.

Ryan, Michael. "Deconstruction and Radical Teaching." Johnson, *Pedagogical Imperative* 45–58.

Said, Edward. Interview. *Diacritics* 6 (1976): 30–47.

Sammons, Jeffrey L. "Squaring the Circle: Observations on Core Curriculum and the Plight of the Humanities." *Profession 86*. New York: MLA, 1986. 14–21.

Scholes, Robert. *Semiotics and Interpretation*. New Haven: Yale UP, 1982.

Schön, Daniel. *The Reflective Practitioner*. New York: Basic, 1983.

Terdiman, Richard. "Structures of Initiation: On Semiotic Education and Its Contradictions in Balzac." Johnson, *Pedagogical Imperative* 198–226.

Weber, Samuel. *Institution and Interpretation*. Minneapolis: U of Minnesota P, 1987.

Winkler, Karen J. "Interdisciplinary Research: How Big a Challenge to Traditional Fields?" *Chronicle of Higher Education* 7 Oct. 1987: 1+.

Chapter 15

1. I would reply that histories create their own exclusions not because they are misrepresentations—a word that requires for its intelligibility the possibility of a representation that is not one—but because as narratives that tell one story rather

than all stories they will always seem partial and inaccurate from the vantage point of other narratives, themselves no less, but differently, exclusionary.

2. From a talk delivered at a meeting of the English Institute, August 27, 1988.

3. See my "Consequences," *Critical Inquiry* 11 (March 1985): 440–441.

4. But one could be strongly committed to Marxism or psychoanalysis at one level and still practice history (or literary criticism or pedagogy) in a way that was free of Marxist or psychoanalytic assumptions (although the practice would flow from some other assumptions). This might hold true even if in answer to a direct question about your practice you declared that it was Marxist or psychoanalytic. Your theory of what you do is logically independent of what you in fact do (although, as I acknowledge above, a relationship, at least in some cases, is always possible). Thus someone might reasonably disagree with your account of your practice and agree wholeheartedly with its assertions. To demand a perfect homology between practice and theory and between theory and politics is, as Catherine Gallagher says, to surrender "to the myth of a self-consistent subject impervious to divisions of disciplinary boundaries and outside the constraints of disciplinary standards" (46).

5. See Roberto Unger, "The Critical Legal Studies Movement, *Harvard Law Review* 96 (1983), and see my "Unger and Milton," in *Doing What Comes Naturally: Change, Rhetoric, and the Practice of Theory in Legal and Literary Studies* (Durham, N.C.: Duke University Press, 1989).

6. In fact, were I to adopt such a focus I would no longer be doing literary criticism, I would be doing something else, sociology or anthropology or systems analysis, etc. It is once again a question of forgetting and remembering: one cannot keep in mind everything at once and still perform specific tasks (except perhaps the specific—and impossible—task of fully enumerating everything). The very possibility of performing a specific task depends on *not* attending to (i.e. forgetting) concerns that would, if they were given their due, involve one in the performing of a different specific task.

7. Cf. Gallagher: "There may be no political impulse whatsoever behind [the] desire to historicize literature. This is not to claim that the desire for historical knowledge is itself historically unplaced or 'objective'; it is, rather, to insist that the impulse, norms, and standards of a discipline called history, which has achieved a high level of autonomy in the late twentieth century, are a profound part of the subjectivity of some scholars and do not in all cases require political ignition" (46). I would only add that the antiprofessionalism displayed by Pecora and Montrose is yet another indication of the idealizing and ahistorical vision that generates their complaints and anxieties.

Index

319